NANOTECHNOLOGY SCIENCE AND TECHNOLOGY

SUPERPARAMAGNETIC IRON OXIDE NANOPARTICLES: SYNTHESIS, SURFACE ENGINEERING, CYTOTOXICITY AND BIOMEDICAL APPLICATIONS

NANOTECHNOLOGY SCIENCE AND TECHNOLOGY

Additional books in this series can be found on Nova's website under the Series tab.

Additional E-books in this series can be found on Nova's website under the E-books tab.

NANOTECHNOLOGY SCIENCE AND TECHNOLOGY

SUPERPARAMAGNETIC IRON OXIDE NANOPARTICLES: SYNTHESIS, SURFACE ENGINEERING, CYTOTOXICITY AND BIOMEDICAL APPLICATIONS

MORTEZA MAHMOUDI
PIETER STROEVE
ABBAS S. MILANI
AND
ALI S. ARBAB

Nova Science Publishers, Inc.
New York

Copyright © 2011 by Nova Science Publishers, Inc.

All rights reserved. No part of this book may be reproduced, stored in a retrieval system or transmitted in any form or by any means: electronic, electrostatic, magnetic, tape, mechanical photocopying, recording or otherwise without the written permission of the Publisher.

For permission to use material from this book please contact us:
Telephone 631-231-7269; Fax 631-231-8175
Web Site: http://www.novapublishers.com

NOTICE TO THE READER

The Publisher has taken reasonable care in the preparation of this book, but makes no expressed or implied warranty of any kind and assumes no responsibility for any errors or omissions. No liability is assumed for incidental or consequential damages in connection with or arising out of information contained in this book. The Publisher shall not be liable for any special, consequential, or exemplary damages resulting, in whole or in part, from the readers' use of, or reliance upon, this material. Any parts of this book based on government reports are so indicated and copyright is claimed for those parts to the extent applicable to compilations of such works.

Independent verification should be sought for any data, advice or recommendations contained in this book. In addition, no responsibility is assumed by the publisher for any injury and/or damage to persons or property arising from any methods, products, instructions, ideas or otherwise contained in this publication.

This publication is designed to provide accurate and authoritative information with regard to the subject matter covered herein. It is sold with the clear understanding that the Publisher is not engaged in rendering legal or any other professional services. If legal or any other expert assistance is required, the services of a competent person should be sought. FROM A DECLARATION OF PARTICIPANTS JOINTLY ADOPTED BY A COMMITTEE OF THE AMERICAN BAR ASSOCIATION AND A COMMITTEE OF PUBLISHERS.

Additional color graphics may be available in the e-book version of this book.

LIBRARY OF CONGRESS CATALOGING-IN-PUBLICATION DATA

Superparamagnetic iron oxide nanoparticles : synthesis, surface engineering, cytotoxicity, and biomedical applications / Morteza Mahmoudi ... [et al.].
    p. cm.
 Includes index.
 ISBN 978-1-61668-964-3 (hardcover)
 1. Nanomedicine. 2. Ferric oxide--Magnetic properties. 3. Nanoparticles. I. Mahmoudi, Morteza, 1979-
 R857.N34S87 2009
 610.28'4--dc22
                              2010036438

Published by Nova Science Publishers, Inc. ✢ *New York*

Dedication

Morteza Mahmoudi: To My Wife, Haniyeh Aghaverdi

ABOUT THE AUTHORS

Morteza Mahmoudi obtained his PhD in 2009 from Sharif University of Technology with specialisation on the cytotoxicity of super-paramagnetic iron oxide nanoparticles (SPIONs). He has authored over 30 publications and book chapters, 20 abstracts, and 2 issued (National) and pending (US) patents on the use of magnetic nanoscale technologies in biomedical applications. His current research involves the magic SPIONs for simultaneous diagnosis and therapeutic applications (www.biospion.com).

Dr. Mahmoudi is a referee of many prestigious ISI journals such as The Journal of Physical Chemistry, Chemical Reviews, ACS Applied Materials & Interfaces, Analytical Chemistry, Journal of Nanobiotechnology, Chemistry of Materials, Nanomedicine and Journal of Colloid and Interface Science.

He has received many awards such as Kharazmi Young Festival Award (2009; the highest national award), 2010 Dr. Mojtahedi Innovation Award for Distinguished Innovation in Research and Education at Sharif University of Technology, introduced as a Scientific Elite by the National Elite Institute  (2009), and Exceptional Talents Award from Dean of Graduate Studies at Sharif University of Technology (2008 & 2009). He was a visiting scientist at Laboratory of Powder Technology (LTP) at Swiss Federal Institute of Technology (EPFL) under supervision of Professor Heinrich Hofmann.

Professor Pieter Stroeve conducts research in mass transfer with chemical reaction, nanotechnology, colloid science and biomass conversion. He has extensively published on mass transfer with non-equilibrium chemical reaction. His focus in nanotechnology is the use of porous templates to make nanoporous membranes, nanotubes, nanowires, nanocables, nanostructured surfaces and their use for biosensors, nanostructured solar cells, detectors and molecular separations. In colloid science he is working on the nanostructure of polyion-surfactant complexes, polycation-polyanion complexes, self-assembled monolayers, and nanoparticle-polymer composites. In biomass conversion, Prof. Stroeve has been working on using pulsed electrical field to

make plants more permeable to catalysts and acid molecules by creating nanopores in the plant membranes. He obtained his B.Sc. from University of California, Berkeley, and his M.Sc and Sc.D. in from Massachusetts Institute of Technology (MIT) in Chemical Engineering.

Dr. Abbas Milani is currently Assistant Professor at the School of Engineering, the University of British Columbia Okanagan. Formerly, he was an NSERC (Natural Sciences and Engineering Research Council of Canada) fellow at the Massachusetts Institute of Technology. His expertise falls in multi-scale modeling, simulation, and multi-criteria optimization of composite materials, structures, and forming processes. He is the recipient of numerous scholarships and awards and his research has been recognized by national and international scientific communities. Most recently, he was a winner of the 2010 International Intelligent Optimal Design Prize by CADLM.

Dr. Ali S. Arbab is a physician research scientist with broad experience in imaging sciences and cell biology. He has extensive experiences in nuclear medicine scanning (SPECT), magnetic resonance imaging (MRI), X-ray computed tomography (CT), ultrasonography, and molecular imaging such as optical imaging, fluorescent and confocal microscopy. He is also experienced in cell biology (including stem cells), cell transfection, making of primed dendritic cells and cytotoxic T-cells, cellular labeling of MR contrast agents, biodistribution of radiopharmaceuticals, radio pharmacokinetics, cell membrane ion channels modulation, tumor model preparation, flowcytometry and immunochemistry. The valuable experiences including troubleshooting has been gathered during his post-doc training at NIH in cellular and molecular imaging and the experiences are further improved when he completed both NIH and department of defense grants. Currently, he has two ongoing grants from NIH.

Contents

Preface		xi
Chapter 1	Introductory to the Book	1
Chapter 2	Synthesis of SPIONs	11
Chapter 3	Application of Statistical Techniques	51
Chapter 4	Coatings of SPIONs	71
Chapter 5	Cytotoxicity of SPIONs	125
Chapter 6	Application of SPIONs	163
Index		213

PREFACE

Nanoscience has been a subject of study for at least a century, although fields such as colloid science and cellular biology were not known by this name. Nanotechnology started in the early 1980s due to the advances made in integrated circuits, and has gained drastic growth and development over the last decade. Today, nanotechnology and nanoscience are now extreme topics of interest, not only in academic communities, but also in industrial sectors due to the high potential for commercial product developments. The rapid launch of new products incorporating nanotechnology is showing a clear trend across a wide spectrum of fields from manufacturing and materials applications, including nanocomposites and coatings, to electronics and IT applications including advanced memory chips and displays. Healthcare and life science applications of nanotechnologies such as nanostructured medical devices and nanotherapeutics are receiving increased attention by industry; although these have had the longest time-to-market.

A distinguishing feature of nanotechnology and nanoscience is the design of new physiochemical properties of nanostructured materials that cannot be attained by using bulk materials. Designed properties of nanomaterials have great potential to enhance, or even transform, many conventional matters in our modern life. One of the promising subfields of bionanoscience is nanomedicine. It is recognized as a highly interdisciplinary field, involving chemistry, physics, biology, engineering, materials science and medicine, with high potential to provide precise diagnosis of diseases and fast, easy, and low cost treatment of catastrophic syndromes with minimal side effects and lower patient compliance. The most promising nanomaterials that have found widespread usage in biomedicine are superparamagnetic iron oxide nanoparticles (SPIONs). SPIONs with a mean particle diameter of about 10 nm suspended in appropriate carrier liquids are known as ferrofluids. A special feature of ferrofluids is the combination of normal liquid behavior with superparamagnetic properties of nanoparticles, enabling their usage in numerous biomedical applications such as magnetic resonance imaging (MRI) for contrast enhancement, drug/gene delivery, (stem) cell tracking, or magnetic separation technologies (e.g. rapid DNA sequencing) and ultrasensitive diagnostic assays. Some of these applications are already well-established, such as magnetic separation and MRI, while others are under research at this stage. The shape, particle size characteristics and surface properties of SPIONs are crucial factors for their successful applications in biomedicine. These factors are normally controlled by the synthesis method and processing conditions utilized in a particular application. On the other hand, the process yield and its reproducibility are among important factors for scale-up manufacturing of SPION products.

In this book, a wide scope of current and future developments of SPIONs, including synthesis, surface modification, cytotoxicity and applications, are covered by combining contributions from faculty members in materials science, chemical engineering, chemistryand biology. Great emphasis is given to the interdisciplinary nature of the bionanoscience and bionanotechnology fields.

After a general introduction in Chapter 1, the authors dedicate Chapter 2 to various synthesis methods together with their advantages and disadvantages. In Chapter 3, a set of most practical statistical techniques for optimum synthesis and characterization of nano-products is included. Chapter 4 covers coating of SPIONs and targeting matters. Chapter 5 presents a thorough review of available *in vitro* and *in vivo* toxicity evaluation methods together with recent toxicity results of SPIONs. Chapter 6 covers biomedical applications of SPIONs from the laboratory to commercialized product scales. It is important to note that the number of SPION-related publications has drastically increased over the past decade. However, most investigations are related to the synthesis, characterization and surface properties of SPIONs. Biological issues have been addressed only in the last few years and their toxicity has been recognized as a safety issue. The authors hope Chapters 5 and 6 can be a small contribution in promoting researcher' interest in the field to explore more safety aspects, risk assessment methods, and regulations for future applications of SPIONs.

Finally, we would like to thank the production team of Nova Science Publishers for their continuous and dedicated support during the preparation of this book.

Morteza Mahmoudi, Pieter Stroeve, Abbas S. Milani, and Ali S. Arbab
Summer 2010

Chapter 1

INTRODUCTORY TO THE BOOK

In recent years, the fabrication of nanoparticles and exploration of their properties have attracted the attention of physicists, chemists, biologists and engineers. [1-4] Interest in nanoparticles arise from the fact that the mechanical, chemical, electrical, optical, magnetic, electro-optical and magneto-optical properties of these particles are different from their bulk properties and depend on the particle size. [5-9] There are numerous areas where nanoparticulate systems are of scientific and technological interest. To produce these particles in appropriate quantities, with controlled particle size is a significant challenge.

1.1. SEMICONDUCTING NANOPARTICLES

Nanotechnology became technologically popular in the design of miniaturized, ultrahigh density integrated circuits and information storage devices of the future and in some sense an alternate route to overcome the 100 nm barrier in conventional electronics and develop nanodimensional molecular electronic devices. [10-13] Owing to their semiconducting properties, group II-VI and IV-VI compounds have been intensively examined by various authors. [14] The size dependent opto-electronic properties of nanoparticles (diameter 1-100 nm) is attributed to quantum confinement effects. [1-13] Briefly, electronic excitation in semiconductors arise from an exciton (an electron and hole bound pair) localized in a potential well. Theoretical calculations [6-13] have demonstrated that when particle sizes corresponding to the De Broglie wavelength of the free charge carriers is approached, quantum confinement effects become dominant. One manifestation of such an effect is an increase in the optical band gap energy with decreasing particle size that is readily manifested as sharp changes in color visible to the naked eye. Depending on the particle size, the color of CdS colloids may vary from blue to red while PbS nanoparticles may appear pale yellow, orange, red or black.

Nanoparticles has been studied in micelles, [15-17] vesicles, [18,19] sol-gel glasses, [20,21] zeolites, [22,23] Langmuir-Blodgett (LB) films [24,25] and polymers. [26,27] In some cases the clusters have poorly defined surfaces and a broad distribution of particle sizes. Bawendi [28] has demonstrated control in preparing monodisperse CdSe clusters using a synthesis medium consisting of trioctylphosphine and its oxide. While size exclusion chromatography permits a narrow size distribution of particle sizes, only minute quantities of the materials are obtained which is unsuitable for any large-scale applications. On the contrary, the synthetic route used by Bawendi [28] permits the production of gram quantities of nanoparticulate solids. Chemical reactions initiated within the microscopic cavities of zeolites, glasses, polymers and micelles provide another process of preparing nanoparticles. The shape and sizes of the nanoparticles in these "microreactors" are largely controlled by the restricted geometry of the cavities in which nucleation and growth of these particles occur. Thermodynamic and entropic requirements also play a crucial role in determining the size of these clusters.

1.2. IRON OXIDE NANOPARTICLES

In the attempt to improve magnetic recording technology, great effort has been made in the past for obtaining high-density recording media. For this purpose, the production of nanoparticles, uniform, highly dispersible and oriented in a matrix are essential. Controlled coercivity (between 500 and 1500 Oe) and high saturation magnetization are important tasks. Increasing the coercivity to an applicable level has been fulfilled by doping or coating the maghemite particles with Co. [29] The production of nanoparticles has been achieved on the basis of a mimetic approach, i.e. with the utilization of an organic support that plays an essential role in the crystal nucleation and in its growth control.

Nguygen and Diaz reported a simple synthesis of bulk poly(pyrrole-N-propylsulfonate) polymer composites containing nanosized magnetite (γ-Fe_2O_3) particles. [30] Using $FeCl_3$, they polymerized pyrrole-N-propylsulfonate, resulting in a black polymeric powder. Using sulfonate groups as nucleation sites for the growth of nanoparticles was a new technique introduced by Ziolo *et al.* [31,32] when they reported a matrix-mediated synthesis of maghemite (γ-Fe_2O_3). The bulk magnetic material that they synthesized, a γ-Fe_2O_3/polymer nanocomposite, was optically active at room temperature. Using $FeCl_3$ and $FeCl_2$ they ion-exchanged sulfonated polystyrene cross-linked with divinylbenzene which yielded a polymer with iron cations ionically bonded to sulfonate groups. Washing with NaOH, then heating to 60 °C, while adding aqueous H_2O_2 solution, oxidized the iron to γ-Fe_2O_3. Ziolo et al. proposed that the sulfonate groups of the polymer matrix provide spatially located sites for the growth of γ-Fe_2O_3, while the void volume in the cross linked resin imposes a limit on the maximum size of the crystals grown, thereby minimizing aggregation of the iron oxide particles. Particles ranged in size from 5-10 nm, as determined by transmission electron microscopy (TEM).

A similar technique was used to prepare a superparamagnetic form of goethite in pores of sulfonated, highly cross-linked poly(divinylbenzene) micro spheres. [33] The synthesized iron oxide has a calculated magnetic susceptibility, which is about 3 orders magnitude larger than the bulk goethite. Chemically, the ferrous ion oxidation reaction that leads to the formation of goethite in the macro porous particles was identical to the conditions reported to the formation of γ-Fe_2O_3/polymer composites using a commercial ion exchange resin consisting of sulfonated, lightly cross-linked polystyrene. The use of polymers with different pore diameter and grade of cross linkage leads to the oxidation of ferrous chloride to different iron oxides. In this behavior it can be shown that organic matrices have a dramatically influence on the crystallization product.

1.3. BIOMINERALIZATION

In some bacteria the phenomena of biomineralization is observed. Biomineralization occurs naturally in many biological systems including bacteria. In presence of a supersaturated solution, minerals may form on the surface of a living organism. Formed structures are characterized by a high degree of regularity and supramolecular architectures are finely controlled. In the case of bacteria iron oxide nanoparticles are synthesized inside the intracellular space. Although biomineralization has been widely studied, the mechanisms of growth are not fully understood. The supramolecular organization of the organic support appears to play an essential role in the process; organized architectures (such as vesicles, micelles, polymeric networks) provide functionalized surfaces that act as templates and assist the interfacial molecular recognition. In many cases the organic matrix provides precise sites for oriented nucleation. Owing to the immense complexities of these biological systems, such processes are not well understood. Efforts to mimic the natural biomineralization processes have resulted in the development of different experimental strategies that induce the growth of organized crystallites of inorganic salts in membranes, [34] monolayers, [34] Langmuir-Blodgett films, [35] reverse micelles and vesicles. [36] While one standard approach involves the development of chemical methods for synthesizing inorganic solids of desired composition, structure and morphology, an alternative simple and unique approach involves in-situ nucleation of nanoparticles in highly organized supramolecular assemblies [34-36] and their subsequent controlled growth into highly organized crystallites. The controlled growth and the formation of crystallites and even metallic particles of definite morphology and dimensions are possible today. [36] These studies have not only paved the way for a better understanding of the crystallization processes in organized biological systems but have also attracted the attention of the physicist, chemist and materials scientist alike as the optical and electronic properties of these nano-dimensional systems are different from their bulk properties that make them extremely important in designing nanotechnology based optical and electronic devices of the future. [37,38] Recent studies have established that by fine tuning the size and local environment of ordered supramolecular assemblies like reverse micelles and vesicles, size dependent quantization/quantum confinement effects may be realized resulting in significant changes in the optical and electronic properties of these systems and the ability to fine tune them for desired applications make such systems extremely attractive. [37,38]

1.4. TYPES OF MAGNETIC IRON OXIDES

While there are many iron oxides and iron hydroxides there are only two ferrimagnetic iron oxides, namely bulk magnetite and bulk maghemite. Magnetite is ferrimagnetic with a Curie temperature of 858 K. Magnetite has the chemical formula Fe_3O_4 and is a spinel ferrite. The common chemical name is ferrous-ferric oxide and thus the formula for magnetite may also be written as $FeO\cdot Fe_2O_3$. Magnetite is one part wüstite (FeO) and one part hematite

(Fe$_2$O$_3$). Magnetite contains Fe^{+2} (Fe ^{II}) and Fe^{+3} ions (FE ^{III}). Thus there are two different oxidation states of the iron in one structure. The chemical IUPAC name is iron (II,III). The Curie temperature of magnetite is 858 K. Magnetite Fe3O4 has a cubic inverse spinel structure which consists of a cubic close packed array of oxide ions where the Fe+2 ions occupy half of the octahedral sites and the Fe^{+3} ions are split evenly across the remaining octahedral sites and the tetrahedral sites. The ferrimagnetism of Fe$_3$O$_4$ arises because the electron spins of the Fe^{+2} and Fe^{+3} ions in the octahedral sites are coupled and the spins of the Fe^{+3} ions in the tetrahedral sites are coupled but anti-parallel to the former. The net effect is that the magnetic contributions of both sets are not balanced and there is permanent magnetism. [39] Extensive studies have revealed that at least 16 different oxides, hydroxides and oxy-hydroxides of iron are possible and the final products depend on the pH, rate of oxidation, reaction conditions and local microenvironment in which the reaction occurs. [40]

Maghemite is an iron oxide that has the same structure as magnetite. Magnetite is a spinel ferrite and is also ferrimagnetic. It has the formula γ-Fe$_2$O$_3$. Oxidation of Fe$_3$O$_4$ can be used to produce the brown pigment which is maghemite, γ-Fe$_2$O$_3$. Additional oxidation gives a red pigment, hematite α-FeO$_3$, which is not ferrimagnetic.

Nanoparticles of magnetite and maghemite are superparamagnetic at room temperature. Superparamagnetic behavior of iron oxide nanoparticles can be attributed to the size of the nanoparticle. For sufficiently small size of the nanoparticle (<30 nm),the thermal energy is sufficient to change the direction of magnetization of the entire nanoparticle. Random fluctuations in the direction of magnetization cause the magnetic field to average to zero. The magnetic moment of the entire nanoparticle can align with an external magnetic field which is called superparamagnetism. [41] Superparamagnetism in many biomedical applications such as drug delivery is useful because the superparamagnetic iron oxide nanoparticles (SPIONs) can be transported by magnetic field effects to the desired site and once the external magnetic field is removed, magnetization disappears and the SPIONs can remains at the target site for a certain period. Nanoparticles of iron oxide are biocompatible, non-toxic, are surface active. Nanoparticles of iron oxides find wide use in biomedical applications. The nanoparticles can be used as contrast agents in magnetic resonance imaging, in labeling of diseased tissues, in magnetically controlled transport of pharmaceuticals, and in the preparation of ferrofluids. Magnetite and to some extent maghemite are mainly used in biomedicine because they are superparamagnetic, biocompatible and non-toxic. Iron oxide is degradable and therefore useful for in vivo applications.

1.5. SYNTHESIS OF SUPERPARAMAGNETIC IRON OXIDE NANOPARTICLES (SPIONS): CHEMICAL COPRECIPITATION

Several methods such as arc discharge, mechanical grinding, laser ablation, microemulsions and high temperature decomposition of organic precursors have been reported for the synthesis of Fe$_3$O$_4$ nanoparticles. [42] The *chemical coprecipitation method* is the simplest, least costly and the most common technique used to produce dispersed (water-based) Fe$_3$O$_4$ nanoparticles at low temperatures. If a suitable surfactant is employed and the processing parameters (pH, reaction temperature, stirring rate, solute concentration, etc) are controlled, the size, distribution and shape of the particles can be tailored. [43-46] Individual nanoparticles and nanoparticle agglomerates can be characterized by the magnetic core size and the hydrodynamic diameter. The parameters are important for targeting purposes. The first one is responsible for the magnetic response in applied inhomogeneous magnetic fields and the second parameter is important for targeting and cell interactions. [47] A coating layer can prevent the agglomeration of the particles, increase the circulation time, and provide biocompatibility. Monodispersed particles with a high saturation magnetization and functionalized with suitable coatings are required for targeting and imaging in hyperthermia, transfection and MRI (magnetic resonance imaging) applications.

Despite the pros and cons of using nanoscale iron oxides for *in vivo* applications, SPIONs remain the only magnetic nanoparticles that have been approved for clinical use to date. [48] Investigators seeking fast-track developments of magnetic-guided therapy often prefer this tried-and-tested option. One solution to the nanoparticles' weak magnetic responsiveness is to maximize the magnetic field at the target sites. [48] It has been recognized that the core size of iron oxide nanoparticles determines the magnetic properties; however, less work has been conducted to investigate the effect of hydrodynamic size on magnetic properties. [49] Recently, Mahmoudi et al. have reported the effect of both stirring rate and base molarities variations on the characteristics of superparamagnetic nanoparticles with the reaction temperature fixed at 35 °C. [50]

1.6. DESIGN OF EXPERIMENTS (DOE) FOR SYNTHESIS OF SPIONS

To reduce the large number of experiments due to a number of synthesis parameters, or factors, researchers often use a design of experiments (DOE) approach. [52,53] The application of DOE methods in optimization of nanomaterials processing is beneficial, particularly given the high cost of experimentation and complex random error structures. Different types of problems can be realized during DOE practice. [52,53] For the first type, one may be interested in a screening procedure in which a small number of factors (called 'main effects') are extracted from a larger pool of factors. For the second type, one may aim at finding a description of how factors affect the response (the input-output relation). Eventually, using such a relation the goal can be to optimize the response surface function. The third type is when the experiments are tuned to give an estimation of testing errors (i.e., the robustness of the solution is of interest rather than its optimality). The fourth type relies on obtaining a mathematical model for the input-output relation and also to estimate the typical size and structure of errors. Mahmoudi et al. have show such an application for SPIONs synthesis and the stabilization by nanoencapsulation using PVA. [50] Specifically, a uniform design (UD) of experiments, with a type-II objective (i.e., finding a description of how the input factors affect the output response and optimizing the response) was adapted to study the effects of stirring rate and chemical composition of the synthesis media on particle size and distribution, the magnetic properties as well as the morphology of the synthesized particles.

REFERENCES

[1] Nirmal, M., Brus, L., Luminescence Photophysics in Semiconductor Nanocrystals *Acc. Chem. Res.* 1999, *32*, 407.

[2] Alivisatos, A. P., Semiconductor Clusters, Nanocrystals, and Quantum Dots. *Science* 1996, 271, 933.

[3] Chan, W. C. W., Nie, S., Quantum Dot Bioconjugates for Ultrasensitive Nonisotopic Detection. *Science* 1998, *281*, 2016.

[4] Dujardin, E. and Mann, S., Bio-Inspired Materials Chemistry. Adv. Mater. 2002, 11, 775-788.

[5] Brus, L. E., Luminescence of direct and indirect gap quantum semiconductor crystallites. *MRS Symp. Ser.* 1992, 272, 215.

[6] Wang, Y. Nonlinear optical properties of nanometer-sized semiconductor clusters. *Acc. Chem. Res.* 1991, *24*, 133.

[7] Steigerwald, M. L., Brus, L. E. Semiconductor crystallites: a class of large molecules. *Acc. Chem. Res.* 1990, *23*, 183.

[8] Norris, D.J., Bawendi, M.G., Brus, L. E. Optical properties of semiconductor nanocrystals (quantum dots). In Molecular Electronics, Jortner, J., Ratner, M., eds., Blackwell Science, 1997.

[9] Weller, H. Quantisized semiconductor particles –A novel state of matter for Materials Science. *Adv. Mater.* 1993, *5*, 88.

[10] Koyama, H., Araki, M., Yamamoto, Y., Koshida, N., Visible photoluminescence of porous silicon and related optical properties. *Japan. J. Appl. Phys.* 1991, *30*, 3606.

[11] Dabbousi, B. O., Bawendi, M. G., Onitsuka, O., Rubner, M. F. Electroluminescence from CdSe quantum-dot/polymer composites. *Appl. Phys. Lett.* 1995, *66*, 1316.

[12] Nirmal, M.; Dabbousi, B. O., Bawendi, M. G., Macklin, J. J., Trautman, J. K., Harris, T. D., Brus, L. E. Fluorescence intermittency in single cadmium selenide nanocrystals. *Nature* 1996, *383*, 802.

[13] Cassagneau, T., Mallouk, T. E., Fendler, J. H. Layer-by-Layer Assembly of Thin Film Zener Diodes from Conducting Polymers and CdSe Nanoparticles. *J. Am. Chem. Soc.* 1998, *120*, 7848.

[14] *Semiconductor Nanoclusters-Physical, Chemical and Catalytic Aspects*, Kamat, P.V.: Meisel, D.; Eds. Elsevier, Amsterdam 1997.

[15] Pileni, M. P., *"Structure and Reactivity in Reverse Micelles,"* Elsevier, Amsterdam, 1989.

[16] Petit, C., Lixon, P., Pileni, M. P. Synthesis of cadmium sulfide in situ in reverse micelles. 2. Influence of the interface on the growth of the particles. *J. Phys. Chem.* 1990, *94*, 1598.

[17] Fendler, J. H., Atomic and molecular clusters in membrane mimetic chemistry. *Chem. Rev.* 1987, *87*, 877.

[18] Youn, H. C., Baral, S., Fendler, J. H. Dihexadecyl phosphate, vesicle-stabilized and in situ generated mixed cadmium sulfide and zinc sulfide semiconductor particles: preparation and utilization for photosensitized charge separation and hydrogen generation. *J. Phys. Chem.* 1998, *92*, 6320.

[19] Kennedy, M. T., Korgel, B. A., Montbouquette, H. G., Zasadinski, J. A. Cryo-Transmission Electron Microscopy Confirms Controlled Synthesis of Cadmium Sulfide Nanocrystals within Lecithin Vesicles. *Chem. Mater.* 1998, *10*, 2116.

[20] Rajh, T., Micic, O.I., Lawless, D., Serpone, N. Semiconductor photophysics. 7. Photoluminescence and picosecond charge carrier dynamics in cadmium sulfide quantum dots confined in a silicate glass. *J. Phys. Chem.* 1992, *96*, 4633.

[21] Minti, H., Eyal, M., Reisfeld, R.,M. Berkovic, G. *Chem. Phys. Lett.* 1991, *183*, 277.

[22] Wang, Y., Herron, N., Photoluminescence and relaxation dynamics of cadmium sulfide superclusters in zeolites. *J. Phys. Chem* 1988, *92*, 4988.

[23] Ozin, G. A., Steele, M. R., Holmes, A. J. Intrazeolite Topotactic MOCVD. 3-Dimensional Structure-Controlled Synthesis of II-VI Semiconductor Nanoclusters. *Chem. Mater.* 1994, *6*, 999.

[24] Yang, J., Meldrum, F. C., Fendler, J. H. Epitaxial Growth of Size-Quantized Cadmium Sulfide Crystals under Arachidic Acid Monolayers. *J. Phys. Chem.* 1995, *99*, 5500.

[25] Yang, J., Fendler, J. H. Morphology Control of PbS Nanocrystallites, Epitaxially Grown under Mixed Monolayers. *J. Phys. Chem.* 1995, *99*, 5505.

[26] Tassoni, R., Schrock, R. R., Synthesis of PbS Nanoclusters within Microphase-Separated Diblock Copolymer Films. *Chem. Mater.* 1994, *6*, 744.

[27] Kane, R. S.; Cohen, R. E.; Silby, R. J., Synthesis of PbS Nanoclusters within Block Copolymer Nanoreactors. *Chem. Mater.* 1996, *8*, 1919.

[28] Empedocles, S. A., Norris, D. J., Bawendi, M. G., Photoluminescence Spectroscopy of Single CdSe Nanocrystallite Quantum Dots. *Phys. Rev. Lett.* 1996, *77*, 3873.

[29] Koike, Y. Fine ceramics, 1985 Shinoku Saito, ed., Elsevier, Crown House.

[30] Nguyen, M.T.; Diaz, A.F. A novel method for the preparation of magnetic nanoparticles in a polypyrrole powder. *Advanced Materials* 1994, *6(11)*, 858-860.

[31] Ziolo, R.F., Giannelis, E.P. Weinstein, B.A., O'Horo, M.P., Ganguly, B.N., Mehrotra, V., Russell, M.W., Huffman, D.R. Matrix-mediated synthesis of nanocrystalline γ-ferric oxide: a new optically transparent magnetic material. *Science* 1992, *257*, 219-223.

[32] Vassiliou, J.K., Mehrotra, V., Russell, M.W., Giannelis, E.P., McMichael, R.D.,, Shull, R. D.; Ronald F. Ziolo Magnetic and optical properties of γ-Fe$_2$O$_3$ nanocrystals. *J. Appl. Phys.* 1993, *73(10)*, 5109-5115

[33] Winnik, F.M.; Morneau, A.; Ziolo, R.F.; Stoever, H.D.H.; Li, W.H. Template-Controlled Synthesis of Superparamagnetic Goethite within Macroporous Polymeric Microspheres. *Langmuir* 1995, *11*, 3660-3666

[34] (a) Furhop, J.-H.; Koning, J. *Membranes and Molecular Assemblies: The Synkinetic Approach*, The Royal Society of Chemistry, London, 1994. (b) Houslay, M.D.; Stanley, K. K. *Dynamics of Biological Membranes*, John-Wiley & Sons, N.Y. 1982. (c) Tien, H. T., *Bilayer Lipid Membranes. Theory and Practice*, Marcell and Dekker, New York, 1974. (d) Fendler, J. H., *Membrane Mimetic Chemistry*, Wiley-Interscience, N.Y., 1982. (e) Kotov, N.A.; Meldrum, F.C.; Wu, C.; Fendler, J. H. *J. Phys. Chem.* 1994, *98*, 2735. (f) Kotov, N.A.; Meldrum, F.C.; Wu, C.; Fendler, J. H. *J. Phys. Chem.* 1994, *98*, 8827. (g) Yi, K.C.; Sanchez-Mendita, V.; Lopez Castanares, R.; Meldrum, F. C.; Fendler, J. H. *J. Phys. Chem.* 1995, *99*, 9869.

[35] (a) Ulman, A. *An Introduction to Ultrathin Organic Films from Langmuir-Blodgett to Self –Assembly*, Academic Press, Boston, 1991.(B) Roberts, G. *Langmuir-Blodgett Films*, Plenum Press, 1990 (c) Gaines, G. L. *Insoluble Monolayers at Liquid-Gas Interfaces*, 1966, Interscience Publishing, N.Y.

[36] (a) Pileni, M. P.; Eds. *Structure and Reactivity in Reverse Micelles*, Elsevier: Amsterdam, 1989. (b) Pileni, M. P.; Brochett, P.; Zemb, T.; Milhaud, J. *Surfactants in Solution*, Mittal, K. L. Ed. P. Bothorel, Marcel-Dekker, N.Y. 1986. (c) Bawendi, M.G.; Steigerwald, M. L.; Brus, L. E. *Annu. Rev. Phys. Chem.* 1990, *41*, 477.(d) Pileni, M. P. in Handbook of Surface and Colloid Chemistry, Eds. Birdi, K. S.; CRC Press, N.Y. 1997. (e) Fendler, J. H.; Fendler, E. J.; *Catalysis in Micellar and Macromolecular Systems*, Academic Press, NY 1975. (f) Fendler, J.H.; Meldrum, F.C. *Adv. Mater.* 1995, 7, 607.

[37] (a) Birge, R. R. *Molecular and Biomolecular Electronics*, American Chemical Society, Washington, DC., 1994. (b) Molecular Electronics Carter, (c) *Materials and Measurements in Molecular Electronics*, Kajimura, K. and Kuroda, S. (Eds.), Springer Proceedings in Physics *81*, Springer-Verlag, Tokyo 1996. (d) Bosshard, C.; Sutter, K.; Pretre, P.; Hulliger, J.; Florsheimer, M.; Katz, P. *Organic Nonlinear Optical Materials*, Gordon and Breach, Basel, 1995. (e) Schieb. S.; Cava, M.P.; Baldwin, J.W.; Metzger, R.M. *Thin Solid Films* 1998, *327-329*, 100. (f) *Molecular Electronics and Molecular Devices*, ed. Seinicki, K.; CRC Press, NY, 1994

[38] (a) Guenther, L. Phys. World. 1990, *3*, 28. (b) Martin, C. R. *Acc. Chem. Res.* 1995, *28*, 61. (c) Ziolo, R. F.; Giannelis, E. P.; Weinstein, B. A.; O"Horo, M. P.; Ganguly, B. N.; Mehrotra, V.; Russel, M. W.; Huffmann, D. R. *Science* 1992, *257*, 219. (d) Ziolo, R. F.; Vassiliou, J. K.; Mehrotra, V; Russel, M. W.; McMicheal, R.D.; Shull, R. D. *J. Appl. Phys.* 1993, *73*, 5109. (e) McMicheal, R. P.; Schull, R. D.; Swartzendruber, L.; Bennet,L. H.; Watson, R. E. *J Magn. Magn. Mater.* 1990, *85*, 219. (f) Winnik, F. M.; Morneau, A.; Ziolo, R. F.; Stroever, H. D. H.; Li, H. *Langmuir* 1995, *11*, 3660.

[39] Greenwood, Norman N.; Earnshaw, A., Chemistry of the Elements (2nd ed.), 1997, Oxford: Butterworth-Heinemann.

[40] Cornell, R.M.; Schwertmann, U. *The Iron Oxides*, Structure, Properties, Reactions, Occurrence and Uses, 1996, VCH: New York.

[41] Buschow, K.H.G., Editor, Hand Book of Magnetic Materials, 2006, Elsevier Press.

[42] Sun, Y. K.; Ma, M.; Zhang, Y.; Gu, N. *Colloids Surf. A: Physicochem. Eng. Aspects.* 2004, *245*, 15–19.

[43] Cornell, R. M.; Schertmann, U. *Iron Oxides in the Laboratory: Preparation and Characterization*, Weinheim, VCH, 1991.

[44] Chen, S.; Feng, J.; Guo, X.; Hong, J.; Ding, W. Mater Lett. 2005, *59*, 985–988.

[45] Bae, D. S.; Han, K. S.; Cho, S.B.; Choi, S. H. Mater Lett. 1998, *37*, 255–258.

[46] Gupta, A.K.; Gupta, M. *Biomaterials* 2005, *26*, 3995–4021.

[47] Gupta, A. K.; Naregalkar, R. R.; Vaidya, V. D.; Gupta, M. *Nanomedicine* 2007, *2(1)*, 23-29.

[48] Gould, P. *Nanotoday* 2006, *1(4)*, 34-39.

[49] Jun, Y. W.; Huh, Y. M.; Choi, J. S.; Lee, J. H.; Song, H. T.; Kim, S. J.; Yoon, S.; Kim, K. S.; Shin, J. S.; Suh, J. S.; Cheon, J. *J. Am. Chem. Soc.* 2005, *127*, 5732–5733.

[50] Mahmoudi, M.; Simchi, A.; Imani, M.; Milani, A. S.; Stroeve, P. *J. Phys. Chem. B.*, 2008, *112(46)*, 14470-14481.
[51] Fowlkes, W. Y.; Creveling, C. M. Engineering Methods for Robust Product Design. *Prentice Hall: New York.* 1995.
[52] Robinson, G. K. Practical Strategies for Experimenting *Wiley: New York.* 2000.

Chapter 2

SYNTHESIS OF SPIONS

The SPIONs offer an exceptional potential for numerous biomedical applications including targeted drag delivery and imaging, gene therapy, biomolecules separations and early diagnosis of cancer. However, the specific particle sizes together with very narrow size distribution are required for desired applications, for instance hyperthermia which needs very uniform magnetic saturations [1-3], in order to have colloids with very uniform physical and chemical properties. In addition, it is now well recognized that nanoparticles of different sizes and distributions can be exposed to different fluid environments and behave differently, particularly in regards to their velocities as they move through narrow blood capillaries, hence, severe agglomeration may occur with nanoparticles of non-optimized size and distribution for desired sites and applications. [4] Furthermore, cellular uptake of nanoparticles firmly depends on their size, size distribution, shape and surface characteristics.[5, 6] In order to produce SPIONs, there are several synthesis methods; each of which have its advantages and disadvantages and will be suitable for specific bio-application purposes which are fully probed in this chapter after a brief description of SPIONs nucleation. The synthesis methods include co-precipitation[7], microemulsion[8], bio-mineralization[9], hydrothermal and high temperature reactions[10], sol-gel[11], polyol[12], flow injection[13], electrochemical[14], aerosol/vapor[15], sonochemical reactions[16], laser ablation[17, 18], and electro-spray synthesis.[19] Choosing a suitable experimental method should be based on the product reproducibility; desired physical, chemical and magnetic properties; and finally the scale-up capability for industry use.

2.1. NUCLEATION OF NANOPARTICLES

In order to achieve SPIONs with a very narrow size distribution, understanding the concepts of SPIONs nucleation and growth are necessary. The narrow size distribution can be reached via the homogenous precipitation in the aqueous medium, which involves the separation of the nucleation and growth of the nuclei.[20] This homogeneous reaction is occurred by the formation of supersaturated growth species.[21] Upon achievement of supersaturating, the solute concentration in solution surpasses the equilibrium concentration (i.e. $C > C_0$), hence, the new species will be synthesized due to the negative amount of the Gibbs free energy changes according to the following equation:[21]

$$\Delta G_v = \frac{kT}{\Omega} \ln\left(\frac{C}{C_0}\right) = -\frac{kT}{\Omega} \ln(1+\sigma) \qquad (2\text{-}1)$$

$$\sigma = \left(\frac{C-C_0}{C_0}\right) \qquad (2\text{-}2)$$

Where ΔG_v is changes in the Gibbs free energy per unit volume of the solid phase, k is the Boltzman constant, T is the temperature, Ω is the atomic volume, C is the concentration of the solute, C_0 is the equilibrium concentration, and σ is the degree of supersaturation. The crucial problem to achieve a very narrow size distribution of nanoparticles is the short burst of nucleation followed by slow controlled growth in the solution.[22]

In this regard, the critical particle size will be recognized, according to the variation of Gibbs free energy, which is called embryo; the stable nanoparticles (*nuclei*) have the bigger size than embryo. In contrast, the nanoparticles with a lower size will be dissolved in the solution. In order to synthesize SPIONs with very narrow size distribution, the nucleation and growth stages must be separated and nucleation stages should be inhibited during growth stage. In order to control the separation of the mentioned two stages together with the mass transfer, the turbulent flow should be created by an ultrasonic generator (ultrasonic probe or bath) and changing the homogenization rates of the solution. [5] The first explanation of these mechanism were claimed by LaMer and Dinegar.[23] The reduction of Gibbs free energy is recognized as a driving force of the both nucleation and growth stages and can be estimated via the following equations:

$$\Delta G = \Delta G_v + \Delta G_\gamma \qquad (2\text{-}3)$$

$$\Delta G_v = A_1 r^3 \qquad (2\text{-}4)$$

$$\Delta G_\gamma = A_2 r^2 \qquad (2\text{-}5)$$

Where ΔG is the overall free energy changes; ΔG_v and ΔG_γ are the free energy changes associated with the volume and surfaces changes of the nucleus with a radius of r, respectively; finally, A_1 and A_2 are model constants. By assuming the spherical shape for the nucleus following equations are achieved:

$$\Delta \mu_v = \frac{4}{3}\pi r^3 \Delta G_v \qquad (2\text{-}6)$$

$$\Delta \mu_s = 4\pi r^2 \gamma \qquad (2\text{-}7)$$

Where $\Delta \mu_v$ and $\Delta \mu_s$ are the changes in volume chemical potential and surface energy, respectively, and γ is the surface energy per unit area. Hence:

$$\Delta G = \Delta\mu_v + \Delta\mu_s = \frac{4}{3}\pi r^3 \Delta G_v + 4\pi r^2 \gamma \tag{2-8}$$

Figure 2.1 represents a schematic diagram of the Gibbs free energy versus the radius of nuclei, according to the equation 2-8. From the **Figure 2.1**, it can be judged between the importance of volume energy and surface free energy according to the resultant curve. If the radius is smaller than r^*, the surface free energy is dominant energy and if the radius is bigger than r^*, it is the volume energy that will be most important. The radius of embryo, r^*, is calculated according to the equation 2-8 when the derivative function of total Gibbs free energy is equal to zero: [24]

$$4\pi r^{*2}\Delta G_v + 8\pi r^* \gamma = 0 \tag{2-9}$$

Hence,

$$r^* = -2\frac{\gamma}{\Delta G_v} \tag{2-10}$$

In addition, the critical energy (ΔG_c) which is necessitated for the formation of stable nanoparticles can be calculated by inserting the critical radius in equation 2-8 as follow.

$$\Delta G_c = \frac{16\pi\gamma^3}{3\Delta G_v^2} \tag{2-11}$$

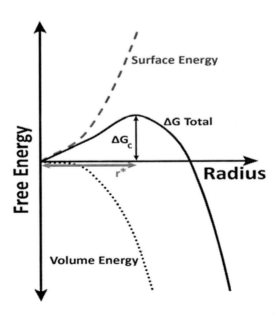

Figure 2.1. Schematic diagram of the Gibbs free energy versus the radius of the nuclei (adapted from [22]).

Finally, the rate of nucleation is:

$$P = \exp\left(-\frac{\Delta G_c}{kT}\right) \tag{2-12}$$

Where P is the probability of the thermodynamic fluctuation of critical energy.

2.2. SYNTHESIS METHODS

As mentioned before, controlling the size, morphology and their magnetic properties are the key factors for SPIONs synthesis and highly dependent on the synthesis methods. Synthesis of SPIONs has been reported by three major routes including chemical, physical, and biological methods.

2.2.1. Chemical Methods

The chemical methods consist of co-precipitation, microemulsions, hydrothermal, solvent and solvent free thermal decomposition, sol-gel, polyol, sonochemical, and electrochemical deposition under oxidized conditions.[25]

2.2.1.1. Co-precipitation
The distinguished features of the co-precipitation method in comparison to other available methods are its facile, efficient chemical pathway and convenient method to achieve both type of iron oxide nanoparticles including magnetite (Fe_3O_4) and maghemite (γ-Fe_2O_3) through the forced hydrolysis of iron salt (i.e. Fe^{2+} and Fe^{3+}) solutions (e.g. in DI water) in the presence of the base.[26, 27] From the thermodynamics point of view, the pH of the base should be in the range of 9-14 in addition to maintaining the molar ratio of iron salts, which should be adjusted to 2:1, in the neutral environment (e.g. nitrogen of argon gas) for optimum magnetite production.[28] The precipitation is achieved via dropwise addition of iron salt solutions to the base solutions or vice versa.[7, 29, 30] It is noteworthy that the SPIONs nucleation will be occurred when the concentration of iron ions exceeds their solubility in solution.[21] The chemical reaction during the synthesis of magnetite nanoparticles can be formulated as follows.[31, 32]

$$Fe^{2+} + 2Fe^{3+} + 8OH^- \rightarrow Fe_3O_4 + 4H_2O \qquad (2\text{-}13)$$

The obtained magnetite has a so-called inverse spinel crystal structure which its layer sequence structure is shown in Figure 2.2. It is noteworthy to mention that the cubic unit cell corresponds to $Fe_{24}O_{32}$; in addition, there are 32 octahedral and 64 tetrahedral holes in the structure. Fe^{3+} ions are placed in both tetrahedral and octahedral positions (1/8 and ¼ of the tetrahedral and octahedral sites, respectively) whereas the Fe^{2+} ions could be just placed on ¼ of the octahedral holes (see Figure 2.2).

The physical and chemical characteristics of obtained magnetite including size, size distribution, shape, magnetic saturation, and composition are significantly dependent on the synthesis parameters such as the type of salts (e.g. chlorides, sulphates, nitrates, and perchlorates), iron salts ratio (Fe^{3+}:Fe^{2+}), type of the base (e.g. Na(OH), NH$_4$(OH) and K(OH)), pH value, ionic strength of the media, homogenization rate, and temperature of the medium.[7, 33-39] It is noteworthy that in the presence of oxygen, magnetite is oxidized to γ-Fe$_2$O$_3$ and/or Fe(OH)$_3$ via the following equations:

$$Fe_3O_4 + 2H^+ \rightarrow \gamma\text{-}Fe_2O_3 + Fe^{2+} + H_2O \quad (2\text{-}14)$$

$$Fe_3O_4 + O_2 + H_2O \rightarrow 3Fe(OH)_3 \quad (2\text{-}15)$$

Beside the synthesis parameters, the maintenance condition of magnetite nanoparticles could have an effect on their both physical and chemical characteristics. For instance, the temperature of storage should be 4°C in order to preserve the colloidal stability up to six months, otherwise, severe agglomeration can be occurred. Interestingly, the pH of the colloid can play a role in transferring electrons/ions, according to equation 2-14. For instance, at pH of less than seven, the surface Fe^{2+} ions is desorbed as hexa-aqua complexes in solution, while at a pH of more than seven, the oxidation of magnetite engages the oxidation-reduction of the surface of magnetite via the migration of ferrous cations through the lattice structure and creating cationic vacancies in order to preserve the charge balance.[40] The formation of these vacancies is fully casual and their ordering are occurs only for nanoparticles with the size more than 5 nm.[40] It is notable that there are various methods to define the size of the obtained SPIONs including transmission electron microscope (TEM), X-ray and dynamic light scattering (DLS). Amoung these methods, X-ray is preferred due to its capability to calculate the average cristallite sizes. The average size of the nanoparticles is calculated from the broadening of the XRD peaks using the Scherrer formula: $d = 0.9\lambda/(w - w_1)\cos(\theta)$, where d is the crystal diameter, w and w_1 are the half-intensity width of the relevant diffraction peak and instrument broadening, respectively, λ is the X-ray wavelength, and θ is the angle of diffraction.

Although, from a long time ago the scientists have known that magnetite could be produced by the mixture of base medium and iron salts (i.e. ferrous and ferric) [41], but the first controlled co-precipitation method for production of SPIONs (magnetite with core size of 12 nm according to Scherrer's equation) was done by Massart. Massart[42] used the bare SPIONs and the severe agglomeration had been observed; in order to prohibit the agglomeration and increase the colloidal stability, nowadays the organic and/or inorganic surfactants/coatings (see Chapter 3) are used.[43-52] In addition, these surfactants can retard the crystal growth by preventing iron ion diffusion to the surfaces of nanoparticles. From that time, enormous investigations were conducted on the co-precipitation method and are still in progress in order to overcome its major disadvantage which is the size polydispersity[53, 54]. This large distribution in size could cause the non-suitable magnetic properties (e.g. blocking temperature, saturation magnetization and surface spin canting effect) and biological responses (very low blood residency time).[40, 55] The main advantage of this technique is the large amount of synthesized SPIONs. Similar to the previous indications, the physical and chemical characteristics including size, size distribution, shape, magnetic saturation, and purity (i.e. composition) of the SPIONs synthesized via co-precipitation are affected by the synthesis process parameters. Hence, several attempts on the synthesis parameters have been made to reach products with very narrow size distributions.[30, 45, 53, 56-59] Synthesis factors have proved to have considerable influence on the average size and phase of the synthesized SPIONs. They include the iron salts ratio (i.e. Fe^{2+}/Fe^{3+})[60], iron concentration in the primary media[61], acidity and the ionic strength of the final media[59, 62], reaction temperature[6, 63], injection flux rates[60], homogenization rates[7, 64], bubbling inert gas thorough the solution[7, 46, 65], etc. Once appropriate synthesis conditions are chosen and used, re-producible SPIONs with narrow size distribution is a secondary but important goal.

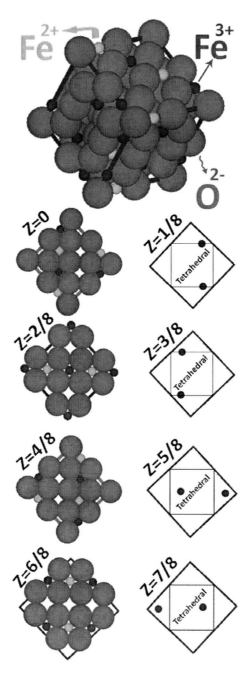

Figure 2.2. 3-D Inverse spinel crystal structure of magnetite and its layer sequence structures according to the Z direction.

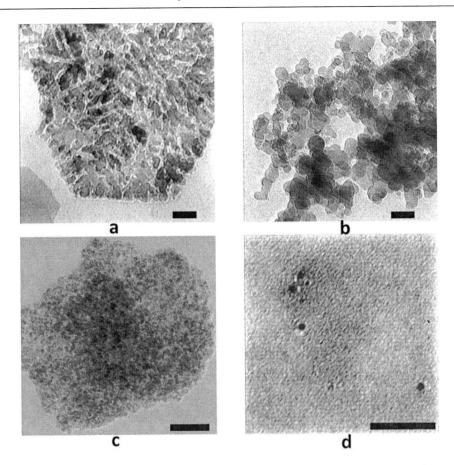

Figure 2.3. SPIONs synthesized via non-optimized co-precipitation method (a) without surfactant and (b) with surfactant. (c) Formation of magnetic beads in optimized synthesis condition (no optimization of polymer amount) and (d) mono-dispersed SPIONs obtained from fully optimized condition. (Scale bar is 50 nm)

Another vital parameter which has proved to have an effect on the size and colloidal dispersity of SPIONs is the amount of the surfactant material. The magnetic saturation of the obtained SPIONs is also related to the coating amount. For instance, it is found that at a fixed stirring rate of the solution, the critical mass of PVA (poly(vinyl alcohol)) should be around 2 times bigger that iron mass in order to maintain a maximum level of magnetic saturation, monodispersity and permeability of SPIONs.[6] Below this amount, magnetic saturation is not significantly affected by the coating agent. Conversely, both the magnetic saturation and permeability are decreased above the mentioned polymer mass ratio due to the formation of magnetic beads (the gel formation properties of PVA become dominant and a nanocomposite of nanoparticles is formed). It is remarkable that the magnetic saturation values of SPIONs are experimentally determined to be in the range of 30-60 emu/g, which is lower than that of the bulk phase, 88 emu/g.[62, 66, 67] Figure 2.3 shows the effect of different synthesis parameter on the characteristics of the obtained nanoparticles.

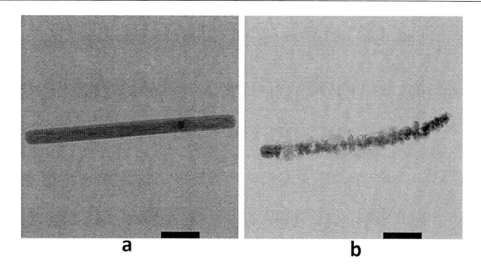

Figure 2.4. TEM micrographs of nanorods for (a) before exposure and (b) after exposure to an electron beam.

Another synthesis parameter that is able to change the shape of the final product is the temperature. For instance, by adjusting the synthesis temperature, the SPIONs nanorods (in the presence of PVA) with tunable sizes which are useful for magnetic sensors, targetted therapy or other applications, can be obtained by co-precipitation.[68] The temperature can also have a significant effect on the polymer molecules shape (i.e. folded or stranded) on the aqueous medium. Although the precise configuration of the polymer chains within the crystalline regions is unresolved at present, the fiber-periodicities suggested a twisted or helical structure in low temperatures and a stranded chain-shape at higher temperatures.[68] By increasing the medium temperature, the SPIONs is arranged on the stranded PVA molecules, hence the nanorods is formed (see Figure 2.4). Figure 2.4(b) reveals multiple magnetic beads, inside the rod, with the magnetic beads wrapped in thin PVA layers adhering to one another in bundles. It is notable that Figure 2.4(b) was obtained after exposure of the rod (Figure 2.4(a)) to the electron beam which caused the degradation of PVA strans on the wall.

2.2.1.2. Microemulsions

Hoar and Schulman[69] reported the oleopathic hydro-micelles for the first time which were subsequently classified as microemulsions[70]. Nowadays, a microemulsion is recognized as a thermodynamically stable transparent isotropic solution of two immiscible liquids, where the micro-domain of either or both liquids is stabilized by the spontaneous self-assembly of the hydrophobic or hydrophilic parts of surfactant molecules.[71, 72] The most important surfactants are anionic (e.g. AOT, and sodium dodecylsulfate (SDS))[73-75], non-anionic (e.g. polyethoxylates, and Lutensol AT50)[43, 47] and cationic (e.g. cetyltrimethyl-ammonium bromide (CTAB), and cetyltrimethylammonium (CTMA))[76-79]. Commonly, microemulsion is the co-precipitation method in a microcavities reactors (restricted space of about 10 nm) which provide a confinement effect that controls nanoparticles sizes and inhibit their aggregations.[80] By introducing iron salts and base medium emulsions, the micro-plates undergo continuously collide, coalesce and break again; resulting the precipitate form in micelles.[81, 82] In this regard, very narrow size SPIONs with the size of 4 nm were prepared by co-precipitation of iron salts with base medium (i.e. ammonium hydroxide) within nano-reactors which were prepared by employing AOT (sodium bis(2-ethylhexylsulfosuccinate)) as surfactant and heptane as continues oil phase.[83] It is notable that the four acceptable theories including the mixed film[84], the solubilization[85], Landau-Ginzburg[86] and the thermodynamic[87] could describe the microemulsion formation. All of these theories confirmed that the crucial matter in the formation of microemulsion is reducing the interfacial total free energy. Comprehensive explanations of these theories are available elsewhere.[70, 84-87] The water-in-oil (W/O) or oil-in-water (O/W) microemulsions, which depend on dispersed and continues phases[88], are extensively employed in order to synthesis nanoparticles with uniform sizes[89-92] including magnetic materials (such as $MnFe_2O_4$, $CoFe_2O_4$, $ZnFe_2O_4$, $MgFe_2O_4$, $CdFe_2O_4$, and $CuFe_2O_4$)[8, 48, 50, 51, 71, 75, 81, 92-99] and more specifically SPIONs[50, 65, 73, 98, 100]. Microemulsions have been employed for numerous applications over a broad range of fields, consisting pharmaceutical, cosmetics, oil recovery, biological membranes (as a model), reaction media, electrochromatography, and food sciences.[101] Beside low scale-up problem which can be important from a technological point of view, another major disadvantage of microemulsion method for preparing SPIONs for biomedical applications is the type of surfactants which are used to facilitate microemulsion formation. Many of the employed surfactants are toxic and needs time consuming and expensive steps for their removal.

In W/O reverse micelles, the aqueous phase is dispersed as nanodrops (typically between 5-50 nm in diameter[102]) encircled by a monolayer of surfactant molecules (10-100 nm)[103] in a continues hydrocarbon phase and vice versa for O/W.[48] Due to their ease of microemulsion formation and purity, the hydrocarbon mineral oils have been employed in the majority microemulsion studies.[104] Other oils have shown difficulties; such as triglycerides which have caused a limitation to penetrate into interfacial film to assist the formation of an optimal curvature. This would be due to the high molecular weight of such oils.[105] It is notable that the first SPIONs which were synthesized in reverse micelles were produced by oxidation of Fe^{2+} salts to maghemite and magnetite.[106] The size of the reverse micelle can have significant effect on the size of obtained SPIONs. The sizes of the micelles and consequently the size and shape of the final product are varied by changing the water-to-surfactant molar ratio, temperature, concentration of both iron salts and base, and somehow by stirring rate;[107, 108] Among these factors, molar ratio of water to surfactant is shown to be a prominent factor in order to determine the diameter of the reverse micelle.[97] In addition to the controlled sizes with a very narrow size distribution, microemulsion method has showed a suitable capability to control the shapes (e.g. nanorods[11] and nanotubes[109]) of obtained nanoparticles.[51, 94, 99] For instance, the formation of iron oxide nanorods by employing oleic acid and benzyl ether has been reported through a sol–gel reaction in reverse micelles.[11]

Finally, the microemulsion seems to be a very suitable method to synthesize the core-shell (SPIONs as core and a protective organic or inorganic shell) structures.[110-112] For instance, silica coated SPIONs (with the core size of 1-2 nm) with uniform size were synthesized by employing W/O microemulsion mediated sonochemical synthesis of SPIONs in nonionic surfactants using two different inorganic bases.[50]

2.2.1.3. Hydrothermal

Hydrothermal method is another promising method to synthesis SPIONs with specific size and morphology.[113-115]. The method employs both elevated pressure (higher than 2000 psi) and temperature (above 200°C) conditions to hydrolyze and dehydrate iron salts, it also benefits from the very low solubility of the resulting iron oxides in water under these conditions to arrive at a supersaturation state.[113] During hydrothermal synthesis, the reaction conditions including solvent and its concentration and temperature together with the reaction time can have crucial effects on performance of the synthesized SPIONs.[10, 116-122] For instance, the sizes of SPIONs were increased by enhancing the reaction temperature; in addition, various phases (e.g. 6-line ferrihydrite and hematite at 250-350°C and pure hematite at >350°C) of SPIONs were synthesized from the same precursor.[123] The generalized hydrothermal method has been used successfully by *wang et al.* [124] for preparing different nano-crystals (e.g. Fe_3O_4, TiO_2, $CoFe_2O_4$ and $BaTiO_3$) via liquid (i.e. ethanol-linoleic acid)-solid (i.e. metal linoleate)-solution (e.g. water-ethanol) reaction according to the phase transfer and separation mechanism which were happening between various interfaces. *Zheng et al.*[125] employed hydrothermal method in order to synthesize ferromagnetic magnetite with a diameter of 27 nm in the presence AOT. The bare[117] and PVA-coated[115] α-Fe_2O_3 were synthesized with narrow size distribution using this method. Many other researchers employed the hydrothermal method for monodisperse SPIONs preparation[10, 118, 119, 121, 123, 126-129]; the results confirm that the optimization of synthesis parameters (e.g. pressure, temperature, reaction time, and type of solvent) is essential for obtaining SPIONs with desired, reproducible and predictable characteristics. *Matson et al.*[130-133] prepared fine iron oxide nanoparticles in a continuous flow reactor, which could rapidly heat the flow solutions by contact with supercritical water in a period of 5-30 s, by hydrothermal method.

A continuous hydrothermal method for preparation of uniform PVA coated SPIONs were proposed by *Teja et al.*[134]. According to the obtained results, both concentration of the PVA and the residue time have an effect on the particle size. More specifically, by increasing the concentration of PVA during low residue time (about 2s), the particle size was decreased, whereas by increasing the residue time to 10s or more, the size was not changed by the variation of PVA concentration.[135]

2.2.1.4. Thermal Decomposition

The decomposition of various iron precursors (e.g. $Fe(Cup)_3$ (Cup: N-nitrosophenylhydroxylamine), $Fe(CO)_5$, or $Fe(acetylacetonate)_3$) [133, 135, 136] in the presence of hot organic surfactants and solvents (e.g. oleylamine[63, 136], fatty acids[137], oleic acid[138-142], hexadecylamine[143], and steric acid[144], and octyl ether[145]) is recognized as high temperature/thermal decomposition method and is used to synthesize SPIONs with excellent size control (e.g. 3, 4, 10, 12, 13, and 20 nm)[145, 146], very narrow size distribution, uniformity and high crystallinity.[82, 147] For instance, the monodispersed maghemite in organic solvents was prepared by injecting the solution of $Fe(Cup)_3$ in octylamine into long chain amine at high temperature (i.e. 250-300°C).[148] Similar results obtained by employing $Fe(CO)_5$ as iron precursor followed by oxidation (with mild oxidant such as trimethylamine $(CH_3)_3NO$).[128] The physical and chemical characteristics of prepared SPIONs by thermal reactions were highly dependent to the reaction times, temperature, both concentration and ratios of the reactants, nature of the solvent and precursors, complexing strength, type of surfactants, boiling temperature of solvent and addition of seeds.[149] It is notable that seed-mediated growth method is advantageous since it is a noninjection or heating-up method with an easily controllable growth process[136], in addition the size of the synthesized nanoparticles is increased through seeded growth (up to 20 nm) and also the distribution of size progressively becomes narrow with time via the intraparticle ripening process together with Oswald ripening process.[63, 146, 150]

According to the literature, there are not numerous reports on detailed studies of the growth process of SPIONs, whereas there are a number of papers on the synthesis route.[151] The comprehensive report on the detailed description of the growth of different SPIONs (with core size of 5 to 10 nm by varying the experimental parameters), which were prepared via seed-mediated method in combination with thermal decomposition growth, was reported by *Huang et al.*[150]. They used two steps consisting the formation of monodispersed SPIONs as seeds and the growth of seeds solution in the presence of both surfactants and precursor under refluxing temperature (i.e. 265°C) for different time intervals (i.e. 30-90 min); the growth process were probed by TEM images (Figure 2.5).[150] The as-synthesized seed nanoparticles showed the average size of 5.4 nm with a narrow size distribution. After 30 min of refluxing, the size distribution became broad; however, the majority of SPIONs still possessed the average size of 5.4 nm. The growth was prohibited after 90 min due to complete consumption of monomers where the 8 nm SPIONs with a very narrow size distribution obtained. There are three stages involved in the formation of the final SPIONs consisting the primary (red balls) and secondary (green balls) nucleation of monomer (Stage I; during the first 30 min), the seed-mediated growth (Stage II), and the intraparticle ripening process (Stage III).

Roca et al.[138] synthesized magnetite with different sizes (5.7-11 nm) with high saturation magnetization (65-72 emu/g) and high initial susceptibility (260-1000 emu/T) values via thermal decomposition method[138]; the high stauration magnetization and susceptibility are required for targeting purposes.[55]

As mentioned before, many biomedical applications of SPIONs such as hyperthermia, magnetic resonance imaging and magnetic cell/biomolecules separation strongly demand a specific particle size; hence, the SPIONs which are prepared by this method could potentially satisfy these bio-applications. However, similar to microemulsion the removal processes of the residual surfactants are time consuming, expensive and cause SPIONs to aggregate.[152, 153] *Park et al.*[151] designed ultra-large-scale synthesis (40g in a single reaction) of monodisperse SPIONs using inexpensive and non-toxic metal salts as reactants, without a size-sorting process (see Figure 2.6). According to their results, the particle size could be controlled simply by varying the experimental conditions.

The effect of oleic acid concentrations on the synthesized SPIONs was explored by *Teng et al.*[154]. According to their results, when the oleic acid:Fe(CO)$_5$ molar ratio was less than 2, only polydispersed nanoparticles were synthesized (Figures 2.7 (c) and (d)) due to inadequate concentration of oleic acid, which makes it ineffective for stabilizing the SPIONs. In contrast, the monodispersed SPIONs were achieved by increasing the oleic acid: Fe(CO)$_5$ molar ratio to 3 or more(Figures 2.7 (a) and (b)). The authors claimed that the oleic acid:Fe(CO)$_5$ molar ratio of 3 is the lower limit for making monodisperse SPIONs.

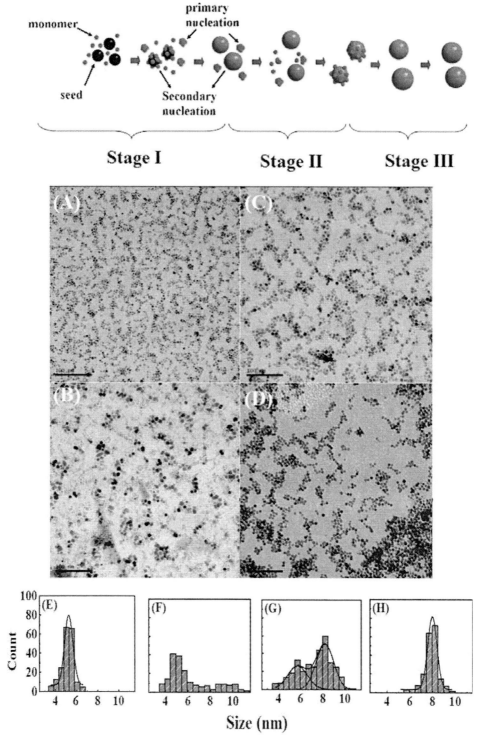

Figure 2.5. TEM bright field images and schematic illustrations of growth particles after (A) 0, (B) 30, (C) 60, and (D) 90 min, and their histograms of size distributions are given in panels E, F, G, and H, respectively. With permission from ref. [150].

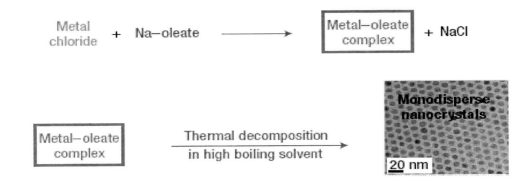

Figure 2.6. The overall scheme for the ultra-large-scale synthesis of monodisperse nanocrystals. Metal–oleate precursors were prepared from the reaction of metal chlorides and sodium oleate. The thermal decomposition of the metal–oleate precursors in high boiling solvent produced monodisperse nanocrystals. With permission from ref. [151].

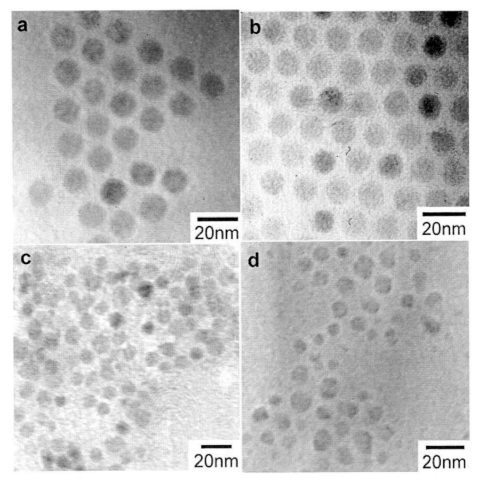

Figure 2.7. Effects of oleic acid:Fe(CO)$_5$ molar ratio of (a) 4, (b) 3, (c) 2 and (d) 1 on the SPIONs sizes and their distributions. With permission from ref. [154].

2.2.1.5. Sol-gel

Another suitable wet-method for metal oxide preparation is sol-gel, which is based on the hydroxylation and condensation of molecular precursors in solution, originating a "sol" of nanoparticles; in addition, the three dimensional metal oxide nanoparticles can be synthesized by further condensation.[155, 156] Using this method, monodisperse nanoparticles with predictable shape are obtained. It is notable that the heat treatment procedure is necessary to obtain a suitable crystallinity.[157] The physical and chemical characteristics of the prepared SPIONs are influenced by the solvent, temperature, nature, concentration and the type of precursors, pH and agitation.[27, 158] Maghemite and maghemite/silica hollow spheres were prepared by repeatedly coating polystyrene lattices with both Fe^{3+} alternately adsorbed with Nafion and combining a sol-gel process based on the hydrolysis of tetraethoxysilane, and subsequently removing the cores upon pyrolysis.[159] *Ennas et al.*[158] probed the effect of temperature on the characteristics of maghemite/silica composite. They used tetraethoxysilane (TEOS) and iron(III) nitrate as starting materials. In a low temperature regime, the amorphous superparamagnetic iron(III) oxide nanoparticles (3-4 nm in diameter) were detected. A small increase in the size of nanoparticles was found by enhancing the temperature (i.e. >700°C), also the nanoparticles were crystalline (γ-Fe_2O_3). Furthermore, the Mössbauer spectra confirmed the large change in the magnetization; magnetic measurements showed the transition from antiferro- to ferrimagnetic behavior. Further increase in temperature gave rise to the formation of α-Fe_2O_3.[158]

Similar results were obtained by using TEOS and $Fe(OC_2H_5)_3$ in alcoholic solution as gelling precursors;[160] the obtained gels were dried and heated in the temperature range of 40-1000°C. The characteristics techniques confirmed the formation of amorphous patterns for the samples which were prepared below 700°C, while heating to higher temperature yielded the formation of α-Fe_2O_3. There are, also, several other researchers which report the synthesis of maghemite/silica nanoparticles via sol-gel method.[161-164] In addition, several groups have reported the preparation of maghemite/aluminum composites.[165-167] *Duraes et al.*[155] reported the formation of the mixture of iron oxide/hydroxide phases via sol-gel method. The existence of hematite (α-Fe_2O_3) and goethite (α-$FeO(OH)$) were confirmed by different characteristics methods including X-ray diffraction, Fourier transform infrared (FTIR) and Mössbauer parameters. The authors also observed the formation of iron hydroxide $Fe(OH)_3$.

Ismail et al.[168] embedded the Y_2O_3 into Fe_2O_3/ TiO_2 mixed oxides by sol-gel method to be used in the photocatalytic property. Since the X-ray diffraction data did not show the presence of the Fe_2O_3 nanoparticles, the authors suggested that Fe^{3+} entered the TiO_2 lattice substitutionally. In addition, FTIR confirmed that the absorption of both Y_2O_3 and Fe_2O_3-doped TiO_2 was similar to pure TiO_2, which indicated that Y_2O_3 and Fe_2O_3, was dissolved into the TiO_2 lattices. Interestingly, either acid or base catalysts were playing an important role in controlling the morphology and the size of mixed oxides.

2.2.1.6. Polyol

Another versatile chemical route for synthesis of the very stable magnetic and monosized nanoparticles with high crystallinity and magnetization is Polyol which refers to the use of polyols (e.g. ethylene glycol, diethylene glycol).[12, 169-174] The main role of polyols is the reduction of the ration of metal salts to metal. In order to have a good potency to dissolve the inorganic compounds, high boiling solvents is used.[12] The prepared nanoparticle showed facile dispersion in either aqueous method or polar solvents because of the formation of hydrophilic polyol ligands in situ on the surface of magnetite nanoparticles.[12] The major advantages of polyol is its kinetically controllable of experimental conditions at both laboratory and manufacturing scales.[12] In addition, according to the existence of polyol medium, the nanoparticles were isolated from the oxidizing atmosphere.[171]

Cai et al.[12] modified the polyol process for preparation of SPIONs where single iron rich precursor was only employed. Since neither reducing agent nor surfactants were used, there were no difficulties for scale-up from the laboratory to manufacturing. Liu et al.[169] obtained monosize sub-5nm polyvinylpyrrolidone-coated SPIONs by polyol synthesis. Their results confirmed the magnetic nanoparticles had high crystallinity with distinct lattices and superparamagnetic behavior at room temperature. Pt/Fe$_3$O$_4$ core-shell nanoparticles have been synthesized by polyol.[170] In the first step, Pt nanoparticles were obtained the reduction of Pt(acetylacetonate)$_2$ by polyethylene glycol, consequently layers of iron oxide were deposited on the surface of Pt nanoparticles via the thermal decomposition of Fe(acetylacetonate)$_3$.

2.2.1.7. Sonochemical

Sonochemical process arises from acoustic cavitation, the formation, growth, and implosive collapse of bubbles in an aqueous media.[175] The localized hot spots (transient temperatures of ~5000 K, pressure of ~1800 atm, and cooling rate of ~10 [10] K/s) are created, due to the collapse of bubbles[176, 177], where the volatilization of the organometallic compounds/volatile solvents decompose inside the cavitating bubble to yield individual metal atoms; hence the agglomeration of prepared atoms have the capability to produce highly porous nanostructured materials such as amorphous metals and alloys.[178-180]

Sonochemical decomposition rates for volatile organometallic compounds/ volatile solvents depend on a variety of experimental parameters including vapor pressure of precursors, solvent vapor pressure, and ambient gas.[181] In order to enhance the yields of the sonochemical route, the volatile ability of the precursors should be increased (since the primary sonochemical reaction site is the vapor inside the cavitating bubbles).[182] Furthermore, the solvent vapor pressure should be low at the sonication temperature, because significant solvent vapor inside the bubble reduces the bubble collapse efficiency.[175]

The major advantage of sonochemical method is its capability to achieve atomic level mixing of the constituent ions in the amorphous phase so that the crystalline phase can be obtained by annealing at relatively low temperatures.[183] It is noteworthy that the cavitation is a quenching process, so the composition of the synthesized materials is identical to the composition of the vapor in the bubbles, without phase separation.[183]

The results of a sonochemical process involving an inorganic/organometallic *volatile* solute showed consistent formation of amorphous nanoparticles.[184] In the majority of the synthesis methods, the coating materials were accomplished *ex situ* (i.e. the SPIONs were synthesized first and then subjected to the coating solution under agitation).[57, 185-192] *Abu Mukh-Qasem et al.* [184] prepared a stable aqueous dispersion of magnetite nanoparticles by employing high intensity ultrasound in a one-step process. This method was based on the decomposition of a volatile organometallic precursor, Fe(CO)$_5$, in the presence of sodium dodecyl sulfate (SDS) as the stabilizer. According to the another report, SPIONs with high magnetization coated with oleic acid (83 emu/g) and crystallinity were prepared by using a sonochemical method.[193]

The obtained semi-amorphous SPIONs have been served as substrates for the self-assembly of thiols[194], alcohols[195], carboxylic acids[196], SDS and octadecyl-trichlorosilane[197].

Sonochemical decomposition of Fe(CO)$_5$ was carried out for maghemite preparation (dimension of 5-16 nm) in the presence of various surfactants including undecenoate, dodecyl sulfonate, and octyl phosphonate.[183] The ionic bindings of the surfactants to the nanoparticle surfaces were explored using FTIR spectroscopy. According to the results obtained by the electron paramagnetic resonance measurements, magnetization curves, zero-field cooled, and field cooled studies, the phosphonate-coated SPIONs behaved in a noticeably different manner from the other coated SPIONs; the reason is that the existence of extra negative charge on the phosphonate, as compared to the carboxylate and sulfonate groups, makes it a strong bridging bidentate ligand. This in turn results in the formation of strong ionic bonds to the surface Fe^{3+} ions, which decreases the number of unpaired spins, possibly through a double super-exchange mechanism through a Fe^{3+}-O-P-O-Fe^{3+} pathway.[183]

Transition metal oxides (i.e. ZnO, CuO, Co$_3$O$_4$, and Fe$_3$O$_4$) were synthesized from the metal acetates by the sonochemical method.[198] Two mechanisms for the formation of the metal oxides from aqueous metal acetates were claimed by the authors including (a) sonochemical oxidation for the preparation of Co$_3$O$_4$, and Fe$_3$O$_4$ and (b) sonochemical hydrolysis for the CuO and ZnO.[198] In order to confirm the idea, the pH values of the reaction mixtures were measured before and after sonication. The results confirmed that during the sonication H$^+$ ions were generated in the formation of CuO and ZnO, whereas OH$^-$ ions were formed in the synthesis of Co$_3$O$_4$ and Fe$_3$O$_4$. Furthermore, it has been showed that the sizes, morphologies, and yields of the obtained nanoparticles were significantly dependent to the composition of the solvent.

Suslick et al. [199] prepared stable nanosized iron (3 to 8 nm) colloid in the polyvinylpyrrolidone (PVP) matrix using high-intensity ultrasound (sonochemical decomposition of Fe(CO)$_5$).

2.2.1.8. Electrochemical Deposition

The era of the electrochemical method for synthesis of metal nano-clusters was developed by *Reetz et al.*[200]; the original method was only used for metals. Following, there are several groups which have employed the method for the preparation of different quantum dots, the magnetic oxide and magnetic mixed oxides (e.g. Sr and Fe).[201-207] *Pascal et al.*[208] used the same type of generation of metal cations in organic medium from a sacrificial anode in order to synthesize Fe$_2$O$_3$. According to their results, the electrolysis potential value was adequate to allow the anodic oxidation of water be present in small amounts in solution; hence, the resulting reactive oxygen complexes the metal cations to give the corresponding metal oxide. A tetraalkylammonium type surfactant was used in the role of both surfactants and electrolyte. The setup of the experiment is presented in Figure 2.8. The maghemite with the size of 3 to 8 nm were prepared via this method. X-ray analysis confirmed the formation of amorphous structure; however, both Raman spectroscopy and electron diffraction showed the crystalline structure of maghemite. It is notable that both magnetite and maghemite were prepared by the electrochemical deposition method under oxidizing conditions.[209]

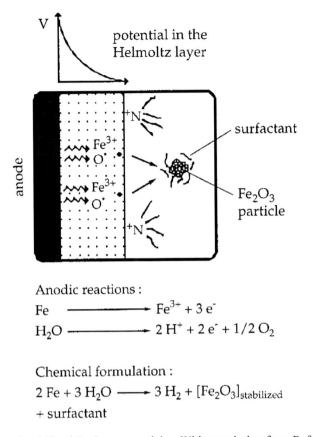

Figure 2.8. Formation of stabilized Fe$_2$O$_3$ nanoparticles. With permission from Ref. [208].

Franger et al.[204] employed the electrochemical synthesis for the preparation of pure and homogeneous magnetite nanoparticles (with the core size of 45 nm). The various physicochemical parameters were studied to understand the mechanism of the reaction and the optimization of the electrosynthesis conditions. According to their results, the benefits of using complexing agents in the aqueous electrolyte can be demonstrated for both the enhancement of the charge transfer kinetics and the decrease of the grain size to the nanometer level. The authors showed that the anodic polarization of a stainless steel electrode, during 30 min under a constant current of 50 mA, can yield the formation of pure, fine and homogeneous magnetite particles, provided the alkaline electrolyte contained complexing thiosulfate ions (i.e. 0.02 mol/L).

2.2.2. Physical Methods

SPIONs prepared by majority of the chemical methods are certainly associated with difficulties caused by the need in freeing surface of the material from chemically and/or physically bound toxic surfactants, in order to make the synthesized particles suitable for biomedical applications. This may be more specifically claimed for polyol and microemulsions.[210] In comparison, the SPIONs which are achieved by the physical methods are not associated with the above mentioned surface impurities. The physical methods consist of flow injection, aerosol, pulsed laser ablation, laser induced pyrolysis, and powder ball milling.[25]

2.2.2.1. Flow Injection

The major Advantage of the majority of physical routs is their capability to be considered for industrial applications. For instance, the use of a segmented flow tubular reactor (SFTR), which includes a premixing of two streams of reagents followed by segmentation of the resulting stream with a separating fluid during its moving to a tubular reactor, was reported for the continuous preparation of different high quality ceramic powders with a narrow size distribution.[211-213] The major disadvantage of the SFTR method is its long synthesis time due to the use of separated steps of premixing and segmentation; hence, this reactor could be inefficient for reactions with fast kinetics. In order to decrease the reaction time, the flow injection synthesis (FIS) were employed by *Wang et al.*[214]. In addition, this method contained the advantage of the flow-injection analytical technique, which was proposed by *Ruzicka et al.*[215], consisting the high reproducibility due to the plug-flow and laminar conditions, a high mixing homogeneity, and an opportunity for a precise external control of the process.[215, 216] It is noteworthy that the plug-flow was detected in a capillary when one liquid is followed by another liquid without observable longitudinal mixing. Laminar flow conditions are observed at low Reynolds number (i.e. ≤200).[13]

Salazar-Alvarez et al.[13] employed the FIS method for the preparation of magnetite nanoparticles. The designed system (Figure 2.9) showed fast precipitation kinetics that resulted in SPIONs with diameter of less than 10 nm. In addition, the influences of both chemical parameters and conditions on the physical and chemical characteristics of the obtained SPIONs were probed. According to the results, the variation of reagent concentrations together with the flow rates allowed the formation of SPIONs with specific size and narrow size distribution; furthermore, authors claimed that the flow segmentation was not required for the synthesis of SPIONs where continuous mode was found to be more advantageous without decreasing the quality of the SPIONs or altering their morphology.

2.2.2.2. Aerosol/Vapor

The nanoparticles with the size of about 20 nm were synthesized via the electrospray aerosol spray pyrolysis methods.[217-220] *Hogan et al.*[221] employed the electrospraying and in-flight heating of ferritin, which is an iron(inner core: FeOOH)–protein (outer shell) complex[222, 223], in order to obtain monodisperse aerosol nanoparticles. According to the Monte Carlo simulation, the authors claimed that the monodisperse nanoparticles (with the size of less than 20 nm) can be produced by the formation of monodispersed droplets which created by the optimized electrospray method.[224-226] It is notable that the role of protein-coat its catalyze potential for the nucleation of FeOOH nanoparticle within its interior and then encapsulates the nanoparticle. [221] Ferritin has been also employed for the liquid-phase synthesis of monodisperse protein-coated SPIONs[223, 227, 228] In addition after heating at 800 °C in a furnace aerosol reactor, the formation of the monodispersed SPIONs it is reported via electrospraying of ferritin.[229] *Shafranovsky et al.*[230] synthesized the iron nanoparticles with iron oxide shell using aerosol method.

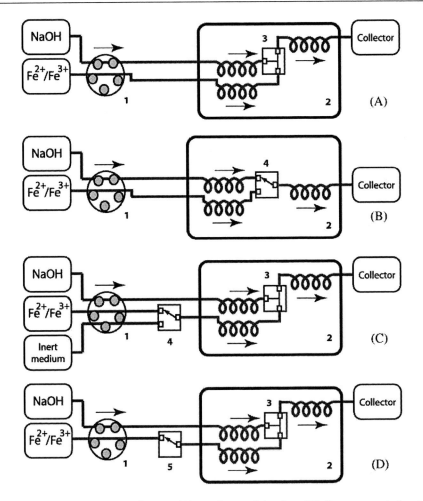

Figure 2.9. Various flow-injection schemes: (A) continuous injection; (B) flow segmentation by injection of reagents in turn; (c) flow segmentation with an inert medium; (D) flow segmentation by injection of one reagent in continuous flow of second reagent. (1) Multi-channel peristaltic pump; (2) thermostat; (3) T-injector; (4) computer-operated switching valve; (5) computer-operated on–off valve. Helixes indicate thermostated capillaries. With permission from ref. [13].

2.2.2.3. Pulsed Laser Ablation

Due to its suppression of nanoparticles contamination, the pulsed laser ablation method have been considered as a safe method among researchers and applied for several types of materials such as metals, intermetallic compounds, simple oxides, complex oxides, and polymers.[231-235] In addition, the ablation process can be easily controlled via optimization of laser parameters such as fluence and wavelength.[234] A typical example of the system setup is illustrated in Figure 2.10. More specifically for the preparation of SPIONs, an iron wire was sent through the pipe furnace to the reaction area by an automatic wire-feeding system. Consequently, a pulsed Nd:YAG laser beam (wavelength 1064 nm, laser pulse width 0.3–20 ms, repetition rate 5–150 Hz, and average laser power 400 W) was focused by a plano-convex lens to the tip of the wire at normal incidence with a diameter of 0.5 mm. As the SPIONs were synthesized by the ablation process, a mixture of N_2 (99.99%) and O_2 was used to transfer the prepared nanoparticles to the collector.[233] Using this method, *Wang et* al.[233] prepared pure γ-Fe$_2$O$_3$ nanoparticles with spherical shape and the size range of 5 to 90 nm. According to the results, both laser power density and carrier gas pressure had significant effect on the size of the prepared SPIONs. In particular, the mean particle sizes were decreased by increasing the laser power density, total mixed gas pressure and the oxygen ratio, respectively.

Kawaguchi et al.[17] synthesized the gold and magnetite composite nanoparticles via a combination of unique laser processes in water. The authors employed pulsed laser ablation in liquid in order to obtain high pure gold nanoparticles and the formation of gold/magnetite composite nanoparticles was confirmed by a nano-soldering effect which was introduced by *Mafune et al.*[232] for the connection of Au and Pt nanoparticles in water medium. Furthermore, the effects of irradiation time on the magnetic properties of the obtained materials were explored. According to the results, both coercivity and temperature hysteresis were qualitatively equal to the source magnetite for the irradiation time of up to 20 min. It is notable that the small decrease in blocking temperature together with little trace of residual coercivity was detected.

Interestingly, *Zbroniec et al.*[236] prepared SPIONs aggregated films using the excimer laser ablation technique by adopting an off-axis configuration and a gas condensation process. ArF excimer (193 nm) laser were employed for ablation of the sintered iron oxide targets (i.e. maghemite) in different carrier gases including Ar, O$_2$, He, N$_2$, Ne and Xe under various pressures. The results confirmed the strong effect of both the ambient gas and its pressure on the composition (i.e. maghemite or its mixture with FeO) of the deposition product.

It was shown that both oxides (i.e. Fe$_2$O$_3$ and FeO) were prepared by using argon as ambient gas at the lowest fluence of 4.9 J/cm^2. In addition, the pure maghemite was achieved by reducing the Ar pressure. The authors claimed that diatomic gases (e.g. N$_2$ and O$_2$) were more efficient in cooling down the evaporated species, causing the non-stoichiometric deposition even at lower pressures.

Some reports claimed the formation of nanoparticle chain aggregates (NCA) after employing the laser ablation method due to the relaxation of the induced tension.[237] *Ogawa et al.*[238] explored the mechanism of this phenomenon. They synthesized the NCA by laser ablation of rotating metal (alumina, iron-oxide, and titania) foils in an oxygen stream. In addition, the operating conditions (e.g. oxygen flow rate and laser frequency) were explored. The results confirmed that the NCA elasticity is driven by the reduction in free energy of the aggregates.

The magnetic characterization of SPIONs prepared by pulsed laser ablation was also carried out. The results confirmed the SPIONs formation and growth in the carrier gas before reaching the collector. The author observed that the temperature variation of coercive field and remanence of the SPIONs were considerably different from those of the corresponding bulk iron oxides.

The current excel experimental setup, which is shown in Figure 2.11, consists of a pulsed excimer laser whose irradiation is focused on a rotating target (e.g. aluminum, titanium, iron, niobium, tungsten and silicon)[234, 238-244] mounted in a cylindrical chamber.[245] This system is eligible to produce different nanoparticles such as SPIONs in the range of 4.9 nm to 13 nm. The critical operating parameters of the system including laser power (100 mJ/pulse) and frequency (2 Hz), and carrier gas flow rate (1 l/min) were optimized and subsequently held constant.[245] It was recognized that the solid-state diffusion coefficient influenced the size of the aerosol in flame reactors, and its effect was reduced for aerosols, which are formed by laser ablation due to the collision-coalescence mechanism and the very fast quenching of the aerosols.[246]

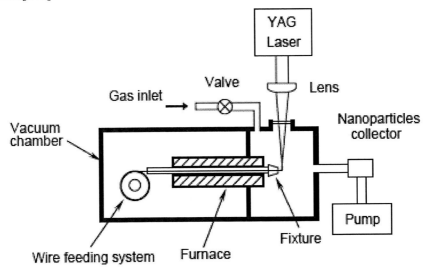

Figure 2.10. Example of laser ablation setup for the preparation of nanoparticle. With permission from ref. [233].

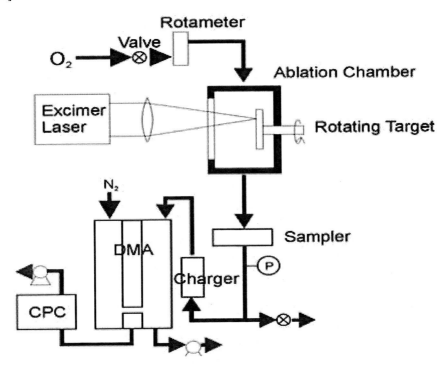

Figure 2.11. Schematic diagram of the excel laser ablation setup.
DMA: differential mobility analyzer; CPC: condensation particle counter; with permission from ref. [245].

2.2.2.4. Laser Induced Pyrolysis

Another continues method, which is used for the preparation of pure and non-aggregated SPIONs without requiring the subsequent size-selection process, is the laser induced pyrolysis under gas phase[247] or aerosol[248] precursors. The basic mechanisms for the formation of particles in gas-phase processes were reported by Pratsinis[249] and Kodas[250] and their co-workers. The CO_2 laser is implemented a new chemical vapor deposition laser technique which is called laser induced pyrolysis.[235] The numerous nanostructured materials were synthesized by laser pyrolysis method including various Si-based compounds (e.g. SiC, Si3N4, Si/C/N)[235, 251-253] and nanocrystalline silicon[254-256], SPIONs and iron carbides nanoparticles[257-259], and carbon-based compounds (e.g. fullerenes[260, 261], carbonaceous nanoparticles[262, 263], carbon nitrides[264] and carbon nanotubes[265, 266]).

Veintemillas-Verdaguer et al.[267] prepared pure maghemite nanoparticles with sizes smaller than 10 nm and a narrow particle size distribution via laser induced pyrolysis of iron pentacarbonyl vapours. The same authors synthesized the pure γ-Fe_2O_3 (with sizes in the range 4 to 17 nm) nanoparticles by a continuous wave CO_2 laser-induced pyrolysis of iron pentacarbonyl vapour in an oxidizing atmospheres including air-argon mixtures and air-ethylene mixtures which were employed to fill the apparatus and as carrier gas.[268] According to the results, both various sizes and degrees of structural order were strongly dependent on the synthesis conditions. In addition, the productivities attained were typically 0.05gh^{-1}. In addition, the effect of laser power and oxygen proportion on the structural and magnetic properties of maghemite nanoparticles was probed.[269] More specifically, very uniform maghemite nanoparticles (with the diameter of 4-7 nm) were synthesized by laser pyrolysis of a mixture of $Fe(CO)_5$–C_2H_4–Air with different degrees of crystallinity and different magnetic properties as a function of the ethylene/Air proportion. [269] According to the results, small polycrystalline SPIONs were obtained under Fe/O_2 molar ratio of 0.5, independently of the laser power used in the experiment. Furthermore, under molar ration of 1 and laser power higher than 53 W single crystal SPIONs were formed. The authors claimed that the crystallinity and particle size could be controlled by optimizing the temperature of the process. The results (i.e. adjusting the experimental conditions) were used by *Morales et al.*[270] for the preparation of dextran coated SPIONs (core size of 2-7nm and hydrodynamic size of 50 nm) with an application as contrast agents in magnetic resonance imaging. It was also found that an important enhancement of the magnetic properties of the product and the suspensions (i.e. saturation magnetization and susceptibility values) takes place as the particle and the crystallite sizes increase. Similarly, the preparation of stable colloidal dispersions of SPIONs with very narrow size distribution using dispersion and simultaneous coating of the powders with dextran in a strong alkaline medium was reported using the same method.[271]

2.2.2.5. Powder Ball Milling

It has been recognized that by the high-energy ball milling (i.e. planetary mill and horizontal ball mill), the grain size of micrometer-sized metal and ceramic particles can easily be reduced to the nanometer size range.[272-275] It is noteworthy that the planetary ball mill is usually used for mechanical alloying and consists of one turn disc (turntable) and two or four bowls where the direction of the turn disc and the bowls are in opposite and causes the centrifugal force on both powder mixture and milling balls in the bowl. Consequently, the powder mixture is fractured and cold-welded under high-energy impact through mechanical alloying process.[276] The major disadvantage of ball milling technique is its very limited yield; hence, this method is mostly suitable for laboratory. *Tan et al.*[276] prepared maghemite nanoparticles with average size of 8 nm using the high-energy ball milling of Fe$_2$O$_3$-based solid solutions which are mixed with different mole percents of SnO$_2$, ZrO$_2$ and TiO$_2$ for ethanol gas sensing applications. Interestingly, the reported results on Fe-oxide showed that, besides controlling crystallite size, high-energy ball milling can induce phase transformations of α-Fe$_2$O$_3$ to γ-Fe$_2$O$_3$ or vice versa depending on both particle size and changes in the chemical state of powder.[277, 278] *Meillon et al.*[279] have probed the transformation mechanism of α-Fe$_2$O$_3$ to γ-Fe$_2$O$_3$, based on the crystallographic concept including both variation of the orientation relationship (between α-Fe$_2$O$_3$ and γ-Fe$_2$O$_3$) and the movement of the oxygen planes which occurs during grinding. The authors suggested that the theory might be valid for the inverse transformation; however, *Senna*[280] claimed that this transformation is initiated by the nucleation of α-Fe$_2$O$_3$ that is stimulated by the high extent of shear deformation due to stored energy via prolonged high energy ball milling. Interestingly, *Hadena et al.*[281] revealed that in nanoparticles with the size of less than 20 nm, transformation of γ-Fe$_2$O$_3$ to α-Fe$_2$O$_3$ can be induced by the disappearance of the vacancy ordering in γ-Fe$_2$O$_3$. *Lee et al.*[282] confirmed that the occurrence of α-γ-α-Fe$_2$O$_3$ transformations during the ball milling process is dependent on the crystallographic concept of ball milling, which in turn relates to both mechanical energy exerted on powders and the particle size reduction.

2.2.3. Biological/ Bio-mineralization

The SPIONs are produced by biological species such as bacteria, fungus, proteins.[25] The bio-mineralization, which is recognized as a new source for preparing super-biocompatible SPIONs[9], is a process at the interface between inorganic and organic phases leading to the nucleation and growth of crystals in both natural (e.g. magnetotactic bacteria)[283, 284] and synthetic systems (e.g. encapsulation of SPIONs in liposomes or phospholipid composition)[285-287]. Intracellular magnetite particles synthesized by several morphological types of magnetotactic bacteria including the spirillar (helical) freshwater species, *Magnetospirillum magnetotacticum,* and four incompletely characterized marine strains: MV-1, a curved rod shaped bacterium; MC-1 and MC-2, two coccoid (spherical) microorganisms; and MV-4, a spirillum have been comprehensively studied. *[288]* Various strains produce crystals with characteristic shapes. Comparing the size and shape distributions of crystals from magnetotactic bacteria with those of synthetic magnetite grains of similar size revealed that the biogenic and synthetic distributions can be statistically distinguishable. In particular, the size distributions of the bacterial magnetite crystals were narrower and had a distribution asymmetry that was the opposite of the non biogenic samples. The only deviation from ideal structure in the bacterial magnetite seems to be the occurrence of spinel-law twins. The major problem of this method is the very limited quantity of the final product.

REFERENCES

[1] Moroz P, Jones SK, Gray BN. Magnetically mediated hyperthermia: current status and future directions. *International Journal of Hyperthermia* 2002;18: 267-284.
[2] Gonzales-Weimuller M, Zeisberger M, Krishnan KM. *Journal of Magnetism and Magnetic Materials* 2009;321: 4.
[3] Gazeau F, Levy M, Wilhelm C. *Nanomedicine* 2008;3: 14.
[4] Hong R, Li HH, Wang H, Li HZ. *China Particuology* 2007;5: 6.
[5] Gupta AK, Gupta M. *Handbook of particulate drug deliVery;* American Scientific Publishers: Steven Ranch, CA 2007.
[6] Mahmoudi M, Simchi A, Milani AS, Stroeve P. Cell toxicity of superparamagnetic iron oxide nanoparticles. *Journal of Colloidal and Interface Science* 2009;336: 510-518.
[7] Mahmoudi M, Simchi A, Imani M, Milani AS, Stroeve P. Optimal Design and Characterization of Superparamagnetic Iron Oxide Nanoparticles Coated with Polyvinyl Alcohol for Targeted Delivery and Imaging. *Journal of Physical Chemistry B* 2008;112: 14470-14481.
[8] Zhi J, Wang YJ, Lu YC, Ma JY, Luo GS. In situ preparation of magnetic chitosan/Fe3O4 composite nanoparticles in tiny pools of water-in-oil microemulsion. *Reactive & Functional Polymers* 2006;66: 1552-1558.
[9] Lisy MR, Hartung A, Lang C, Schuler D, Richter W, Reichenbach JR, Kaiser WA, Hilger I. Fluorescent bacterial magnetic nanoparticles as bimodal contrast agents. *Investigative Radiology* 2007;42: 235-241.
[10] Zhu HL, Yang DR, Zhu LM. Hydrothermal growth and characterization of magnetite (Fe3O4) thin films. *Surface & Coatings Technology* 2007;201: 5870-5874.
[11] Woo K, Lee HJ, Ahn JP, Park YS. Sol-gel mediated synthesis of Fe2O3 nanorods. *Advanced Materials* 2003;15: 1761-+.
[12] Cai W, Wan JQ. Facile synthesis of superparamagnetic magnetite nanoparticles in liquid polyols. *Journal of Colloid and Interface Science* 2007;305: 366-370.
[13] Salazar-Alvarez G, Muhammed M, Zagorodni AA. Novel flow injection synthesis of iron oxide nanoparticles with narrow size distribution. *Chemical Engineering Science* 2006;61: 4625-4633.
[14] Santos FJ, Varanda LC, Ferracin LC, Jafelicci M. Synthesis and electrochemical behavior of single-crystal magnetite nanoparticles. *Journal of Physical Chemistry C* 2008;112: 5301-5306.
[15] Tartaj P, Gonzalez-Carreno T, Rebolledo AF, Bomati-Miguel O, Serna CJ. Direct aerosol synthesis of carboxy-functionalized iron oxide colloids displaying reversible magnetic behavior. *Journal of Colloid and Interface Science* 2007;309: 68-71.
[16] Kim EH, Lee HS, Kwak BK, Kim BK. *10th International Conference on Magnetic Fluids*; Elsevier Science Bv: Guaruja, BRAZIL 2004: 3.
[17] Kawaguchi K, Jaworski J, Ishikawa Y, Sasaki T, Koshizaki N. Preparation of gold/iron-oxide composite nanoparticles by a unique laser process in water. *Journal of Magnetism and Magnetic Materials* 2007;310: 2369-2371.
[18] Pereira A, Cros A, Delaporte P, Itina T, Sentis M, Marine W, Thomann AL, Boulmer-Leborgne C. Formation of iron oxide nanoparticles by pulsed laser ablation. In: *Proceedings of SPIE - The International Society for Optical Engineering*; 2006.

[19] Basak S, Chen DR, Biswas P. Electrospray of ionic precursor solutions to synthesize iron oxide nanoparticles: Modified scaling law. *Chemical Engineering Science* 2007;62: 1263-1268.

[20] Sugimoto T. *Fine Particles: Synthesis, characterization and mechanism of growth.* New York: Marcel Dekker 2000.

[21] Cao G. *Nanostructure and Nanomaterials: Synthesis, properties and applications.* 1 ed. 2004, London: imperial college press.

[22] Tartaj P, Morales MP, Veintemillas-Verdaguer S, Gonzalez-Carreno T, Serna CJ. Synthesis, properties and biomedical applications of magnetic nanoparticles. *Handbook of Magnetic Materials*; Elsevier: Amsterdam, The Netherlands 2006.

[23] Lamer VK, Dinegar RH. *Journal of the American Chemical Society* 1950;72: 4847-4854.

[24] Reed-Hill R, Abbaschian R. *Physical Metallurgy Principles.* 3rd ed. 1994, Boston: PWS Publishing Company.

[25] Mahmoudi M, Sen T, Sant S, Wang B, Khademhosseini A. *Superparamagnetic Iron Oxide Nanoparticles for Drug Delivery.* Expert Opinion in Drug Delivery 2010.

[26] Lu AH, Salabas EL, Schuth F. Magnetic nanoparticles: Synthesis, protection, functionalization, and application. *Angewandte Chemie-International Edition* 2007;46: 1222-1244.

[27] Laurent S, Forge D, Port M, Roch A, Robic C, Elst LV, Muller RN. Magnetic iron oxide nanoparticles: Synthesis, stabilization, vectorization, physicochemical characterizations, and biological applications. *Chemical Reviews* 2008;108: 2064-2110.

[28] Jolivet JP, Cassaignon S, Chaneac C, Chiche D, Tronc E. Design of oxide nanoparticles by aqueous chemistry. In: *14th International Sol-Gel Conference.* Montpellier, FRANCE: Springer; 2007. p. 299-305.

[29] Kang YS, Risbud S, Rabolt JF, Stroeve P. Synthesis and characterization of nanometer-size Fe_3O_4 and γ-Fe_2O_3 particles. *Chemistry of Materials* 1996;8: 2209-2211.

[30] Valenzuela R, Fuentes MC, Parra C, Baeza J, Duran N, Sharma SK, Knobel M, Freer J. Influence of stirring velocity on the synthesis of magnetite nanoparticles (Fe_3O_4) by the co-precipitation method. *Journal of Alloys and Compounds* 2009;488: 227-231.

[31] Cornell RM, Schertmann U. *Iron oxides in the laboratory; preparation and characterization.* Weinheim: VCH 1991.

[32] Cotton FA, Wilkinson G. *Advanced inorganic chemistry.* New York: Wiley Interscience 1988.

[33] Mahmoudi M, Shokrgozar MA, Simchi A, Imani M, Milani AS, Stroeve P, Vali H, Hafeli UO, Bonakdar S. Multiphysics Flow Modeling and in Vitro Toxicity of Iron Oxide Nanoparticles Coated with Poly(vinyl alcohol). *Journal of Physical Chemistry C* 2009;113: 2322-2331.

[34] Khalafalla SE, Reimers GW. *IEEE Trans. Magn.* 1980;16.

[35] Hadjipanayis GC, Siegel RW. Nanophase materials: synthesis, properties and applications. *NATO ASI Series, Applied Sciences,* vol. E260. Dordrecht: Kluwer; 1993.

[36] Sjogren CE, Briley-Saebo K, Hanson M, Johansson C. Magnetic characterization of iron oxides for magnetic resonance imaging. *Magn Reson Med* 1994;31: 5.

[37] Tominaga M, Matsumoto M, Soejima K, Taniguchi I. Size control for two-dimensional iron oxide nanodots derived from biological molecules. *Journal of Colloid and Interface Science* 2006;299: 761-765.

[38] Thapa D, Palkar VR, Kurup MB, Malik SK. Properties of magnetite nanoparticles synthesized through a novel chemical route. *Materials Letters* 2004;58: 2692-2694.
[39] Weissleder R. *US Patent* 1996;5,492,814.
[40] Morales MP, Veintemillas-Verdaguer S, Montero MI, Serna CJ, Roig A, Casas L, Martinez B, Sandiumenge F. Surface and internal spin canting in gamma-Fe2O3 nanoparticles. *Chemistry of Materials* 1999;11: 3058-3064.
[41] J. LeFort CR. *Acad. Sci. Paris* 1852;34.
[42] Massart R. *IEEE Trans. Magn.* 1981;17.
[43] Dimitrova GT, Tadros TF, Luckham PF, Kipps MR. *Investigations into the phase behavior of nonionic ethoxylated surfactants using H-2 NMR spectroscopy.* Langmuir 1996;12: 315-318.
[44] Hu SH, Liu DM, Tung WL, Liao CF, Chen SY. Surfactant-Free, Self-Assembled PVA-Iron Oxide/Silica Core-Shell Nanocarriers for Highly Sensitive, Magnetically Controlled Drug Release and Ultrahigh Cancer Cell Uptake Efficiency. *Advanced Functional Materials* 2008;18: 2946-2955.
[45] Kandori K, Horii I, Yasukawa A, Ishikawa T. Effects of surfactants on the precipitation and properties of colloidal particles from forced hydrolysis of fecl3-hcl solution. *Journal of Materials Science* 1995;30: 2145-2152.
[46] Kim DK, Zhang Y, Voit W, Rao KV, Muhammed M. Synthesis and characterization of surfactant-coated superparamagnetic monodispersed iron oxide nanoparticles. In: 3rd International Conference on Scientific and Clinical Applications of Magnetic Carriers. Rostock, Germany: *Elsevier Science Bv;* 2000. p. 30-36.
[47] Landfester K, Bechthold N, Tiarks F, Antonietti M. Miniemulsion polymerization with cationic and nonionic surfactants: A very efficient use of surfactants for heterophase polymerization. *Macromolecules* 1999;32: 2679-2683.
[48] Lawrence MJ. Surfactant systems - microemulsions and vesicles as vehicles for drug-delivery. In: *1st International Meeting on the Scientific Basis of Modern Pharmacy.* Athens, Greece: Medecine Et Hygiene; 1994. p. 257-269.
[49] Leem G, Sarangi S, Zhang SS, Rusakova I, Brazdeikis A, Litvinov D, Lee TR. *Surfactant-Controlled Size and Shape Evolution of Magnetic Nanoparticles.* Crystal Growth & Design 2009;9: 32-34.
[50] Santra S, Tapec R, Theodoropoulou N, Dobson J, Hebard A, Tan WH. *Synthesis and characterization of silica-coated iron oxide nanoparticles in microemulsion: The effect of nonionic surfactants.* Langmuir 2001;17: 2900-2906.
[51] Sicoli F, Langevin D. Shape fluctuations of microemulsions droplets - role of surfactant film bending elasticity. In: Ottewill RH, Rennie AR, editors. *VII Conference of the European-Colloid-and-Interface-Society - Trends in Colloid and Interface Science VIII.* Bristol, England: Dr Dietrich Steinkopff Verlag; 1993. p. 223-225.
[52] Wang XM, Zhang CN, Wang XL, Gu HC. The study on magnetite particles coated with bilayer surfactants. *Applied Surface Science* 2007;253: 7516-7521.
[53] Schwarzer HC, Peukert W. Combined experimental/numerical study on the precipitation of nanoparticles. *Aiche Journal* 2004;50: 3234-3247.
[54] Gribanow NM, Bibik EE, Buzunov OV, Naumov VN. *Journal of Magnetism and Magnetic Materials* 1990;85: 4.
[55] Mahmoudi M, Milani AS, Stroeve P. Synthesis and Surface Architecture of Superparamagnetic Iron Oxide Nanoparticles for Application in Drug Delivery and

Their Biological Response: A Review. *International Journal of Biomedical Nanoscience and Nanotechnology 2010;*in press.

[56] Bautista MC, Bomati-Miguel O, Morales MD, Serna CJ, Veintemillas-Verdaguer S. Surface characterisation of dextran-coated iron oxide nanoparticles prepared by laser pyrolysis and coprecipitation. In: 5th International Conference on Scientific and Clinical Applications of Magnetic Carriers. Lyon, FRANCE: Elsevier *Science Bv;* 2004. p. 20-27.

[57] Lee J, Isobe T, Senna M. Preparation of ultrafine Fe3O4 particles by precipitation in the presence of PVA at high pH. *Journal of Colloid and Interface Science* 1996;177: 490-494.

[58] Peng J, Zou F, Liu L, Tang L, Yu L, Chen W, Liu H, Tang JB, Wu LX. Preparation and characterization of PEG-PEI/Fe3O4 nano-magnetic fluid by co-precipitation method. *Transactions of Nonferrous Metals Society of China* 2008;18: 393-398.

[59] Vayssieres L, Chaneac C, Tronc E, Jolivet JP. Size tailoring of magnetite particles formed by aqueous precipitation: An example of thermodynamic stability of nanometric oxide particles. *Journal of Colloid and Interface Science* 1998;205: 205-212.

[60] Jolivet JP. *De la Solution a` l'Oxyde*. InterEditions et CNRS Editions: Paris, France 1994.

[61] Babes L, Denizot B, Tanguy G, Le Jeune JJ, Jallet P. Synthesis of iron oxide nanoparticles used as MRI contrast agents: A parametric study. *Journal of Colloid and Interface Science* 1999;212: 474-482.

[62] Jiang WQ, Yang HC, Yang SY, Horng HE, Hung JC, Chen YC, Hong CY. Preparation and properties of superparamagnetic nanoparticles with narrow size distribution and biocompatible. *Journal of Magnetism and Magnetic Materials* 2004;283: 210-214.

[63] Sun SH, Zeng H, Robinson DB, Raoux S, Rice PM, Wang SX, Li GX. Monodisperse MFe2O4 (M = Fe, Co, Mn) nanoparticles. *Journal of the American Chemical Society* 2004;126: 273-279.

[64] Massart R, Cabuil V. *J. Chem. Phys.* 1987;84.

[65] Gupta AK, Wells S. *IEEE Trans. Nanobiosci.* 2004;3.

[66] Gomez-Lopera SA, Plaza RC, Delgado AV. Synthesis and characterization of spherical magnetite/biodegradable polymer composite particles. *Journal of Colloid and Interface Science* 2001;240: 40-47.

[67] Lan Q, Liu C, Yang F, Liu SY, Xu J, Sun DJ. Synthesis of bilayer oleic acid-coated Fe3O4 nanoparticles and their application in pH-responsive Pickering emulsions. *Journal of Colloid and Interface Science* 2007;310: 260-269.

[68] Mahmoudi M, Simchi A, Imani M, Stroeve P, Sohrabi A. Templated growth of superparamagnetic iron oxide nanoparticles by temperature programming in the presence of poly(vinyl alcohol). *Thin Solid Films 2010*;in press.

[69] Hoar TP, Schulman JH. Transparent water-in-oil dispersions: the oleopathic hydro-micelle. *Nature* 1943;152: 102-103.

[70] Schulman JH, Stoeckenius W, Prince LM. Mechanism of formation and structure of micro emulsions by electron microscopy. *Journal of Physical Chemistry* 1959;63: 1677-1680.

[71] Bagwe RP, Kanicky JR, Palla BJ, Patanjali PK, Shah DO. Improved drug delivery using microemulsions: Rationale, recent progress, and new horizons. *Critical Reviews in Therapeutic Drug Carrier Systems* 2001;18: 77-140.
[72] Langevin D. *Annu. Rev. Phys. Chem.* 1992;43.
[73] Lee KM, Sorensen CM, Klabunde KJ, Hadjipanayis GC. *IEEE Trans. Magn.* 1992;28.
[74] O'Connor CJ, Seip C, Sangregorio C, Carpenter E, Li S, Irvin G, John VT. *Mol. Cryst. Liq. Cryst.* 1999;335.
[75] Liz L, Quintela MAL, Mira J, Rivas J. Preparation of colloidal fe3o4 ultrafine particles in microemulsions. *Journal of Materials Science* 1994;29: 3797-3801.
[76] Perez-Luna VH, Puig JE, Castano VM, Rodriguez BE, Murthy AK, Kaler EW. Langmuir 1990;6: 5.
[77] Jayakrishnan J, Shah DO. *J. Polym. Sci., Polym. Lett.* Ed. 1984;22.
[78] Ferrick MR, Murtagh J, Thomas JK. *Macromolecules* 1989;22: 3.
[79] Antonietti M, Hentze HP. *Advanced Materials* 1996;8: 5.
[80] Pileni MP. *Journal of Physical Chemistry* 1993;97: 6961.
[81] Lawrence MJ, Rees GD. Microemulsion-based media as novel drug delivery systems. *Advanced Drug Delivery Reviews* 2000;45: 89-121.
[82] Gupta AK, Gupta M. Synthesis and surface engineering of iron oxide nanoparticles for biomedical applications. *Biomaterials* 2005;26: 3995-4021.
[83] Lopez-Quintela MA, Rivas J. *Journal of Colloids and Interface Science* 1993;158: 446.
[84] Prince LM. The mixed film theory. In: *Microemulsions: Theory and Practise*. Prince, L. M., Ed., Academic Press, London 1977: 91-132.
[85] Friberg SE. Microemulsions and micellar solutions. In: *Microemulsions:Theory and Practise*. Prince, L. M., Ed., Academic Press, London. 1977: 133-148.
[86] Chen K, Jayaprakash C, Pandit R, Wenzel W. Microemulsions: a Landau-Ginzburg theory. *Physical Review Letters* 1990;65: 2736-2739.
[87] Ruckenstein E, Chi JC. Stability of microemulsions. *J. Chem. Soc. Farad.* T 1975;271: 1690-1707.
[88] Pillai V, Shah DO. Synthesis of high-coercivity cobalt ferrite particles using water-in-oil microemulsions. *Journal of Magnetism and Magnetic Materials* 1996;163: 243-248.
[89] Li T, Moon J, Morrone AA, Mecholsky JJ, Talham DR, Adair JH. Preparation of Ag/SiO$_2$ Nanosize Composites by a Reverse Micelle and Sol−Gel Technique. *Langmuir* 1999;15: 7.
[90] Stathatos E, Lianos P, DelMonte F, Levy D, Tsiourvas D. *Formation of TiO2 nanoparticles in reverse micelles and their deposition as thin films on glass substrates.* Langmuir 1997;13: 4295-4300.
[91] Shiojiri S, Hirai T, Komasawa I. Immobilization of semiconductor nanoparticles formed in reverse micelles into polyurea via in situ polymerization of diisocyanates. Chemical Communications 1998: 1439-1440.
[92] Shah DO. *Microemulsions, and Monolayers-Science and Technology*. Marcel Dekker Inc.: New York 1998.
[93] Kosak A, Makovec D, Drofenik M, Znidarsic A. In situ synthesis of magnetic MnZn-ferrite nanoparticles using reverse microemulsions. In: *International Conference on Magnetism (ICM 2003)*. Rome, ITALY: Elsevier Science Bv; 2003. p. 1542-1544.

[94] Lopez-Quintela MA. Synthesis of nanomaterials in microemulsions: formation mechanisms and growth control. *Current Opinion in Colloid & Interface Science* 2003;8: 137-144.
[95] Munshi N, De TK, Maitra A. Size modulation of polymeric nanoparticles under controlled dynamics of microemulsion droplets. *Journal of Colloid and Interface Science* 1997;190: 387-391.
[96] Pang YX, Bao XJ. Aluminium oxide nanoparticles prepared by water-in-oil microemulsions. *Journal of Materials Chemistry* 2002;12: 3699-3704.
[97] Paul BK, Moulik SP. Uses and applications of microemulsions. *Current Science* 2001;80: 990-1001.
[98] Perez JAL, Quintela MAL, Mira J, Rivas J, Charles SW. Advances in the preparation of magnetic nanoparticles by the microemulsion method. *Journal of Physical Chemistry* B 1997;101: 8045-8047.
[99] Xiang JH, Yu SH, Liu BH, Xu Y, Gen X, Ren L. Shape controlled synthesis of PbS nanocrystals by a solvothermal-microemulsion approach. *Inorganic Chemistry Communications* 2004;7: 572-575.
[100] Liu C, Zou BS, Rondinone AJ, Zhang ZJ. Reverse micelle synthesis and characterization of superparamagnetic MnFe2O4 spinel ferrite nanocrystallites. *Journal of Physical Chemistry* B 2000;104: 1141-1145.
[101] Flanagan J, Singh H. Microemulsions: A Potential Delivery System for Bioactives in Food. *Critical Reviews in Food Science and Nutrition* 2006;46: 221-237.
[102] Sharma MK, Shah DO. *Macro- and Microemulsions.* ed. D.O. Shah (American Chemical Society, Washington, DC, 1985) p. 1.
[103] Pileni M, Duxin N. *Chem. Innovation* 2000;30.
[104] Kabir H, Aramaki K, Ishitobi M, Kunieda H. Cloud point and formation of microemulsions in sucrose dodecanoate systems. *Colloids and Surfaces A* 2003;216: 65-74.
[105] Gaonkar AG, Bagwe RP. Microemulsions in Foods: Challenges and Applications. *Surfactant Science Series* 2003;109: 407-430.
[106] Inouye K, Endo R, Otsuka Y, Miyashiro K, Kaneko K, Ishikawa T. *J. Phys. Chem. B* 1982;86: 5.
[107] Feltin N, Pileni MP. *Langmuir* 1997;13: 3927.
[108] Pileni MP. *Nature Materials* 2003;2: 6.
[109] Tan W, Santra S, Zhang P, Tapec R, Dobson J. *US Patent* 6548 264 2003.
[110] Atarashi T, Kim YS, Fujita T, Nakatsuka K. *Journal of Magnetism and Magnetic Materials* 1999;201: 7.
[111] Wang G, Harrison A. *Journal of Colloids and Interface Science* 1999;217: 203.
[112] Ohmori M, Matijevic E. *Journal of Colloids and Interface Science* 1992;150: 594.
[113] Tavakoli A, Sohrabi M, Kargari A. A review of methods for synthesis of nanostructured metals with emphasis on iron compounds. *Chemical Papers* 2007;61: 151-170.
[114] Teja AS, Holm LJ. Production of magnetic oxide nanoparticles. In: Y.P. Sun, Editor, *Supercritical Fluid Technology in Materials Science and Engineering: Synthesis, Properties, and Applications*, Elsevier 2002: 23.
[115] Xu CB, Teja AS. Continuous hydrothermal synthesis of iron oxide and PVA-protected iron oxide nanoparticles. *Journal of Supercritical Fluids* 2008;44: 85-91.

[116] Chen D, Xu R. *Mater. Res. Bull.* 1998;33: 1015.
[117] Cote LJ, Teja AS, Wilkinson AP, Zhang ZJ. Continuous hydrothermal synthesis and crystallization of magnetic oxide nanoparticles. *Journal of Materials Research* 2002;17: 2410-2416.
[118] Giri S, Samanta S, Maji S, Ganguli S, Bhaumik A. Magnetic properties of alpha-Fe2O3 nanoparticle synthesized by a new hydrothermal method. *Journal of Magnetism and Magnetic Materials* 2005;285: 296-302.
[119] Mao BD, Kang ZH, Wang EB, Lian SY, Gao L, Tian CG, Wang CL. Synthesis of magnetite octahedrons from iron powders through a mild hydrothermal method. *Materials Research Bulletin* 2006;41: 2226-2231.
[120] Sreeja V, Joy PA. Microwave-hydrothermal synthesis of gamma-Fe2O3 nanoparticles and their magnetic properties. *Materials Research Bulletin* 2007;42: 1570-1576.
[121] Wang J, Sun JJ, Sun Q, Chen QW. One-step hydrothermal process to prepare highly crystalline Fe3O4 nanoparticles with improved magnetic properties. *Materials Research Bulletin* 2003;38: 1113-1118.
[122] Xu CB, Lee J, Teja AS. Continuous hydrothermal synthesis of lithium iron phosphate particles in subcritical and supercritical water. *Journal of Supercritical Fluids* 2008;44: 92-97.
[123] Teja AS, Koh PY. Synthesis, properties, and applications of magnetic iron oxide nanoparticles. *Progress in Crystal Growth and Characterization of Materials* 2009;55: 22-45.
[124] Wang X, Zhuang J, Peng Q, Li Y. *Nature* 2005;437: 121.
[125] Zheng Y-H, Cheng Y, Bao F, Wang Y-S. *Mater. Res. Bull.* 2006;41: 525.
[126] Burda C, Chen XB, Narayanan R, El-Sayed MA. Chemistry and properties of nanocrystals of different shapes. *Chemical Reviews* 2005;105: 1025-1102.
[127] Shaw RW, Brill TB, Clifford AA, Eckert CA, Franck EU. Supercritical water - a medium for chemistry. *Chemical & Engineering News* 1991;69: 26-39.
[128] Hyeon T, Lee SS, Park J, Chung Y, Bin Na H. Synthesis of highly crystalline and monodisperse maghemite nanocrystallites without a size-selection process. *Journal of the American Chemical Society* 2001;123: 12798-12801.
[129] Butter K, Kassapidou K, Vroege GJ, Philipse AP. Preparation and properties of colloidal iron dispersions. Journal of Colloid and Interface Science 2005;287: 485-495.
[130] Darab J, Matson D. *Journal of Electronic Materials* 1998;27: 1068-1072.
[131] Matson DW, Linehan JC, Bean RM. *Materials Letters* 1992;14: 222-226.
[132] Matson DW, Linehan JC, Darab JG, Buehler MF. *Energy & Fuels* 1994;8: 10-18.
[133] Zhong Z, Ho J, Teo J, Shen S, Gedanken A. *Chemistry of Materials* 2007;19: 4776-4782.
[134] Xu C, Teja AS. *Journal of Supercritical Fluids* 2008;44: 85-91.
[135] Xu C, Lee J, Teja AS. *Journal of Supercritical Fluids* 2008;44: 92-97.
[136] Kwon SG, Piao Y, Park J, Angappane S, Jo Y, Hwang NM, Park JG, Hyeon T. Kinetics of monodisperse iron oxide nanocrystal formation by "heating-up" process. *Journal of the American Chemical Society* 2007;129: 12571-12584.
[137] Jana NR, Chen YF, Peng XG. Size- and shape-controlled magnetic (Cr, Mn, Fe, Co, Ni) oxide nanocrystals via a simple and general approach. *Chemistry of Materials* 2004;16: 3931-3935.

[138] Roca AG, Morales MP, O'Grady K, Serna CJ. Structural and magnetic properties of uniform magnetite nanoparticles prepared by high temperature decomposition of organic precursors. *Nanotechnology* 2006;17: 2783-2788.

[139] Roca AG, Morales MP, Serna CJ. Synthesis of monodispersed magnetite particles from different organometallic precursors. In: *41st IEEE International Magnetics Conference* (Intermag 2006). San Diego, CA: Ieee-Inst Electrical Electronics Engineers Inc; 2006. p. 3025-3029.

[140] Bronstein LM, Huang XL, Retrum J, Schmucker A, Pink M, Stein BD, Dragnea B. Influence of iron oleate complex structure on iron oxide nanoparticle formation. *Chemistry of Materials* 2007;19: 3624-3632.

[141] Jung H, Park H, Kim J, Lee JH, Hur HG, Myung NV, Choi H. Preparation of riotic and abiotic iron oxide nanoparticles (IOnPs) and their properties and applications in heterogeneous catalytic oxidation. *Environmental Science & Technology* 2007;41: 4741-4747.

[142] Samia ACS, Hyzer K, Schlueter JA, Qin CJ, Jiang JS, Bader SD, Lin XM. Ligand effect on the growth and the digestion of co nanocrystals. *Journal of the American Chemical Society* 2005;127: 4126-4127.

[143] Li Y, Afzaal M, O'Brien P. The synthesis of amine-capped magnetic (Fe, Mn, Co, Ni) oxide nanocrystals and their surface modification for aqueous dispersibility. *Journal of Materials Chemistry* 2006;16: 2175-2180.

[144] Majewski P, Thierry B. Functionalized magnetite nanoparticles - Synthesis, properties, and bio-applications. *Critical Reviews in Solid State and Materials Sciences* 2007;32: 203-215.

[145] Tartaj P, Morales MP, Veintemillas-Verdaguer S, Gonzalez-Carreno T, Serna CJ. The preparation of magnetic nanoparticles for applications in biomedicine. *J. Phys. D: Appl. Phys.* 2003;36: 182-197.

[146] Sun S, Zeng H. Size-Controlled Synthesis of Magnetite Nanoparticles. *Journal of American Chemical Society* 2002;124: 2.

[147] Zhao S, Wu HY, Song L, Tegus O, Asuha S. Preparation of gamma-Fe2O3 nanopowders by direct thermal decomposition of Fe-urea complex: reaction mechanism and magnetic properties. *Journal of Materials Science* 2009;44: 926-930.

[148] Rockenberger J, Scher EC, Alivisatos AP. *J. Am. Chem. Soc.* 1999;121: 2.

[149] Sato S, Murakata T, Yanagi H, Miyasaka F, Iwaya S. *Journal of Materials Science* 1994;29: 5657.

[150] Huang JH, Parab HJ, Liu RS, Lai TC, Hsiao M, Chen CH, Sheu HS, Chen JM, Tsai DP, Hwu YK. Investigation of the Growth Mechanism of Iron Oxide Nanoparticles via a Seed-Mediated Method and Its Cytotoxicity Studies. *Journal of Physical Chemistry C* 2008;112: 15684–15690.

[151] Park J, An KJ, Hwang YS, Park JG, Noh HJ, Kim JY, Park JH, Hwang NM, Hyeon T. Ultra-large-scale syntheses of monodisperse nanocrystals. *Nature Materials* 2004;3: 891-895.

[152] Pinna N, Grancharov S, Beato P, Bonville P, Antonietti M, Niederberger M. *Chemistry of Materials* 2005;17: 3044-3049.

[153] Majewski P, Thierry B. *Critical Reviews in Solid State and Material Sciences* 2007;32: 203-215.

[154] Teng X, Yang H. Effects of surfactants and synthetic conditions on the sizes and self-assembly of monodisperse iron oxide nanoparticles. *Journal of Materials Chemistry* 2004;14: 774-779.
[155] Duraesa L, Costab BFO, Vasquesa J, Camposc J, Portugal A. Phase investigation of as-prepared iron oxide/hydroxide produced by sol–gel synthesis. *Materials Letters* 2005;59: 859– 863.
[156] Roy RA, Roy R. *Mater. Res. Bull.* 1984;19: 169.
[157] Kojima K, Miyazaki M, Mizukami F, Maeda K. J. *Sol-Gel Sci. Technol.* 1997;8: 77.
[158] Ennas G, Musinu A, Piccaluga G, Zedda D, Gatteschi D, Sangregorio C, Stanger JL, Concas G, Spano G. Characterization of Iron Oxide Nanoparticles in an Fe2O3-SiO2 Composite Prepared by a Sol-Gel Method. *Chemistry of Materials* 1998;10: 495-502.
[159] Dai Z, Meiser F, Mohwald H. Nanoengineering of iron oxide and iron oxide/silica hollow spheres by sequential layering combined with a sol–gel process. *Journal of Colloidal and Interface Science* 2005;288: 298-300.
[160] Guglielmi M, Principi G. J. *Non-Cryst. Solids* 1982;48: 161.
[161] Raileanu M, Crisan M, Petrache C, Crisan D, Zaharescu M. Fe2O3-SiO2 Nanocomposites obtained by different sol-gel routes. *Journal of Optoelectronics and Advanced Materials* 2003;5: 693-698.
[162] del Monte F, Morales MP, Levy D, Fernandez A, Ocana M, Roig A, Molins A, O'Grady K, Serna CJ. Formation of ç-Fe2O3 Isolated Nanoparticles in a Silica Matrix. *Langmuir* 1997;13: 3627-3634.
[163] Niznansky D, Rehspringer JL, Drillon M. Preparation of Magnetic Nanoparticles (y-Fe2O3) in the Silica Matrix. *IEEE transactions on Magnetics* 1994;30: 821-823.
[164] Bentivegna F, Ferre J, Nyvlt M, Jamet JP, Imhoff D, Canva M, Brun A, Veillet P, Visnovsky S, Chaput F, Boilot JP. Magnetically textured g-Fe2O3 nanoparticles in a silica gel matrix: Structural and magnetic properties. *Journal of Applied Physics* 1998;83: 7776-7788.
[165] Gash AE, Simpson RL, Tillotson TM, Satcher JH, Hrubesh LW. in: J. Kennedy (Ed.), *Proc. of the 27th International Pyrothechnics Seminar,* IIT Research Institute, Chicago 2000: 41-53.
[166] Tillotson TM, Gash AE, Simpson RL, Hrubesh LW, Satcher JH, Poco JF. J. *Non-Cryst. Solids* 2001;285: 338– 345.
[167] Gash AE, Tillotson TM, Satcher JH, Poco JF, Hrubesh LW, Simpson RL. *Chemistry of Materials* 2001;13: 999-1007.
[168] Ismail AA. Synthesis and characterization of Y2O3/Fe2O3/TiO2 nanoparticles by sol–gel method. *Applied Catalysis B: Environmental* 2005;58: 115-121.
[169] Liu H-L, Ko SP, Wu J-H, Jung MH, Min JH, Lee JH, An BH, Kim YK. One-pot polyol synthesis of monosize PVP-coated sub-5nm Fe3O4 nanoparticles for biomedical applications. *Journal of Magnetism and Magnetic Materials* 2007;310: 815-817.
[170] Tzitzios VK, Petridis D, Zafiropoulou I, Hadjipanayis G, Niarchos D. Synthesis and characterization of L10 FePt nanoparticles from Pt–Fe3O4 core-shell nanoparticles. *Journal of Magnetism and Magnetic Materials* 2005;294: 95-98.
[171] Joseyphus RJ, Kodama D, Matsumoto T, Sato Y, Jeyadevan B, Tohji K. Role of polyol in the synthesis of Fe particles. *Journal of Magnetism and Magnetic Materials* 2007;310: 2393-2395.
[172] Fievet F, Lagier JP, Blin B, Beaudoin B, Figlarz M. *Solid State Ionics* 1989;198: 32-33.

[173] Feldmann C, Jungk G-O. *Angew. Chem. Int. Ed.* 2001;40: 359.
[174] Kitamoto Y, Kimura K, Yamazaki Y. Chemical synthesis of FePt/Fe-oxide core/shell nanoparticles. *Funtai Oyobi Fummatsu Yakin/Journal of the Japan Society of Powder and Powder Metallurgy* 2009;56: 121-126.
[175] Suslick KS. *Ultrasound: Its Chemical, Physical, and Biological Effects,* VCH, New York 1988.
[176] Suslick KS. Sonochemistry. *Science* 1990;247: 1439-1445.
[177] Flint EB, Suslick KS. The temperature of cavitation. Science 1991;253: 1397-1399.
[178] Suslick KS, Flint EB, Grinstaff MW, Kemper KA. Sonoluminescence from metal carbonyls. *Journal of Physical Chemistry* 1993;97: 3098-3099.
[179] Suslick KS, Choe SB, Cichowlas AA, Grinstaff MW. Sonochemical synthesis of amorphous iron. *Nature* 1991;353: 414-416.
[180] Suslick KS, Hyeon T, Fang M, Cichowlas AA. Sonochemical synthesis of nanostructured catalysts. *Materials Science and Engineering A* 1995;204: 186-192.
[181] Suslick KS, Hyeon T, Fang M. Nanostructured Materials Generated by High-Intensity Ultrasound: Sonochemical Synthesis and Catalytic Studies. *Chemistry of Materials* 1996;8: 2172-2179.
[182] Suslick KS, Hammerton DA, Cline Jr RE. The sonochemical hot spot. *Journal of the American Chemical Society* 1986;108: 5641-5642.
[183] Shafi K, Ulman A, Yan XZ, Yang NL, Estournes C, White H, Rafailovich M. Sonochemical synthesis of functionalized amorphous iron oxide nanoparticles. *Langmuir* 2001;17: 5093-5097.
[184] Abu Mukh-Qasem R, Gedanken A. Sonochemical synthesis of stable hydrosol of Fe3O4 nanoparticles. *Journal of Colloid and Interface Science* 2005;284: 489-494.
[185] Khalafalla SE, Reimers GW. Preparation of dilution-stable aqueous magnetic fluids. *IEEE Transactions on Magnetics* 1980;MAG-16: 178-183.
[186] Shen L, Stachowiak A, Fateen SEK, Laibinis PE, Hatton TA. Structure of alkanoic acid stabilized magnetic fluids. A small-angle neutron and light scattering analysis. *Langmuir* 2001;17: 288-299.
[187] Underhill RS, Liu G. Triblock nanospheres and their use as templates for inorganic nanoparticle preparation. *Chemistry of Materials* 2000;12: 2082-2091.
[188] Shimoiizaka J, Nakatsuka K, Fujita T, Kounosu A. Sink-float separators using permanent magnets and water based magnetic fluid. *IEEE Transactions on Magnetics* 1980;MAG-16: 368-371.
[189] Wormuth K. Superparamagnetic latex via inverse emulsion polymerization. *Journal of Colloid and Interface Science* 2001;241: 366-377.
[190] Mendenhall GD, Geng Y, Hwang J. Optimization of long-term stability of magnetic fluids from magnetite and synthetic polyelectrolytes. *Journal of Colloid and Interface Science* 1996;184: 519-526.
[191] Ding XB, Sun ZH, Wan GX, Jiang YY. Preparation of thermosensitive magnetic particles by dispersion polymerization. *Reactive and Functional Polymers* 1998;38: 11-15.
[192] Bacri JC, Perzynski R, Salin D, Cabuil V, Massart R. Ionic ferrofluids: A crossing of chemistry and physics. *Journal of Magnetism and Magnetic Materials* 1990;85: 27-32.

[193] Kim EH, Lee HS, Kwak BK, Kim BK. Synthesis of ferrofluid with magnetic nanoparticles by sonochemical method for MRI contrast agent. *Journal of Magnetism and Magnetic Materials* 2005;289: 328-330.

[194] Kataby G, Prozorov T, Koltypin Y, Cohen H, Sukenik CN, Ulman A, Gedanken A. Self-assembled monolayer coatings on amorphous iron and iron oxide nanoparticles: Thermal stability and chemical reactivity studies. *Langmuir* 1997;13: 6151-6156.

[195] Kataby G, Ulman A, Prozorov R, Gedanken A. Coating of amorphous iron nanoparticles by long-chain alcohols. *Langmuir* 1998;14: 1512-1515.

[196] Kataby G, Cojocaru M, Prozorov R, Gedanken A. Coating carboxylic acids on amorphous iron nanoparticles. *Langmuir* 1999;15: 1703-1708.

[197] Rozenfeld O, Koltypin Y, Bamnolker H, Margel S, Gedanken A. *Langmuir* 1994;10: 627.

[198] Kumar RV, Diamant Y, Gedanken A. Sonochemical Synthesis and Characterization of Nanometer-Size Transition Metal Oxides from Metal Acetates. *Chemistry of Materials* 2000;12: 2301-2305.

[199] Suslick KS, Fang M, Hyeon T. Sonochemical Synthesis of Iron Colloids. *Journal of the American Chemical Society* 1996;118: 11960-11961.

[200] Reetz MT, Helbig W, Quaiser SA. In *Active Metals: Preparation, Characterization, Applications; Furstner,* A., Ed.; VCH: Weinheim 1996.

[201] Amigo R, Asenjo J, Krotenko E, Torres F, Tejada J, Brillas E. Electrochemical synthesis of new magnetic mixed oxides of Sr and Fe: Composition, magnetic properties, and microstructure. *Chemistry of Materials* 2000;12: 573-579.

[202] Penner RM. Hybrid electrochemical/chemical synthesis of quantum dots. *Accounts of Chemical Research* 2000;33: 78-86.

[203] Torres F, Amigo R, Asenjo J, Krotenko E, Tejada J, Brillas E. Electrochemical route for the synthesis of new nanostructured magnetic mixed oxides of Mn, Zn, and Fe from an acidic chloride and nitrate medium. *Chemistry of Materials* 2000;12: 3060-3067.

[204] Franger S, Berthet P, Berthon J. Electrochemical synthesis of Fe3O4 nanoparticles in alkaline aqueous solutions containing complexing agents. *Journal of Solid State Electrochemistry* 2004;8: 218-223.

[205] Kim JW, Choi SH, Lillehei PT, King GC, Chu SH, Park Y. Biologically derived nanoparticle arrays via an electrochemical reconstitution of ferritin and their applications. In: *Meeting Abstracts;* 2004. p. 2627.

[206] Marken F, Patel D, Madden CE, Millward RC, Fletcher S. The direct electrochemistry of ferritin compared with the direct electrochemistry of nanoparticulate hydrous ferric oxide. *New Journal of Chemistry* 2002;26: 259-263.

[207] Khan HR, Petrikowski K. Magnetic and structural properties of the electrochemically deposited arrays of Co and CoFe nanowires. *Journal of Magnetism and Magnetic Materials* 2002;249: 458-461.

[208] Pascal C, Pascal JL, Favier F, Moubtassim MLE, Payen C. Electrochemical synthesis for the control of gamma-Fe2O3 nanoparticle size. Morphology, microstructure, and magnetic behavior. *Chemistry of Materials* 1999;11: 141-147.

[209] Khan HR, Petrikowski K. Anisotropic structural and magnetic properties of arrays of Fe26Ni74 nanowires electrodeposited in the pores of anodic alumina. *Journal of Magnetism and Magnetic Materials* 2000;215: 526-528.

[210] Stein A, Schroden RC. Colloidal crystal templating of three-dimensionally ordered macroporous solids: Materials for photonics and beyond. *Current Opinion in Solid State and Materials Science* 2001;5: 553-564.

[211] Jongen N, Donnet M, Bowen P, Lemaî®tre J, Hofmann H, Schenk R, Hofmann C, Aoun-Habbache M, Guillemet-Fritsch S, Sarrias J, Rousset A, Viviani M, Buscaglia MT, Buscaglia V, Nanni P, Testino A, Herguijuela R. Development of a continuous segmented flow tubular reactor and the "Scale-out" concept - In search of perfect powders. *Chemical Engineering and Technology* 2003;26: 304-306.

[212] Donnet M, Jongen N, Lemaî®tre J, Bowen P. New morphology of calcium oxalate trihydrate precipitated in a segmented flow tubular reactor. *Journal of Materials Science Letters* 2000;19: 749-750.

[213] Jongen N, Lemaî®tre J, Bowen P, Hofmann H. Aqueous synthesis of mixed yttrium-barium oxalates. *Chemistry of Materials* 1999;11: 712-718.

[214] Wang L, Muhammed M. Synthesis of zinc oxide nanoparticles with controlled morphology. *Journal of Materials Chemistry* 1999;9: 2871-2878.

[215] Ruzicka J, Hansen EH. Flow-injection analysis and its early history. *Talanta* 1982;29: 157.

[216] Harries N, Burns JR, Barrow DA, Ramshaw C. A numerical model for segmented flow in a microreactor. *International Journal of Heat and Mass Transfer* 2003;46: 3313-3322.

[217] Ude S, Fernandez De La Mora J, Alexander Iv JN, Saucy DA. Aerosol size standards in the nanometer size range: II. Narrow size distributions of polystyrene 3-11 nm in diameter. *Journal of Colloid and Interface Science* 2006;293: 384-393.

[218] Kim JH, Germer TA, Mulholland GW, Ehrman SH. Size-monodisperse metal nanoparticles via hydrogen-free spray pyrolysis. *Advanced Materials* 2002;14: 518-521.

[219] Dahl JK, Weimer AW, Lewandowski A, Bingham C, Bruetsch F, Steinfeld A. Dry reforming of methane using a solar-thermal aerosol flow reactor. *Industrial and Engineering Chemistry Research* 2004;43: 5489-5495.

[220] Prin EM, Glikin MA, Kutakova DA, Kauffeldt T. Catalyst aerosol flow reactor for industrial application. *Journal of Aerosol Science* 1998;29.

[221] J. Hogan Jr C, Biswas P. Narrow size distribution nanoparticle production by electrospray processing of ferritin. *Journal of Aerosol Science* 2008;39: 432-440.

[222] Liu G, Debnath S, Paul KW, Han W, Hausner DB, Hosein HA, Michel FM, Parise JB, Sparks DL, Strongin DR. Characterization and surface reactivity of ferrihydrite nanoparticles assembled in ferritin. *Langmuir* 2006;22: 9313-9321.

[223] Meldrum FC, Heywood BR, Mann S. Magnetoferritin: In vitro synthesis of a novel magnetic protein. *Science* 1992;257: 522-523.

[224] Chen DR, Pui DYH, Kaufman SL. Electrospraying of conducting liquids for monodisperse aerosol generation in the 4 nm to 1.8 m diameter range. *Journal of Aerosol Science* 1995;26: 963-977.

[225] Hogan Jr CJ, Yun KM, Chen DR, Lenggoro IW, Biswas P, Okuyama K. Controlled size polymer particle production via electrohydrodynamic atomization. *Colloids and Surfaces A: Physicochemical and Engineering Aspects* 2007;311: 67-76.

[226] Rosell-Llompart J, Fernández de la Mora J. Generation of monodisperse droplets 0.3 to 4 μm in diameter from electrified cone-jets of highly conducting and viscous liquids. *Journal of Aerosol Science* 1994;25: 1093-1119.

[227] Douglas T, Dickson DPE, Betteridge S, Charnock J, Garner CD, Mann S. Synthesis and structure of an iron(III) sulfide-ferritin bioinorganic nanocomposite. *Science* 1995;269: 54-57.

[228] Polanams J, Ray AD, Watt RK. Nanophase iron phosphate, iron arsenate, iron vanadate, and iron molybdate minerals synthesized within the protein cage of ferritin. *Inorganic Chemistry* 2005;44: 3203-3209.

[229] Gonzalez D, Nasibulin AG, Jiang H, Queipo P, Kauppinen EI. Electrospraying of ferritin solutions for the production of monodisperse iron oxide nanoparticles. *Chemical Engineering Communications* 2007;194: 901-912.

[230] Shafranovsky EA, Petrov YI. Aerosol Fe nanoparticles with the passivating oxide shell. *Journal of Nanoparticle Research* 2004;6: 71-90.

[231] Usui H, Shimizu Y, Sasaki T, Koshizaki N. Photoluminescence of ZnO nanoparticles prepared by laser ablation in different surfactant solutions. *Journal of Physical Chemistry* B 2005;109: 120-124.

[232] Mafune F, Kohno JY, Takeda Y, Kondow T. Nanoscale soldering of metal nanoparticles for construction of higher-order structures. *Journal of the American Chemical Society* 2003;125: 1686-1687.

[233] Wang Z, Liu Y, Zeng X. One-step synthesis of -Fe2O3 nanoparticles by laser ablation. *Powder Technology* 2006;161: 65-68.

[234] Sasaki T, Terauchi S, Koshizaki N, Umehara H. The preparation of iron complex oxide nanoparticles by pulsed-laser ablation. *Applied Surface Science* 1998;127-129: 398-402.

[235] Cannon WR, Danforth SC, Flint JH, Haggerty JS, Marra RA. Sinterable ceramic powders from laser-driven reactions - 1. process description and modeling. In: *Journal of the American Ceramic Society.* 7 ed; 1982. p. 324-330.

[236] Zbroniec L, Sasaki T, Koshizaki N. Effects of ambient gas and laser fluence on the compositional changes in iron oxide particle aggregated films prepared by laser ablation. *Applied Physics A: Materials Science and Processing* 2004;79: 1783–1787.

[237] Friedlander SK, Jang HD, Ryu KH. *Applied Physics Letter* 1998;72: 1.

[238] Ogawa K, Vogt T, Ullmann M, Johnson S, Friedlander SK. Elastic properties of nanoparticle chain aggregates of TiO2, Al2O3, and Fe2O3 generated by laser ablation. *Journal of Applied Physics* 2000;87: 63-73.

[239] Karch J, Birringer R, Gleiter H. Ceramics ductile at low temperature. *Nature* 1987;330: 556-558.

[240] Karch J, Birringer R. Nanocrystalline ceramics: Possible candidates for net-shape forming. *Ceramics International* 1990;16: 291-294.

[241] Kresse M, Wagner S, Pfefferer D, Lawaczeck R, Elste V, Semmler W. Targeting of ultrasmall superparamagnetic iron oxide (USPIO) particles to tumor cells in vivo by using transferrin receptor pathways. *Magnetic Resonance in Medicine* 1998;40: 236-242.

[242] Nishide T, Sato M, Hara H. Crystal structure and optical property of TiO2 gels and films prepared from Ti-edta complexes as titania precursors. *Journal of Materials Science* 2000;35: 465-469.

[243] Remillard JT, McBride JR, Nietering KE, Drews AR, Zhang X. Real Time in Situ Spectroscopic Ellipsometry Studies of the Photocatalytic Oxidation of Stearic Acid on Titania Films. *Journal of Physical Chemistry* B 2000;104: 4440-4447.

[244] Windeler RS, Lehtinen KEJ, Friedlander SK. Production of nanometer-sized metal oxide particles by gas phase reaction in a free jet. II: Particle size and neck formation - Comparison with theory. *Aerosol Science and Technology* 1997;27: 191-205.

[245] Ullmann M, Friedlander SK, Schmidt-Ott A. Nanoparticle formation by laser ablation. *Journal of Nanoparticle Research* 2002;4: 499–509.

[246] Friedlander SK, Ogawa K, Ullmann M. *Journal of Polymer Science* 2000;38: 2658–2665.

[247] Dumitrache F, Morjan I, Alexandrescu R, Morjan RE, Voicu I, Sandu I, Soare I, Ploscaru M, Fleaca C, Ciupina V, Prodan G, Rand B, Brydson R, Woodword A. Nearly monodispersed carbon coated iron nanoparticles for the catalytic growth of nanotubes/nanofibres. *Diamond and Related Materials* 2004;13: 362-370.

[248] Veintemillas-Verdaguer S, Bomatí-O, Morales MP, Di Nunzio PE, Martelli S. Iron ultrafine nanoparticles prepared by aerosol laser pyrolysis. *Materials Letters* 2003;57: 1184-1189.

[249] Pratsinis SE, Vemury S. Particle formation in gases: A review. *Powder Technology* 1996;88: 267-273.

[250] Kodas TT, Hampden-Smith M. *Aerosol Processing of Materials,* Wiley/VCH New York 1999.

[251] Fantoni R, Borsella E, Piccirillo S, Ceccato R, Enzo S. Laser synthesis and crystallographic characterization of ultrafine SiC powders. *Journal of Materials Research* 1990;5: 143-150.

[252] Rice GW, Woodin RL. Kinetics and mechanism of laser-driven powder synthesis from organosilane precursors. *Journal of Materials Research* 1989;4: 1538-1548.

[253] Alexandrescu R, Morjan I, Borsella E, Botti S, Fantoni R, Dikonimos-Makris T, Giorgi R, Enzo S. Composite ceramic powders obtained by laser induced reactions of silane and amines. *Journal of Materials Research* 1991;6: 2442-2451.

[254] Ehbrecht M, Kohn B, Huisken F, Laguna MA, Paillard V. Photoluminescence and resonant Raman spectra of silicon films produced by size-selected cluster beam deposition. *Physical Review B - Condensed Matter and Materials Physics* 1997;56: 6958-6964.

[255] Ledoux G, Guillois O, Porterat D, Reynaud C, Huisken F, Kohn B, Paillard V. Photoluminescence properties of silicon nanocrystals as a function of their size. *Physical Review B - Condensed Matter and Materials Physics* 2000;62: 15942-15951.

[256] Huisken F, Ledoux G, Guillois O, Reynaud C. Light-emitting silicon nanocrystals from laser pyrolysis. *Advanced Materials* 2002;14: 1861-1865.

[257] Huisken F, Kohn B, Alexandrescu R, Morjan I. Mass spectrometric characterization of iron clusters produced by laser pyrolysis and photolysis of Fe(CO)5 in a flow reactor. *European Physical Journal D* 1999;9: 141-144.

[258] Hofmeister H, Huisken F, Kohn B, Alexandrescu R, Cojocaru S, Crunteanu A, Morjan I, Diamandescu L. Filamentary iron nanostructures from laser-induced pyrolysis of iron pentacarbonyl and ethylene mixtures. *Applied Physics A: Materials Science and Processing* 2001;72: 7-11.

[259] Majima T, Miyahara T, Haneda K, Ishii T, Takami M. Preparation of iron ultrafine particles by the dielectric breakdown of Fe(CO)5 using a transversely excited atmospheric CO2 laser and their characteristics. *Japanese Journal of Applied Physics, Part 1: Regular Papers and Short Notes and Review Papers* 1994;33: 4759-4763.

[260] Ehbrecht M, Faerber M, Rohmund F, Smirnov VV, Stelmakh O, Huisken F. CO2-laser-driven production of carbon clusters and fullerenes from the gas phase. *Chemical Physics Letters* 1993;214: 34-38.

[261] Voicu I, Armand X, Cauchetier M, Herlin N, Bourcier S. Laser synthesis of fullerenes from benzene-oxygen mixtures. *Chemical Physics Letters* 1996;256: 261-268.

[262] Galvez A, Herlin-Boime N, Reynaud C, Clinard C, Rouzaud JN. Carbon nanoparticles from laser pyrolysis. *Carbon* 2002;40: 2775-2789.

[263] Morjan I, Voicu I, Dumitrache F, Sandu I, Soare I, Alexandrescu R, Vasile E, Pasuk I, Brydson RMD, Daniels H, Rand B. Carbon nanopowders from the continuous-wave CO2 laser-induced pyrolysis of ethylene. *Carbon* 2003;41: 2913-2921.

[264] Alexandrescu R, Cojocaru S, Crunteanu A, Petcu S, Cireasa R, Voicu I, Morjan I, Fatu D, Huisken F. Laser pyrolysis of carbon-nitrogen gas-phase compounds: An attempted approach to carbon nitride formation. *Carbon* 1998;36: 795-800.

[265] Alexandrescu R, Crunteanu A, Morjan RE, Morjan I, Rohmund F, Falk LKL, Ledoux G, Huisken F. Synthesis of carbon nanotubes by CO2-laser-assisted chemical vapour deposition. *Infrared Physics and Technology* 2003;44: 43-50.

[266] Rohmund F, Morjan RE, Ledoux G, Huisken F, Alexandrescu R. Carbon nanotube films grown by laser-assisted chemical vapor deposition. *Journal of Vacuum Science and Technology B: Microelectronics and Nanometer Structures* 2002;20: 802-811.

[267] Veintemillas-Verdaguer S, Morales MP, Sema CJ. *Materials Letters* 1998: 35227-35231.

[268] Veintemillas-Verdaguer S, Morales MP, Sema CJ. Effect of the oxidation conditions on the maghemites produced by laser pyrolysis. *Applied Organometallic Chemistry* 2001;15: 365-372.

[269] Veintemillas-Verdaguer S, Bomati-Miguel O, Morales MP. Effect of the process conditions on the structural and magnetic properties of $^3\zeta$-Fe2O3 nanoparticles produced by laser pyrolysis. *Scripta Materialia* 2002;47: 589-593.

[270] Morales MP, Bomati-Miguel O, de Alejo RP, Ruiz-Cabello J, Veintemillas-Verdaguer S, O'Grady K. Contrast agents for MRI based on iron oxide nanoparticles prepared by laser pyrolysis. *Journal of Magnetism and Magnetic Materials* 2003;266: 102-109.

[271] Veintemillas-Verdaguer S, Del Puerto Morales M, Bomati-Miguel O, Bautista C, Zhao X, Bonville P, De Alejo RP, Ruiz-Cabello J, Santos M, Tendillo-Cortijo FJ, Ferreir$^{3|}$s J. Colloidal dispersions of maghemite nanoparticles produced by laser pyrolysis with application as NMR contrast agents. *Journal of Physics D: Applied Physics* 2004;37: 2054-2059.

[272] Tan OK, Cao W, Zhu W. Alcohol sensor based on a non-equilibrium nanostructured xZrO2-(1-x)$\zeta\pm$-Fe2O3 solid solution system. *Sensors and Actuators, B: Chemical* 2000;63: 129-134.

[273] Cao W, Tan OK, Zhu W, Jiang B. Mechanical alloying and thermal decomposition of (ZrO2)0.8-($\zeta\pm$-Fe2O3)0.2 powder for gas sensing applications. *Journal of Solid State Chemistry* 2000;155: 320-325.

[274] Jiang JZ, Lu SW, Zhou YX, Mǿrup S, Nielsen K, Poulsen FW, Berry FJ, McMannus J. Correlation of gas sensitive properties with Fe2O3-SnO2 ceramic microstructure prepared by high energy ball milling. *Materials Science Forum* 1997;235-238: 941-946.

[275] Jiang JZ, Lin R, Lin W, Nielsen K, Mǿrup S, Dam-Johansen K, Clasen R. Gas-sensitive properties and structure of nanostructured (ż±-Fe2O3)x-(SnO2) 1-x materials prepared by mechanical alloying. *Journal of Physics D: Applied Physics* 1997;30: 1459-1467.

[276] Tan OK, Cao W, Hu Y, Zhu W. Nanostructured oxides by high-energy ball milling technique: Application as gas sensing materials. Solid State Ionics 2004;172: 309-316.

[277] Senna M. Difference in the transformation processes of pressed and vibro-milled ³ż-Fe2O3 to ż±-Fe2O3. *Journal of Applied Physics* 1978;49: 4580-4582.

[278] Zdujic M, Jovalekic Œ, Karanovic L, Mitric M, Poleti D, Skala D. Mechanochemical treatment of Fe2O3 powder in air atmosphere. *Materials Science and Engineering A* 1998;245: 109-117.

[279] Meillon S, Dammak H, Flavin E, Pascard H. Existence of a direct phase transformation from haematite to maghemite *Philosophical Magazine Letters* 1995;72: 105-110.

[280] Senna M. *Journal of Applied Physics* 1978;49: 4580.

[281] Hadena K, Morish AH. *Solid State Communication* 1977;22: 779.

[282] Lee JS, Lee CS, Oh ST, Kim JG. Phase evolution of Fe2O3 nanoparticle during high energy ball milling. *Scripta Materialia* 2001;44: 2023-2026.

[283] Blakemore RP. Magnetotactic Bacteria. Annual Review of Microbiology 1982;36: 20.

[284] Matsunaga T, Kamiya S. Use of magnetic particles isolated from magnetotactic bacteria for enzyme immobilization. *Applied Microbiology and Biotechnology* 1987;26: 328-332.

[285] Rocha FM, de Pinho SC, Zollner RL, Santana MHA. Preparation and characterization of affinity magnetoliposomes useful for the detection of antiphospholipid antibodies. In: 3rd International Conference on Scientific and Clinical Applications of Magnetic Carriers. Rostock, Germany: *Elsevier Science Bv;* 2000. p. 101-108.

[286] Ito A, Ino K, Kobayashi T, Honda H. The effect of RGD peptide-conjugated magnetite cationic liposomes on cell growth and cell sheet harvesting. *Biomaterials* 2005;26: 6185-6193.

[287] Gonzales M, Krishnan KM. Synthesis of magnetoliposomes with monodisperse iron oxide nanocrystal cores for hyperthermia. In: *5th International Conference on Scientific and Clinical Applications of Magnetic Carriers.* Lyon, FRANCE: Elsevier Science Bv; 2004. p. 265-270.

[288] Devouard B, Posfai M, Hua X, Bazylinski DA, Frankel RB, Buseck PR. Magnetite from magnetotactic bacteria: Size distributions and twinning. *American Mineralogist 1998;83: 1387-1398.*

Chapter 3

APPLICATION OF STATISTICAL TECHNIQUES

As evident from the preceding chapters, SPIONs offer a unique potential for numerous applications in biomedicine. However, it is also evident that SPIONs' characteristics with regards to their (narrow) size, (uniform) size distribution, (high) magnetic saturations, (engineered) surface architectures and their impurity or structural imperfections are among factors that can influence their ultimate success during in vivo applications [1-3]. These factors, in turn, are linked to the process parameters used during the synthesis and characterization of the particles. For instance, using coprecipitation techniques, a number of synthesis parameters are reported to have a direct effect on the ensuing average particle size and the structure of the synthesized magnetic solutions. They include the iron-to-salt ratio [4], iron concentration in the primary media [5], acidity and the ionic strength of the final media [6,7], homogenization rate [8,9], etc. Furthermore, when polymeric materials are employed as coatings of SPION in drug delivery, each polymer type can yield different sets of performance characteristics with regards to the drug release time at desired sites, stabilization, and biocompatibility/toxicity level, etc. The coatings may be added to the nanoparticles during [10,11] or after [9,12] particles' syntheses.

The use of statistical analysis and optimization techniques in the nano-science field is rather new (see, e.g., [9]). A systematic analysis of the synthesis process parameters and the application of optimization techniques can improve the functionlization of SPION products. This chapter is intended to review basic hypothesis testing and design of experiment (DOE) techniques that can be beneficial for optimum design and characterization of SPIONs. To gain a more in-depth understanding of the underlying formulations, the reader is encouraged to consult the original texts in the field such as [13-15]. Also, the application of several multiple criteria decision making/MCDM models [16] in material selection can be found in other studies such as [17]. Recently, Jahan et al. [18] also reported a comprehensive review on material screening and selection methods, and some of the reported techniques may be adapted for optimum material selection of coatings on nano-particles. Although the examples in this chapter will focus on mono-objective problems, the underlying procedure can be conveniently extended to multiple-criteria problems as reviewed in the study [18]. Finally, it is worth noting that for particular physical quantities at the nano-level, measurements can occasionally be too expensive or even infeasible to run. In such situations, physical quantities of interest are normally predicted by computer simulation models. Recently, Mahmoudi et al. [19] developed a multi-physics computational model to predict the velocity field of ferrofluids containing SPIONs for drug delivery in a blood vessel. The DOE techniques can be equally helpful during the analyses of physical and/or computer experiments.

3.1. THE TWO SAMPLE T-TEST

This test is used to evaluate whether or not the means of two sample groups are *statistically* different. Such hypothesis tests during data analysis can help analysts avoid the risks associated with reaching wrong conclusions; e.g., with respect to the effect of treatment factors on particular objective functions such as such the size of synthesized particles and their magnetization. Experimenters in materials science often call one of the material sample groups "control group" and the other one, "treatment group". It is assumed that the measured data from both groups fall on a normal distribution.

Statistically, the 'null hypothesis' (H_0) and the so-called 'alternative hypothesis' (H_1) for a two sample t-test is defined as:

$$H_0 : \mu_1 = \mu_2$$
$$H_1 : \mu_1 \neq \mu_2$$

We would like to test and see which of these hypotheses are statistically true. If we cannot reject H_0, that means that there is no difference between the control and treatment samples and thus the treatment factor has no significant effect on the response.

Let us first remind the sample mean and standard deviation formulas as follows.

$$\bar{y} = \frac{1}{n}\sum_{i=1}^{n} y_i$$

$$S^2 = \frac{1}{n-1}\sum_{i=1}^{n}(y_i - \bar{y})^2$$

y_i are the observations and for both samples of the test, the mean and variance can be calculated using the above equations. Note that $\bar{y}$ and S^2 'estimate' the population/actual mean and variance, respectively. For example, the expected value of $\bar{y}_1$ is μ_1). As the sample size n increases, more reliable estimations can be made.

Following, the formula for a two-sample t-test is defined as a signal-to-noise ration. It takes the difference between the two sample means and divides it by the standard error of that difference. This is shown graphically in Figure 3.1 and written formally as:

$$t_0 = \frac{\bar{y}_1 - \bar{y}_2}{S_p\sqrt{\frac{1}{n_1} + \frac{1}{n_2}}}$$

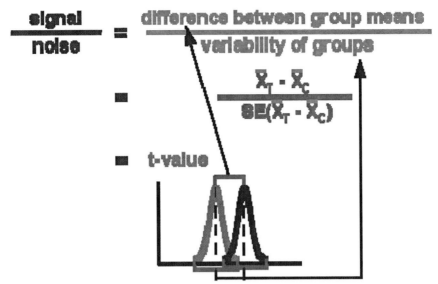

Source: http://www.socialresearchmethods.net/kb/stat_t.php

Figure 3.1. The concept of signal-to-noise in the definition of the value used for the two sample t-test.

$$S_p^2 = \frac{(n_1-1)S_1^2 + (n_2-1)S_2^2}{n_1 + n_2 - 2}$$

n_1 and n_2 are the sample size of the control and treatment groups. S_p^2 is a pooled variance of the population, estimated form the individual variance of samples.

The t-value can be positive or negative depending on which mean sample is larger than the other. Once a t-value is calculated, tables of statistical significance can be used to test whether or not the value is large enough to claim a meaningful difference between the two sample groups. To test the significance, however, the analyst should accept a risk level (often called α). In most materials research, this significance level is taken to be 0.05. It means the analysis takes only 5% risk on rejecting H_0 (i.e., accepting H_1) while in fact H_0 would have been true. In other words, the analyst's confidence level in interpreting the results is 95%. If the absolute value of the calculated t-ration is greater than the table value, it is concluded that H_0 is rejected; otherwise H_0 is true and cannot be rejected. In mathematical terms, the criterion for the rejection of the null hypothesis is $|t| > t_{\alpha/2,\nu}$, where $t_{\alpha/2,\nu}$ refers to a value from the t-distribution table with a significance level of α and the degree of freedom of ν. In the case of two-sample t-test, assuming an equal (pooled) population variance for both samples, ν is calculated as [15]:

$$\nu = n_1 + n_2 - 2$$

The t-distribution percentage point for $\alpha = 0.05$ has been included in Table 3.1.

Table 3.1. t-distribution percentage points for $\alpha = 0.05$

v	$t_{\alpha/2,v}$	v	$t_{\alpha/2,v}$	v	$t_{\alpha/2,v}$
2	4.3027	11	2.2010	21	2.0796
3	3.1824	12	2.1788	22	2.0739
4	2.7765	13	2.1604	23	2.0687
5	2.5706	14	2.1448	24	2.0639
6	2.4469	15	2.1315	25	2.0595
7	2.3646	16	2.1199	26	2.0555
8	2.3060	17	2.1098	27	2.0518
9	2.2622	18	2.1009	28	2.0484
10	2.2281	19	2.0930	29	2.0452
		20	2.0860	30	2.0423

Example 1:

Consider two samples of SPIONs prepared for a magnetization saturation (emu/g) test by means of Vibrating Sample Magnetometry (VSM). The first sample contains bare particles whereas in the second sample, SPIONs are coated with PVA. The polymer-to-iron mass ratio for particles in the second sample group is 1.5. Each sample group contains 10 repeats of the test and sample statistics have been found to be as follows.

Control sample group

$\bar{y}_1 = 16.76 \; (\text{emu}/\text{g})$
$S_1^2 = 0.1 \; (\text{emu}/\text{g})^2$
$n_1 = 10$

Treatment sample group

$\bar{y}_1 = 17.92 \; (\text{emu}/\text{g})$
$S_1^2 = 0.061 \; (\text{emu}/\text{g})^2$
$n_1 = 10$

It is desired to know whether or not the effect of polymer coating on magnetization saturation is statistically significant under the condition tested.

Solution:

Assuming that the population standard deviations (or variances) are equal (i.e., the treatment factor has no effect on the standard deviation), we can write our hypothesis as:

$H_0 : \mu_1 = \mu_2$
$H_1 : \mu_1 \neq \mu_2$

Also, $v = n_1 + n_2 - 2 = 18$. With $\alpha = 0.05$, the reject criterion for H_0 becomes: $|t| > t_{0.025,18}$. From Table 3.1, $t_{0.025,18} = 2.1009$. The pooled variance, S_p^2, with the given information is found to be 0.081, which yields $S_p = 0.284 \; \text{emu}/\text{g}$. Subsequently, the test statistic is found to be

$$t_0 = \frac{\bar{y}_1 - \bar{y}_2}{S_p\sqrt{\frac{1}{n_1} + \frac{1}{n_2}}} = \frac{16.76 - 17.92}{0.284\sqrt{\frac{1}{10} + \frac{1}{10}}} = -9.13$$

Comparing this t-ratio to the table (critical) value, $|-9.13| > 2.1009$, and thus H_0 should be rejected. This means, with a minimum confidence level of 95%, one can conclude the treatment factor (polymer coating) has a significant effect on the mean response. Note that without such a statistical test, an analyst would mistakenly assume that the difference between the two mean values (16.76 emu / g and 17.92 emu / g) is insignificant and no effect of the treatment factor exists. The high effect of polymer coating has become evident for several types of SPION as reviewed in [20].

3.2. Hypothesis Testing for Outlier Detection – Grubbs' Test

A similar two sample t-test procedure discussed in Section 3.1 may be used for detecting outlier (unusual) points within experimental datasets. For instance, assume that you are given *n* data points from an MTT assay experiment. One sample for the t-test can be formed by including all the *n* data points. A second sample can be formed by including (*n*-1) points where a suspected outlier point has been excluded from the original dataset. A t-test between these two samples can assess the likelihood of the suspected point being actually an outlier. This procedure is formally known as Grubbs' Test.

In Grubbs' outlier detection method, a G-ratio (may also be shown as a t-ratio) is defined as follows.

$$G = \frac{|y - \bar{y}|}{S}$$

y is the suspected outlier point; $\bar{y}$ is the sample mean, and S is the sample standard deviation (which would be an estimation of the population standard deviation). After calculating the G-ratio for the suspected outlying value, it is compared to the significance critical values for the test (Table 3.2). If the calculated value of G is equal to or exceeds the critical value, the suspected point is considered an outlier and it should be removed from the dataset. If G is smaller than the table value, the suspected point is not an outlier and should be retained in the dataset. Statistically, such test hypotheses are written as follows.

$$\begin{cases} H_0 : y \text{ is not an outlier in the data set} \\ H_1 : y \text{ is an outlier in the data set} \end{cases}$$

It can be shown that for a sample of size n, and assuming an α-level significance, the critical (tabulated) value of the G-test is found by:

$$G_{critical} = \frac{(n-1)}{\sqrt{n}} \sqrt{\frac{t^2_{\alpha/(2n),n-2}}{n-2+t^2_{\alpha/(2n),n-2}}}$$

Finally, it is to note that the outlier test can be done/repeated on any suspected point with a value such as y; however, in practice, the smallest or the largest value in a sample would be more suspected.

Table 3.2. Critical values for the two-sided test of outliers based on a level of significance [13]

Sample size, n	Critical values for Grubbs' test ($\alpha = 0.05$)
	1.155
3	1.481
4	1.715
5	1.887
6	2.020
7	2.126
8	2.215
9	2.290
10	2.355
11	2.412
12	2.462
13	2.507
14	2.549
15	2.585
16	2.620
17	2.651
18	2.681
19	2.709
20	2.822
25	2.908
30	2.979
35	3.036
40	3.128
50	3.383
100	

3.3. DESIGN OF EXPERIMENTS

Experiments are often conducted by scientists to understand the effect of different factors on materials' (e.g., mechanical, chemical, thermal) behaviour. Can we improve the response of a material by controlling/varying the treatment factors during a particular process/application? The best tool for analyzing the experimental factors in a systematic way is designing our experiments ahead of time. Using a proper design of experiment (DOE), we may even be able reduce the number of necessary experiments to complete our analysis, as opposed to the trial and error approaches that can be time consuming and/or costly. A DOE analysis starts with a complete recognition of the problem at hand, followed by the selection of the number of factors and their levels for the analysis. One should also be able to select the response variable(s) of interest. He/she can then choose one of the pre-designed experimental tables from the literature and conduct experiments at different combination of factor levels as suggested by the DOE tables. In some cases, analysts prefer to design their own tables to meet particular constraints that they may encounter during their application. After conducting the experiments, the analyst can perform different statistical analyses such as hypothesis testing or the analysis of variance (ANOVA) to draw statistical conclusions on the effect of different factors on the mean response. In fact, it is based on these effects that one would make his/her recommendations to improve a product/material/process quality.

Several experimental strategies have been established in the literature. These include full fractional, fractional factorial, central composite, Taguchi, D-optimality, and uniform designs (see, e.g., [14,15, 21]). To show a few examples, Table 3.3 includes a uniform design and a Taguchi design with different number of factors, levels, and constraints. In this table, levels (e.g., 1, 2,...,) refer to the actual values of a factor. For instance, level 1 of a stirring rate factor-$x1$ means 1750 rpm, whereas level 2 of the factor means 2000 rpm. A main advantage of fractional factorial designs, such as the uniform and Taguchi designs, is that they can significantly reduce the total number of experiments (especially when the total number of runs should not exceed a threshold due to time/cost constraints). On the other hand, full factorial designs are advantageous in that they give more thorough information on the effect of factors. In a full factorial design, all possible combinations of factor levels are experimented. When more complex information regarding the effect of factors, e.g., their interaction is required, full factorial designs are chosen with lower number of levels. Factor interactions are normally analyzed by means of interaction plots. A sample of such plot is shown in Figure 3.2. Each line in the plot indicates the variation of the mean response with factor A when it changes from low (-) to high (+), while keeping the B factor at a fixed level (+/-). Here '+' and '-'levels are identical to level numbers 1 and 2 used in Table 3.1.

Recalling Figure 3.2, assume that we are interested in maximizing the response. If factor B is fixed at its low level (B-), then a higher value of A yields a better response. On the other hand, if B is fixed at its higher level (B+), an opposite effect of A on the response is seen. This is a severe case of two-factor interaction. In practice interactions may happen to be low or medium. Such information can be extremely important for an analyst in optimizing the response. For instance, in the case of an interaction similar to Figure 3.2, it may not be sensible to examine either of the main effects (A or B) separately and their effects must be assessed simultaneously. A experimenter may be misled by fixing fix one of these factors and looking at the effect of the other factor to achieve an optimum response. Finally, it is worth noting that while interaction plots are helpful in interpreting experimental results, if the random error/non-repeatability during an experiment is very large, the interaction effect can become insignificant relative to the measurement error, even if it appears significant in the plot. This information can be formally quantified by means of ANOVA as shown through an illustrative example in the following section.

Table 3.3. Examples of designed experiments

2 factors, 6 levels, Constraint: maximum of 6 experiments can be conducted	2 factors, 7 levels Constraint: maximum of 9 experiments can be conducted
Uniform design (centered L2): x1 x2 ------------- 5 5 4 1 2 2 3 6 1 4 6 3 Note: the total of possible experiments would be 2^6 and it has been reduced to 6.	Taguchi L8 design: | Experiment Number	Column
	1 2 3 4 5 6 7
1	1 1 1 1 1 1 1
2	1 1 1 2 2 2 2
3	1 2 2 1 1 2 2
4	1 2 2 2 2 1 1
5	2 1 2 1 2 1 2
6	2 1 2 2 1 2 1
7	2 2 1 1 2 2 1
8	2 2 1 2 1 1 2 | Note: the total number of experiments would be 2^7 and it has been reduced to 8.

See also: http://www.stat.psu.edu/~rli/DMCE/ UniformDesign/ & http://www.york.ac.uk/depts/maths/tables/taguchi_table.htm

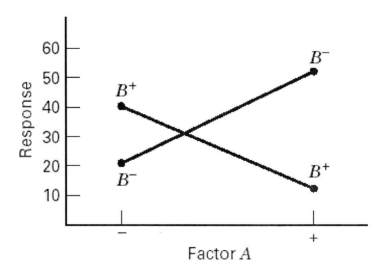

Figure 3.2. Example of a severe interaction between factors A and B.

Application of Statistical Techniques

Example 2:

During synthesis of a set of SPIONs via the coprecipitation technique, the effect of two different molarities of an Iron chloride (NaOH) solution was examined. The two chosen levels are 1.2 and 1.4. The homogenization rate was also varied at two levels (3600 rpm and 9000 rpm). To design the experiment, a 2^2 factorial design was used and at each factor combination (rpm, molarity), two replicates of the test (the synthesis) was performed. Using the XRD technique, the measured particles size values (in nm) are reported in the following table. A factor level -1/+1 refers to the actual low/high value of that factor (e.g., -1 for factor A means 3600 rpm). The column of AB interaction factor is shown in the table for subsequent calculation purposes (it is simply formed by multiplying the A and B columns). The numbers in bracket show the order of runs during experimentation.

It is desired to:
1. It is desred to:calculate the factor effects.
2. pPlot the main effects as well the AB interaction effect.
3. perform a two-way ANOVA and draw conclusions regarding the significance of treatment factors (rpm and morality) on minimizing the mean particle size. Which factor is more significant?
4. write out a prediction model.

	Test Matrix Coded Factor Levels			Measured response (y)		
				Two Replicates		
Factor combination #	A: RPM	B: Molarity	AB	I	II	Total
1	-1	-1	+1	5.2 (3)	5.1 (6)	10.3
2	+1	-1	-1	4.4 (4)	4.5 (7)	8.9
3	-1	+1	-1	4.0 (1)	4.0 (2)	8.0
4	+1	+1	+1	3.8 (5)	3.9 (8)	7.7

Solution:

The effect estimate for each factor can be done by subtracting the average response at that factor's high and low levels. This is shown below. Note that each row of the experimental matrix represents a sample of size two (two replicates).

$$\text{eff}A = \bar{y}_{A^+} - \bar{y}_{A^-} = \frac{8.9+7.7}{4} - \frac{10.3+8.0}{4} = -0.425$$

$$\text{eff}B = \bar{y}_{B^+} - \bar{y}_{B^-} = \frac{8.0+7.7}{4} - \frac{10.3+8.9}{4} = -0.875$$

$$\text{eff}AB = \bar{y}_{AB^+} - \bar{y}_{AB^-} = \frac{10.3+7.7}{4} - \frac{8.9+8.0}{4} = 0.28$$

From the factor effects, one can readily conclude that considering A alone, by increasing this factor, a lower response may be obtained (i.e., negative effect). A similar trend but with a sharper slop is expected for B. Also, the interaction effect between A and B is not zero. To visualize these effects, the main factor and interaction plots are shown in Figures 3.3 (a) to (c). Each square point corresponds to the average value of the response at the corresponding factor level. For the sake of completeness, the error bars at each point (resulting from the replicates of the test) are included in the interaction plot. The calculated effects are proportional to the slopes of these plots.

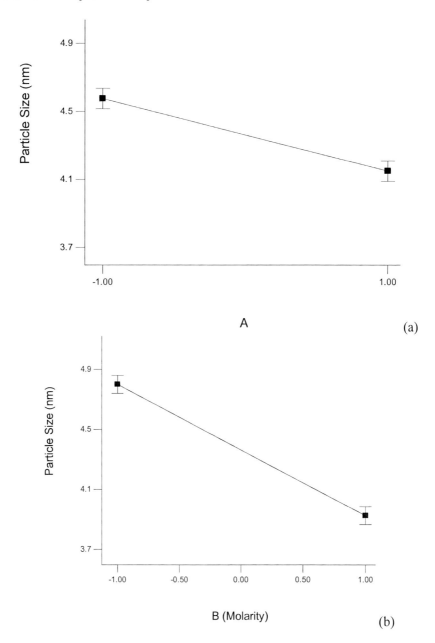

Figure 3.3 Continued

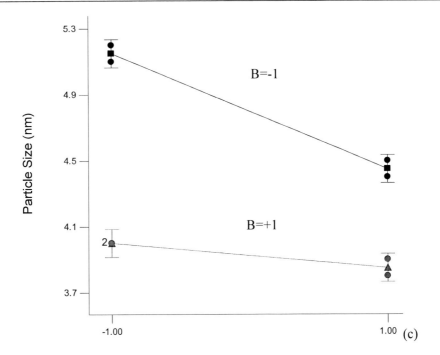

Figure 3.3. (a and b) Main factor plots, (c) the interaction plot.

To perform a two-way analysis of variance with the two controlled (non-random) effects A and B, a standard ANOVA display can be used as shown in Figure 3.3. In this ANOVA, a, r, and n refer to the number of levels of A, number of levels of B, number of the test replicates at each point, and the total number of experiments, respectively. By using the computational formulas in the bottom row of the display, the goal is to find the ratio values in the last column of ANOVA table. y_{ijt} denotes a data point with the i-th level of factor A, j-th level of factor B, and the t-th replicate. The 'dot' subscripts refer to a summation. For example, $y_{1..}$ means the sum of all observations when A is set at level 1 (i.e., 8.9+7.7). Thus, the calculation of formula for ssA reads

$$ssA = \frac{(8.9+7.7)^2 + (10.3+8.0)^2}{2\times 2} - \frac{(8.9+7.7+10.3+8.0)^2}{2\times 2\times 2} = 0.36.$$

The ratios in the last column of the display specify the magnitude of sum of squares due to factors relative to the sum of squares due to the random (experimental) error. A higher ratio means a higher factor significance. More precisely, the ratios are F-statistics that can be compared to the table values from an F-distribution. The result of the comparison is often presented by DOE packages (such as Minitab and Design-Expert) by a term called "Prob > F". A probability (Prob) value less than 0.05 (i.e., more than 95% confidence level) indicate that model terms are statistically significant. Finally, the percentage contribution of significant factors to the total variance can be found by dividing factor's sum of squares to the total sum of squares (sstot). For instance in Table 3.4,

$$A\% = \frac{ssA}{sstot} = \frac{0.36}{2.06} \times 100 = 17.47\%.$$

Table 3.4. Two-way ANOVA display including interaction effect and test replications [14]

Source of Variation	Degrees of Freedom	Sum of Squares	Mean Square	Ratio
Factor A	$a-1$	ssA	$\frac{ssA}{a-1}$	$\frac{msA}{msE}$
Factor B	$b-1$	ssB	$\frac{ssB}{b-1}$	$\frac{msB}{msE}$
AB	$(a-1)(b-1)$	$ssAB$	$\frac{ssAB}{(a-1)(b-1)}$	$\frac{msAB}{msE}$
Error	$n-ab$	ssE	$\frac{ssE}{n-ab}$	
Total	$n-1$	$sstot$		

Computational Formulae for Equal Sample Sizes

$$ssE = \sum_i \sum_j \sum_t y_{ijt}^2 - \sum_i \sum_j y_{ij.}^2/r$$

$$sstot = \sum_i \sum_j \sum_t y_{ijt}^2 - y_{...}^2/n$$

$$n = abr$$

$$ssA = \sum_i y_{i..}^2/(br) - y_{...}^2/n$$

$$ssB = \sum_j y_{.j.}^2/(ar) - y_{...}^2/n$$

$$ssAB = \sum_i \sum_j y_{ij.}^2/r - \sum_i y_{i..}^2/(br) - \sum_j y_{.j.}^2/(ar) + y_{...}^2/n$$

Table 3.5. ANOVA results in the given example

Source of Variation	Degrees of Freedom	Sum of Squares (ss)	Mean Square	F-Ratio	Prob > F	Percentage Contributions
A	1	0.36	0.36	96.33	0.0006 (significant)	17.47%
B	1	1.53	1.53	408.33	< 0.0001 (significant)	74.27%
AB	1	0.15	0.15	40.33	0.0031 (significant)	7.28%
Error	4	0.015	3.750E-003			0.72%
Total	7	2.06				

The ANOVA results (Table 3.5) imply both the homogenization rate and the (NaOH) molarity and their interactions are significant. The more significant factor in minimizing the mean particle size of SPIONs is the molarity (with a percentage contribution of 74.25%). The interaction effect is present (with a contribution of ~7.28%), but according to Figure 3.3(c), it does not signify a severe type (notice that both lines have a negative slope, as opposed to the severe case shown in Figure 3.2). This means, by increasing the molarity value a lower size may be expected regardless of the homogenization rate level; of course this conclusion is only reliable at the experimented factor levels; more levels may be chosen to give more detailed information on the factor effects. The response surface shown in Figure 3.4 shows the contours of particle size response in the factors space. Curved contours indicate some interaction effects between the treatment factors. It is also notable that there are several combinations of the rate and molarity values that give the same particle size. The smallest particle size (~3.85 nm) falls at the top right corner of the graph where both the homogenization rate and molarity values are at their high levels: A=+1 and B=+1.

Once the significant factors and their interactions are identified, a simple 'effect model' can be used to predict the response in the factors' space. In the present example, given that all the three effects are significant, the prediction model reads

$$\hat{y} = \bar{\bar{y}} + \frac{effA}{2} \times A + \frac{effB}{2} \times B + \frac{effAB}{2} \times AB$$

where, $\bar{\bar{y}}$ is the average of all data points. Substituting the calculated effects from the first part of the example, the model follows

$$\hat{y} = 4.36250 - 0.21250\,A - 0.43750\,B + 0.13750\,AB$$

The factor values in this prediction equation are coded and correspond to their levels. For instance, at a stirring rate 0f 900 rpm and a molarity of 1.4, the particle size is predicted by

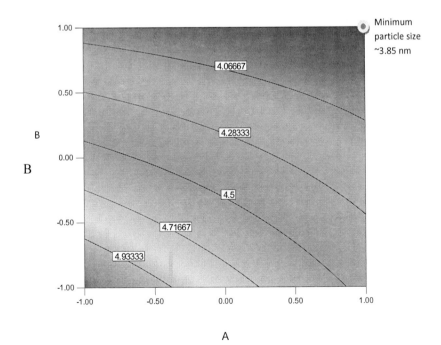

Figure 3.4. Particle size response variation in the space of design factors.

$$\hat{y}\Big|_{A=1,B=1} = 4.3625 - 0.2125(1) - 0.4375(1) + 0.1375(1)(1) = 3.85 \, \text{nm}$$

Such prediction models can be particularly useful to find the response value at untried points. In that case, the factor levels may be found by interpolation. For instance, a stirring rate of 6300 RPM would mean a '0.5' level for factor A.

Alternatively, one can select a particular form of prediction function (polynomial, exponential, etc) and curve fit a model. In that case, the model is referred to as the 'regression model'. In the present example, if one chooses the regression model to be of a linear form, a least square estimation via curve fitting reads

$$\text{Minimize } RSS = \sum_{k=1}^{n} \left(y_k^{observation} - y_k^{prediction} \right)^2$$

where n is the total number of data points (here $n=8$) and $y^{prediction}$ is assumed to be:

$$y^{prediction} = \hat{y}(A,B) = \alpha_0 + \alpha_1 \times A + \alpha_2 \times B + \alpha_3 \times AB.$$

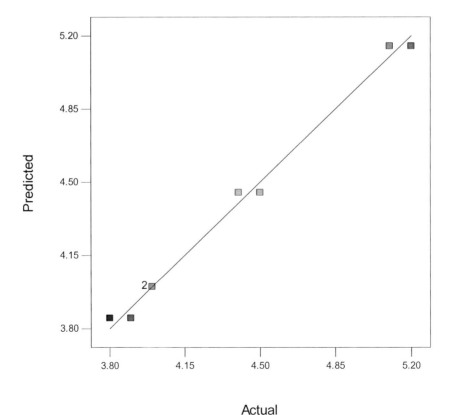

Figure 3.5. Predicted values versus actual/observed values in Example 2.

$(\alpha_0, \alpha_1, \alpha_2, \alpha_3)$ are model constants and are found upon the minimization of the RSS function. The regression model coefficients in this example will be identical to those of the effect model reported above. The R^2-value is found to be 99.27% for the linear model (suggesting a good quality of fit). This would be expected by noting a low error percentage in Table 3.5. If a poor fit is found, one can choose models with higher order effects such as A^2, B^2. The quality/ *lack of fit* can also be judged from a plot of predicted values versus the actual (observed) values (Figure 3.5).

Remark: How to Perform the Analysis with More than Two Factors, Two Levels?

When there are more than two main t factors and/or the numbers of levels for all or some of the factors are more than two, the DOE/ANOVA procedure can be similarly performed, though over higer dimesional factor spcaes. Practially, as the number of treatment factors and their levels increase, fractional factorial tables would be more preferred over the full factorial designs. Also, the dimension of interaction (cross product) terms among treatment factors will become larger. For instance, if one deals with three factors A, B, and C, the interaction terms of interest can be AB, AC, BC, and ABC. If the experimenter has a priorknowledge of interaction effects, cross product terms may be assumed to be minimalduring analysis. In Table 3.3, this would mean that no row for the SS contribution of the interaction term may be included (hence potentially making the random error contribution larger). Regardless, the formulation of SS for single-factor effects remains unchanged. Using powerful statistical and DOE tools such as SAS, Minitab, and Design Expert, high-dimensional experimental design and optimization problems are handled with no significant computational time. However, it is of outmost importance to be aware of assumptions and principles underlying the model calculations.

3.3.1. Checking Model Assumptions: Effect of Potentially Outlying Observations

After performing an ANOVA, there are four main fundamental assumptions that need to be checked before making final recommendations regarding optimum factor levels, significance of parameters, etc. This step of analysis is often referred to as post-ANOVA or *diagnostics* analysis.

- Checking for the independence/randomness of errors.
- Checking for the constant variance assumption of errors.
- Checking for normality of errors (are the sample data at each factor combination taken from a normal distribution?)
- Checking for outliers (unusual, very large or very low, observations).

The bulk of these diagnostics is based on the so-called 'residuals' which is calculated from the prediction model and observed data. Formally, a residual at a given point is defined as the difference between the actual observation and the model prediction. For instance, in Example 2, for a particular observation at the *t*-th replication of a test with factor A and B, respectively, fixed at the ith and j^{th} levels, the residual is defined as $e_{ijt} = y_{ijt} - \bar{y}_{ij.}$, where, $\bar{y}_{ij.}$ is the average of replications at the corresponding factors' combination. For instance, $e_{111} = y_{111} - \bar{y}_{11.} = 3.80 - 3.85 = 0.05$ nm. For an unbiased estimation, the expected value of this residual is zero, meaning that the expected value of the observation and model prediction are the same: $\hat{y}_{ij} = E(y_{ij}) = \bar{y}_{ij}$.

To perform the diagnostic checks, it is often recommended to work with normalized (also called standardized) residuals rather than the non-normalized ones. This is especially useful during the identification of outliers. In general, a standardized residual for the k-th observation can be found from e_k as:

$$z_k = \frac{e_k}{s_{e_k}}$$

where s_{e_k} is the standard deviation of the residual e_k and may be estimated by the square root of the mean square error from ANOVA, s. In fact, the normalized residual is a ratio indicating the number of standard deviations that separates the actual and predicted values. Alternatively, advanced statistical packages use a so-called 'internally Studentized' residual in which the standard deviation of a residual is more accurately estimated by $\frac{e_k}{s\sqrt{1-h_{kk}}}$. The quantity h_{kk} is the *leverage*, a measure of the *influence* of an individual observation point on the model predictions. Leverages range from 0 to 1.0. A leverage of 1 means that the predicted value is equal to the observed value in that particular point and, thus, any potential measurement/bias error associated with that observation is carried into the model and its predicted values.

Having calculated the normalized residuals (standardized or stundentized), the independent error assumption is checked by plotting the residuals against the observation order. The residuals in this plot should be randomly scattered around zero with no particular smooth pattern. For the analysis in Example 2, this diagnostic plot is shown as in Figure 3.6.

Once the error independence check is passed, one should check the equal-variance assumption by plotting the plot of residuals versus the fitted values. In this plot, the residuals should lie within a uniform band and not scale up or down with the variation of mean response. For the analysis of Example 2, this plot is shown in Figure 3.7. If the plot indicates unequal variances, a transformation of data (e.g., natural log, square root, inverse, etc) may be used to remedy the problem; see, e.g., [14,15]. Depending on the type of experimentation, it is possible that that the error variance increases/decreases as the mean response increases (indicating a 'megaphone' pattern). It is also to note that the experience shows that keeping the sample size unchanged during a DOE analysis yields less erroneous conclusions, unless there is a very good knowledge of which treatment factor levels are more sensitive.

For the normality assumption, it is checked whether or not the error terms (residuals) come from a normal distribution. This is done by a 'normal probability plot' in which residuals are compared against their expected normal % probability. If the points in this plot *approximately* follow a linear trend, it can be said that the normality assumption is satisfied (larger sample sizes result in more accurate check of this assumption). For the analysis of Example 2, the normality plot is shown in Figure 3.8.

It is worth ending this chapter with emphasizing that in practice, *one source of deviation from a model assumption may cause other deviations.* Hence, an outlier point may be mistakenly identifed due to the error variables not being normally distributed, or having different variances, or an incorrect choice of a prediction model. As suggested by several textbooks, if all model assumptions hold correctly, including normality, the majority of standardized/stundetized residuals should *approximately* reside between −3 and +3. If some points show standardized/stundentized residuals beyond this range, they are then suspected of being outlier and may be removed from the dataset (or the actual experiment may be repeated at those points). "If there are more outliers than expected under normality, then the true confidence levels are lower than stated and the true significance levels are higher [14]." According to Figure 3.7 no outliers are detected in Example 2.

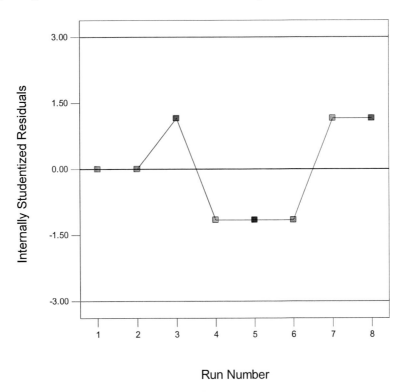

Figure 3.6. A plot of residuals versus the run number (checking the independence of errors).

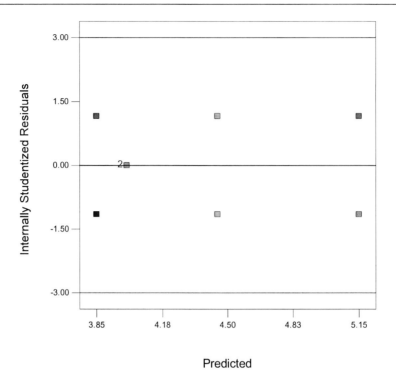

Figure 3.7. A plot of residuals versus the predicted values (checking the constant variance assumption).

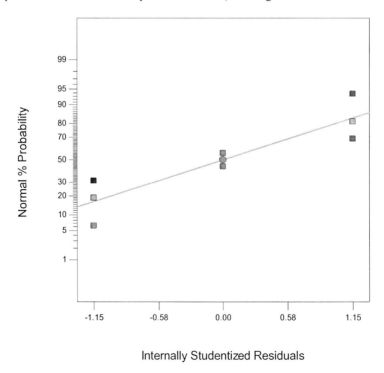

Figure 3.8. A plot of residuals versus their normal % probability (checking the normality assumption).

REFERENCES

[1] Moroz, P., Jones, S.K. and Gray, B.N. (2002) 'Magnetically mediated hyperthermia: current status and future directions', *International Journal of Hyperthermia*, Vol. 18, pp.267–284.

[2] Gonzales-Weimuller, M., Zeisberger, M. and Krishnan, K.M. (2009) *Journal of Magnetism and Magnetic Materials*, Vol. 321, No. 4.

[3] Gazeau, F., Levy, M. and Wilhelm, C. (2008) *Nanomedicine*, Vol. 3, No. 14.

[4] Jolivet, J.P. (1994) De la Solution a l'Oxyde, *Inter Editions et CNRS Editions*, Paris, France.

[5] Babes, L., Denizot, B., Tanguy, G., Le Jeune, J.J. and Jallet, P. (1999) 'Synthesis of iron oxide nanoparticles used as MRI contrast agents: a parametric study', *Journal of Colloid and Interface Science*, Vol. 212, pp.474–482.

[6] Vayssieres, L., Chaneac, C., Tronc, E. and Jolivet, J.P. (1998) 'Size tailoring of magnetite particles formed by aqueous precipitation: an example of thermodynamic stability of nanometric oxide particles', *Journal of Colloid and Interface Science*, Vol. 205, pp.205–212.

[7] Jiang, W.Q., Yang, H.C., Yang, S.Y., Horng, H.E., Hung, J.C., CHEN, Y.C. and HONG, C.Y. (2004) 'Preparation and properties of superparamagnetic nanoparticles with narrow size distribution and biocompatible', *Journal of Magnetism and Magnetic Materials*, Vol. 283, pp.210–214.

[8] Massart, R. and Cabuil, V. (1987) *J. Chem. Phys.*, Vol. 84.

[9] Mahmoudi, M., Simchi, A., Imani, M., Milani, A.S. and Stroeve, P. (2008) 'Optimal design and characterization of superparamagnetic iron oxide nanoparticles coated with polyvinyl alcohol for targeted delivery and imaging', *Journal of Physical Chemistry B*, Vol. 112, pp.14470–14481.

[10] Mahmoudi, M., Simchi, A., Milani, A.S. and Stroeve, P. (2009) 'Cell toxicity of superparamagnetic iron oxide nanoparticles', *Journal of Colloid and Interface Science*.

[11] Merikhi, J., Jungk, H.O. and Feldmann, C. (2000) 'Sub-micrometer CoAl2O4 pigment particles – synthesis and preparation of coatings', *Journal of Materials Chemistry*, Vol. 10, pp.1311–1314.

[12] Miller, M.M., Prinz, G.A., Cheng, S.F. and Bounnak, S. (2002) 'Detection of a micron-sized magnetic sphere using a ring-shaped anisotropic magnetoresistance-based sensor: a model for a magnetoresistance-based biosensor', *Applied Physics Letters*, Vol. 81, pp.2211–2213.

[13] Bolton, S., Dekker, M. Pharmaceutical statistics: *Practical and clinical applications. Second edition.* By: New York. 1990.

[14] Dean, A., Voss, D., *Design and Analysis of Experiments*, NY: Springer, 1999.

[15] Montgomery, D.C., *'Design and Analysis of Experiments'*, NY: Wiley, 2005.

[16] Yoon, K. P., Hwang, C.-L., *'Multiple Attribute Decision Making: An Introduction'*, CA: SAGE Publications, 1995.

[17] Shanian, A., Milani, A.S., Carson, C., Abeyaratne, R.C. (2008) "A new application of ELECTRE III and revised Simos' procedure for group material selection under weighting uncertainty", *Knowledge-Based Systems*, 21(7): 709-720.

[18] Jahan, A., Ismail, M.Y., Sapuan, S.M., Mustapha, F. (2010) *"Material screening and choosing methods – A review"*, Materials and Design, Vol. 31: 696-705.

[19] Mahmoudi, M., Shokrgozar, M.A., Simchi, A., Imani, M., Milani, A.S., Stroeve, P., Vali, H., Hafeli, U.O. and Bonakdar, S. (2009) 'Multiphysics flow modeling and in vitro toxicity of iron oxide nanoparticles coated with poly (vinyl alcohol)', *Journal of Physical Chemistry C*, Vol. 113, pp.2322–2331.

[20] Mahmoudi, M., Milani, A.S., Stroeve, P. (2010) "Surface architecture of superparamagnetic iron oxide nanoparticles for application in drug delivery and their biological response: A review", Special Issue on Advancing Drug Delivery Systems with Nanotechnology, *International Journal of Biomedical Nanoscience and Nanotechnology* (in press).

[21] Bate, S.T., Jones, B. (2008) "A review of uniform cross-over designs", Special Issue on Statistical Design and Analysis in the Health Sciences, *Journal of Statistical Planning and Inference*, Vol. 138 (2): 336-351.

Chapter 4

COATINGS OF SPIONS

Since synthesized bare SPIONs normally tend to reduce their surface energy, clusters and agglomerates are formed; hence, the colloidal stability of magnetic suspensions, which is recognized as a crucial factor for bio-applications of SPIONs due to the very fast detection of unstable colloidal nanoparticles by immune system, is significantly reduced. In principle, the resultant between the attractive and repulsive sides of three fundamental forces including van der Waals, electrostatic and magnetic dipolar forces can control the stability of ferrofluids. It is noteworthy that the steric repulsion is often added to the mentioned forces for polymeric coated SPIONs. The stability of ferrofluids can be tracked via dynamic light scattering (DLS) method. [1-6] It is important to note that the bare SPIONs which are synthesized in chemical routes are covered by hydroxyl groups. The reason is that the iron atoms, which are located at the surface of SPIONs, can act as Lewis acids and coordinate with molecules that donate lone-pair electrons; hence, Fe atoms coordinate with water, which dissociates readily to leave the SPIONs surface. Another problem, which is associated with reducing the colloidal stability, is the isoelectric point of bare SPIONs which is similar to that of the biological environments. [7-9] It is notable that at isoelectric point, the surface charge density of SPIONs is around zero and the nanoparticles have great capability for agglomeration and losing their stability. In order to change the isoelectric point together with controlling the resultant forces in the colloidal ferrofluids, SPIONs should be coated with appropriate materials. These employed materials should have the capability to increase the surface hydrophilicity of SPIONs together with their colloidal stability and dispersability without degradation of their magnetic properties. The ability to escape from reticuloendothelial system (RES) and the consequent increase in the blood circulation half-life may be regarded as another advantage of using coatings for the SPIONs in biomedical applications. [10-13] Furthermore, coatings should be able to isolate the magnetic core from *in vivo* environment, improving the biocompatibility of the magnetic nanomaterials. The type and amount of coating has been recognized as a supercritical matter in the biological applications of magnetic materials. In this chapter, different coating systems including polymeric and non-polymeric coatings are discussed.

4.1. COLLOIDAL STABILITY OF SPIONS

As mentioned earlier, the colloidal stability of the synthesized SPIONs is generally controlled by three main forces including hydrophobic-hydrophilic, magnetic, and van der Waals forces. Due to its hydrophobic interactions, SPIONs have a tendency to form aggregations. The obtained aggregates (e.g. in the form of clusters) can participate in the further aggregation process due to the magnetic dipole-dipole interactions. In addition, in the presence of an external magnetic field, extra magnetization of these clusters can happen and can cause even more aggregation [14]. For minimizing the total surface/interfacial energy, nanometer size SPIONs are aggregated in colloids via attractive van der Waals forces.

Polymeric and non-polymeric coatings are extensively employed on the surface of SPIONs as stabilizing agents during the preparation process of SPIONs. [15-25] In order to use SPIONs in biomedical applications, the employed molecules should be biocompatible, biodegradable, and functionalizable. Wang *et al* [26] have reported a method where hydrophobic surface of coated SPIONs was inverted to hydrophilic surface via alpha cyclodextrin by host-guest interactions and the nanoparticles were able to disperse from the organic to aqueous solution. Inversion of hydrophobic surfaces to hydrophilic surfaces has also been reported by *Pellegrino et al* [27] using an amphiphillic polymer shell. Various polymers such as poly (ethylene-co-vinyl acetate), poly vinylpyrrolidone (PVP), polylactic-co-glycolic acid (PLGA), polyethylene glycol (PEG),: polyvinyl alcohol (PVA) have also been used as coating materials in aqueous suspension [28]. Natural dispersants such as gelatin, dextran, polylactic acids, starch, albumin, liposomes, erythrocytes, chitosan, ethyl cellulose have also been used for dispersion in aqueous medium. Post-synthesis modification of SPIONs has been widely studied as core-shell nanoparticles. Materials used for creation of rigid core-shell SPIONs are mostly silica, metal (e.g. gold and cadmium/selenium) and carbon. The rigid shells used on the surface of SPIONs are known to be biocompatible but not bio-degradable; the distinguished feature of these shells is their protection from the toxic Fe^{2+} ions penetration due to the SPIONs wrapping in biological environments. [29, 30]

4.2. POLYMERIC COATING

Polymers have been used extensively as coatings for SPIONs due to their biocompatibility together with their capability for multi-task role such as colloidal stabilization, delivering biologically active agents with a controlled release profile, and targeting capability to specific tissues via conjugation with specific ligands. [75] In addition, the polymeric coatings can be induced either during [31, 32] or after synthesis [17-20, 33] depending on the desired chemical and physical properties. The well-known polymers, which were employed as coating of SPIONs, are polyethylene glycol (PEG) [34, 35], polyethylene glycol fumarate (PEGF) [18], polyvinyl alcohol (PVA) [17, 19, 25, 36], polyacrylic acid (PAA) [37-39], poly(N-isopropylacrylamide) (PNIPAm) [40-42], dextran [43-55], poly(D,L-lactide) [56-58], poly(D,L-lactide-co-glycolide) (PLGA) [59, 60], alginate [61-84], chitosan [85-98], and polyethylenimine [99-105].

Linear PEG

$$H-(OCH_2CH_2)_n-OH$$

Linear monomethoxy–PEG

$$CH_3-(OCH_2CH_2)_n-OH$$

Figure 4.1. Structural formulae of the polyethylene glycol molecules.

The main disadvantages of polymeric coating is the penetration of toxic ions (e.g. Fe^{2+}) due to the degradation of SPIONs (e.g. via enzyme) in biological environments. The advantages of non-polymeric coatings, such as silica [106-127] and gold [99, 128-141], have been focused on the protecting of SPIONs from biomolecules attack.

4.2.1. Polyethylene Glycol

Polyethylene glycol (PEG), which is one of the most important hydrophilic and water soluble polymers, is recognizing as a member of polyether family. The PEG (see Figure 4.1) ability for creating various functional groups (e.g. amine, carboxylic acid and hetrobifunctional derivatives), due to their one (for monomethoxy–PEG) or two terminal hydroxyl functional groups, has been significantly used to conjugate different biological moieties. The era of PEG usage in biomedical applications were started from its great potential as safe drug carrier. There were many problems in drug delivery systems, for instance polypeptide drugs, such as their susceptibility to destruction via proteolytic enzymes, short circulation time, low solubility, rapid blood clearance via kidneys and even their tendency to produce neutralizing antibodies. [142] In order to overcome the aforementioned problems, scientists have conducted comprehensive investigations on the improvement of clinical properties of polypeptides (e.g. via altering amino-acid sequences) and on employing the vehicles carriers (e.g. liposome). [143, 144] According to the results, for instance by *Davis et al.* [145], the most power full system was the pegylation which is the attachment of PEG to bio-molecules (e.g. protein and drug molecules). [146, 147] It is notable that pegylation could significantly enhance both pharmacokinetic (i.e. the effect of the body fluid on drugs such as absorption, distribution, metabolism and excretion of drugs) and pharmacodynamic (i.e. the effect of drugs on the body such as drug receptor, efficacy, response and adverse effect such as nausea and viral load) properties of the peptide drugs due to its ability to water solubility enhancement, reduction of renal clearance together with biocompatibility improvement. [148]

A suitable characteristic of PEG is its great biocompatibility and toxico-kinetics profile, which has caused its approval in Food and Drug Administration (FDA) of USA. One of the major problems during *in vivo* applications of SPIONs, more specifically for targeted delivery and imaging applications, is their fast removal from the blood stream. A significant decrease in the ferrofluid efficacy is achieved due to the fast removal of SPIONs from the blood. In order to increase their stability in blood stream, PEG has been used as a promising candidate as surface coating. Beside drug maintenance, it has been confirmed that the blood circulation half-life is considerably enhanced by employing the PEG as a coating on the surface of

SPIONs due to its hemocompatible, non-antigenic and non-immunogenic interactions with bloodstream. The latter causes negligible reticuloendothelial system (RES; i.e. an association of body immune cells (e.g. white blood cells such as macrophages, T- and B- cells) with capability of phagocytosis) detection. [12, 13, 34, 149-155]

In addition, the coated SPIONs with engineered PEG complexes (i.e. coupling of 6-hydroxy-chromone-3-carbaldehyde to the PEG-DPA (dopamine)-SPIONs) [156] has shown noteworthy enhancements in the colloidal stability in physiological mediums, which is recognized as a crucial factor for the bio-application of SPIONs. [157, 158] PEG and folic acid coated SPIONs could increase the cell labeling (via intracellular uptake) for the targeting purposes, which have been probed for both mouse macrophage and human breast cancer cell lines. [109]

The PEG-coated SPIONs can enter to the intracellular environments due to their better interactions with lipid bilayers of cell membrane. [159] It is notable that the amphiphilic nature of the PEG could cause its dissolving into both polar and nonpolar solvents. [160] To obtaining very stable bonds between PEG and SPIONs, *Herve et al.* [161] employed a multi-step preparation of aqueous suspensions of PEGylated ferrofluids by functionalization of the SPIONs surfaces with PEG molecules by an intermediate of silanes. The stable initial SPIONs, was prepared and tailored by silanization, followed by anchoring PEG on these NPs through a covalent bond.

SPIONs showed very low efficacy for the brain targeted imaging and drug delivery applications, due to the blood–brain barrier (BBB) which could isolate the brain via its special endothelial cells and sealing location with tight junctions. [162] The BBB has a potential to prohibit the entrance of therapeutic compounds (e.g. for treatment of neurological or psychiatric disorders) to the brain. [146] Direct injection of SPIONs to brain tissue may cause high unpredictable risks for patients. Combination of n-hexadecylcyanoacrylate and PEG [163] nanoparticles could pass the BBB (without any damages) and penetrates to the brain. [164] Hence, using this combination at the surface of SPIONs could significantly increase their efficacy for reaching the brain tissue and treat the diseases by catching into deep areas of the brain (e.g. the striatum, hippocampus, and hypothalamus).

4.2.2. Polyethylene Glycol Fumarate

Let us recall that the stability of the coating materials has been recognized to be a crucial factor for successful bio-applications of SPIONs. The reason is that once the coated nanoparticles are placed in intracellular environment, the coating is likely to be digested. The digestion of coating causes an exposure of necked SPIONs to other cellular components and organelles; thus, the overall integrity of the cells could be controlled by the stability time of the coating. [90]

In order to increase the stability of polymeric coatings, the unsaturated polyesters, which are mainly based on propylene glycol/ethylene glycol and fumaric acid, have been proposed by researchers. [18, 165] The main advantages of these unsaturated aliphatic polyesters is their ability to create a crosslinked network around the SPIONs via curing of their unsaturated double bonds. [166-168] It is notable that these unsaturated polyesters have many other potential biomedical applications, e.g., in bone substitutes, cements or cartilage repair. [169-173] One of the great promising candidates from unsaturated polyesters family is the poly(ethylene glycol-*co*-fumarate) (PEGF) which is a cross-linkable and biodegradable polyester. [174] There are many approaches for the PEGF synthesis such as employing triethyl amine or potassium carbonate (K2CO3) as a proton scavenger (such as propylene oxide (PO)) [169, 171]. Another method includes using chemical reduction/oxidation reaction (RedOx) as initiator systems such as ammonium persulfate/ascorbic acid (APS/AA) or ammonium persulfate/*N,N,N′,N′*-tetramethylethylenediamine (APS/TEMED) which are used in preparing the polymer networks thereof. [175]

These polyesters are not self-crosslinkable, so the crosslinking process should be initiated by employing heat, chemical reaction, and UV or visible light irradiation. [18] It is noteworthy that various monofunctional/difunctional monomers (e.g. polyethylene glycol diacrylates [176, 177] and *N,N′* methylene bisacrylamide [178]) and vinyl monomers (e.g. as *N*-vinyl-2-pyrrolidone and styrene [170, 179]) have been utilized for facilitating the crosslinking reaction.

One of the highly unsaturated macromers is fumaric acid which can be crosslinked with or without employing the multifunctional crosslinking agent (e.g. poly(propylene fumarate)-diacrylate (PPF-DA) and poly(ethylene glycol)-diacrylate (PEG-DA) or dimethacrylate) [180] or reactive diluent to form the 3D polymeric networks. [171] *Mahmoudi et al.* [18] used the PEGF macromers ass a coating of SPIONs and crosslinked them by redox polymerization [18] in the presence of chemical initiator (Ammonium persulphate) for drug delivery applications (Figure 4.2). The specific feature of this novel composition is the reduction of either protein interaction on the surface of nanoparticles or the burst release effect (i.e. quick release of a large drug volume upon injection of the drug-loaded nanoparticles into the body) in drug delivery applications. The burst effect of loaded tamoxifen citrate (TMX, a member of anti estrogen compounds used as first line candidate in estrogen-responsive breast cancer) on the cross-linked PEGF-coated-SPIONs can be reduced by 21% in comparison with the non-crosslinked polymeric network. [18]

He et al. [165] synthesized biodegradable core-shell (a hydrophobic core and hydrophilic shell) polymeric nanoparticles for encapsulation and delivery of Paclitaxel drug (which is a chemotherapeutic drug used for the treatment of different cancer types such as ovarian, breast, brain, colon, lung, and even AIDS-related cancers) [181-186] to tumor cells. In this regard, poly (lactide-*co*-glycolide fumarate) (PLGF) and Poly (lactide-fumarate) (PLAF) were prepared via condensation polymerization technique of poly(L-lactide-co-glycolide) (ULMWPLGA) with fumaryl chloride (FuCl). Poly(lactide-co-ethylene oxide fumarate) (PLEOF) macromer was obtained by reacting poly(L-lactide) (ULMW PLA) and PEG with FuCl. The blend PLGF/PLEOF and PLAF/PLEOF macromers were self-assembled into nanoparticles via dialysis. The prepared microcapsules are illustrated in Figure 4.3. According to the results, PLEOF macromer showed a surface active agent role during self-assembly; hence, the core-shell nanoparticles (i.e. core: PLGF/PLAF macromers and shell: PLEOF macromer) were created. The drug encapsulation yield was 56-70%. Considering degradation process, it was shown that the release profile was controlled by hydrolytic degradation and erosion of the polymeric matrix due to the fact that PLGF and PLAF nanoparticles were degraded within 15 and 28 days, respectively. [165]

4.2.3. Polyvinyl Alcohol

The era of preparation of polyvinyl alcohol (PVA), by hydrolyzing polyvinyl acetate in ethanol in the presence of potassium hydroxide, was introduced by *Hermann and Haehnel* in 1924, which attracted much attention in pharmaceutical and biomedical applications. [162, 187-189] Due to its excellent film forming, hydrophilicity, biocompatibility, emulsifying and adhesive properties, PVA has been widely used as a coating agent for SPIONs; [19, 190] PVA can prevent the agglomeration of particles mostly due to the steric hindrance mechanism, thus giving rise to the amount of monodisperse nanoparticles. [17, 19, 45, 190-194] Similar to PEGF, PVA has a crosslinking potential via either chemical or physical crosslinking agents. [195-197]

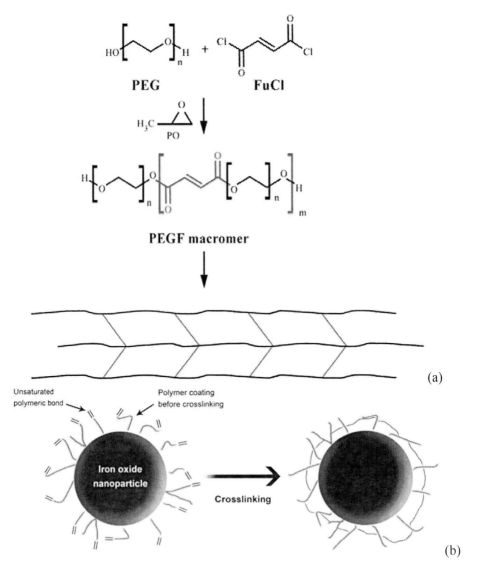

Figure 4.2 Continued

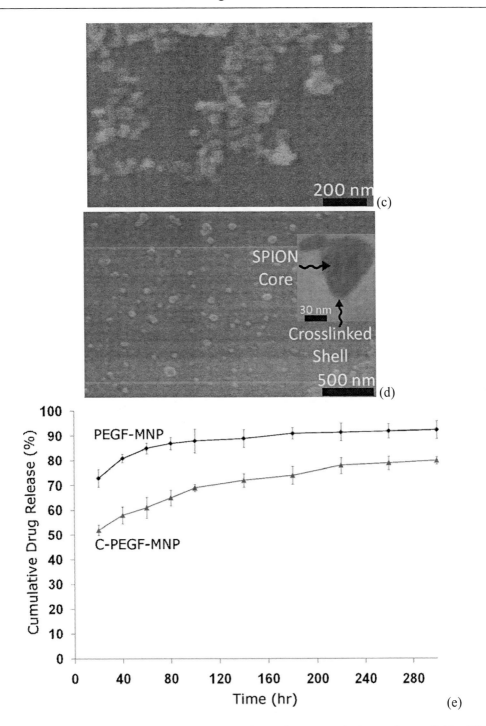

Figure 4.2. (a) Cross-linking of PEGF on SPIONs surfaces; (b) Proposed scheme of cross-linking PEGF that coats the iron oxide nanoparticles; (c) SEM image of C(Cross-linked)-PEGF-MNP(Magnetic NanoParticles (i.e. SPIONs)); (d) SEM and TEM images of selected C-PEGF nanoparticles; (e) Release profile of tamoxifen from PEGF-MNP and C-PEGF-MNP over 12 days. With permission from reference [18].

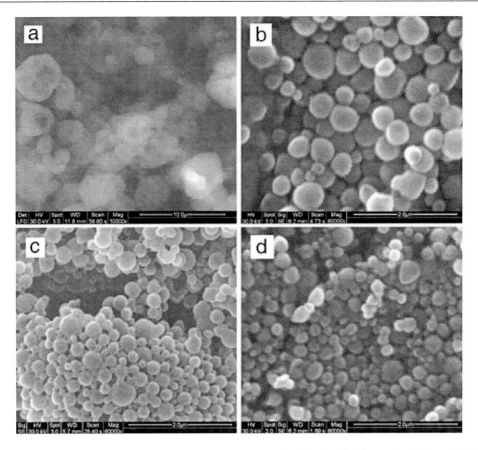

Figure 4.3. SEM images of the nanoparticles which were synthesized with (a) PLEOF, (b), PLGF, (c) mixture of 90 wt% PLGF and 10% PLEOF, and (d) mixture of 90% PLGF and 10% PLEOF and loaded with 6 wt% (based on PLGF+ PLEOF weights) Paclitaxel. (Scale bar is 2 μm). With permission from ref. [165].

As mentioned before, the iron atoms on the surface of SPIONs plays a Lewis acids role and coordinate with water molecules that donate lone-pair electrons; hence, the surface of SPIONs is covered by hydroxyl groups. After interaction with PVA, the coordination covalent bond is created between SPIONs and PVA due to the presence of many hydroxyl groups in the polymer chain. The shift of bond to the higher energy (see Figure 4.4) in the FTIR spectra has confirmed the creation of coordination covalent bond.

Due to the excellent film formation properties of PVA, using an appropriate amount of polymer as a coating on SPIONs is crucial for the shape of the final product. For instance, using PVA/iron mass ratio (r) of 2-3 has caused the formation of single coated SPIONs, in contrast a high amount of PVA ($r>3$) caused the formation of magnetic beads. [24] It is notable that magnetic bead is the random dispersion of SPIONs in the polymer matrix; the PVA matrix is formed because of its excellent film formation properties. [19] *Petri-Fink et al.* [198] synthesized different coating (PVA and its carboxylate, thiol or amino-functionalized derivative) on the surface of SPIONs. The uptake potential of obtained coated SPIONs by human cancer cells was examined and the results confirmed the significant effect of the structure of polymeric shell on the interactions between coated SPIONs and human melanoma tumor cells. [29, 199]

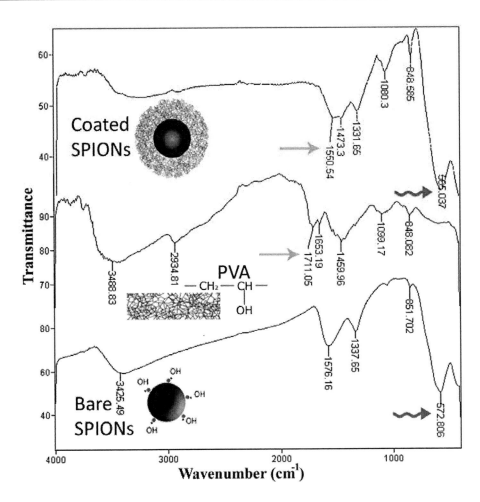

Figure 4.4. FT-IR spectrum of (a) uncoated SPION, (b) PVA and (c) coated SPIONs. Shift in the bonds (red and green vectors) confirmed the formation of chemical bond between SPIONs and PVA. Redrawn with permission from ref. [19].

Interestingly, PVA have a mechano-thermal properties meaning that the physical shape and chemical characteristics of its molecules can be changed via changes in machanical and thermal conditions of the synthesis stage. [19, 22] For instance, it has been recognized that the PVA molecules which are dispersed by high homogenization rates in hot water (i.e. 70 °C) may undergo thermo-physical stretching. [200] *Mahmoudi et al.* [22] used the mentioned unique properties of PVA in order to synthesize the magnetite with nanorod shapes. According to their results, small amounts of PVA act as a template in hot water (70 °C), leading to the oriented growth of SPIONs nanorods. In contrast, PVA coated magnetite SPIONs and magnetic beads were formed at a relatively lower temperature of 30 °C. The reason was the folded shape of polymer molecules rather than stretching shape in the low temperature. It is notable that the PVA can be crystalline via freeze/thaw cycles. For instance, Ricciardi *et al.* [201, 202] have reported X-ray diffraction analysis of PVA samples synthesized by the freeze/thaw technique. According to their results, it was revealed that the percentage of rigid PVA portions and the values of swollen amorphous component are in good agreement and the apparent crystalline dimension along with the preferable lattice direction [10ī] is a function of the number of freeze/thaw cycles for the PVA hydrogel. By employing freeze/thaw technique, stable magnetite nanorods with rigid PVA wall were synthesized. [22] It is noteworthy that stable nanorods have proven to show great potential in imaging and drug delivery applications. [203, 204]

4.2.4. Polyacrylic Acid

Polyacrylic acid (PAA), which contains highly hydrophilic groups, is used extensively as a coating of SPIONs due to its potential to enhance either colloidal stability or biocompatibility of the final nano-product. [160, 205-211] In order to control the properties of prepared SPIONs for biomedical applications, the use of copolymers of PAA (e.g. chitosan-PAA) are preferred over homo-polymers, due to the flexibility of copolymers for designing a material with specific properties. [162-165] Since PAA has a linear structure, it has good potential to control the growth direction of nanoparticles by its molecular structure; for instance, the formation of PAA-coated spindle-shaped nanoparticles was reported by *Iijima et al.* [212].

Ma et al. [38] synthesized PAA-coated SPIONs with the size of 246 nm. The recombinant tissue plasminogen activator was immobilized to the prepared nanoparticles through carbodiimide-mediated amide bond formation and used for targeting thrombolysis *in vivo*, under magnetic guidance. The authors suggested that this system may achieve reproducible and effective target thrombolysis with <20% of common dose of recombinant tissue plasminogen activator. *Mak et al.* [213] employed strong acid cation nano-adsorbent with sulfonic acid groups on the surface via binding and sulfonation of PAA on SPIONs. According to the FTIR results, the authors observed that the sulfonation was accomplished by the reaction between the carboxylic acid groups of PAA and the amino groups of sulfanilic acid via carbodiimide activation.

4.2.5. Poly(N-isopropylacrylamide)

Poly(N-isopropylacrylamide) (PNIPAm), introduced originally in the 1950s [214], is recognized as a thermo-reversible, pH-responsive and water-soluble polymer whose aqueous solutions exhibit a low critical solution temperature (LCST) around 33 °C. [215-217] The LCST, in which the 90% of PNIPAm mass is removed, is a reversible phase transition from a swollen hydrated state to a shrunken dehydrated state. The PNIPAm is completely in the form of liquid in the human body (temperature of 37°C); hence, it has great potential in controlled and targeted drug delivery and imaging applications in conjunction with SPIONs. [218-220]

The pH-responsive systems were prepared by layer-by-layer adsorption technique using magnetic nanoparticles and PNIPAm microgels. [221, 222] *Sun et al.* [223] and *Lai et al.* [224] prepared PNIPAm-coated SPIONs, which were used as contrast agents for magnetic resonance imaging (MRI). The product showed lower critical solution temperature (LCST) at 31 °C. It is notable that LCST, which is pressure dependent, is recognized as a critical temperature below which a mixture is miscible in all proportions. [225]

Associating interactions (e.g. strong polar interactions and hydrogen bonds) and compressibility effects are the main forces creating the LCST. [226]

Li et al. [89] prepared thermo-responsive, pH-responsive, and magnetic-sensitive composite microspheres using emulsion polymerization of NIPAm and chitosan containing oleic acid modified SPIONs as a dispersed phase. It is notable that the role of chitosan was to induce pronounced pH sensitivity upon the existence of ionizable amino functional groups on its backbone. These obtained microspheres demonstrated a LCST of about 31 °C in water. The LCST amount could be varied from 28 °C to 32 °C according to the acidic and basic pH, respectively. Via electromagnetically induced heating, the prepared composite microspheres have the capability to be heated to 45 °C using an alternating electromagnetic field.

4.2.6. Dextran

Dextran, having various molecular weights ranging from 10-150 kDa, is a kind of natural polymers which is made of anhydroglucose units consisting mainly alpha-D(1-6) linkages. However, some unusual 1,3 glucosidic linkages are also present at branching points. Dextran has been used as a coating for SPIONs due to its great biocompatibility and biodegradability properties together with its colloidal stability properties (more stability could be obtained by conjugation of Herceptin to dextran-coated SPIONs. [52]). [43-55] Furthermore, dextran-coated SPIONs have demonstrated a good capability to enhance the blood circulation time which is a crucial factor for targeted delivery and imaging purposes. [49] It is notable that the formation of protein corona (i.e. dynamic protein coating on the surface of unmodified dextran-coated SPIONs) is responsible for the aforementioned decrease in their blood circulation time. [227, 228] The main protein causing the reduction in blood circulation half-time of nanoparticles is the opsonin. [47, 229] The role of other plasma proteins (e.g. complement, fibronectin and fibrinogen) in the unmodified dextran-coated SPIONs clearance was also shown by *Simberg et al.* [43]. The results established the selectivity of plasma proteome towards SPIONs surfaces. Furthermore, by using knockout mice, it was shown that the attached plasma proteins were unlikely to play a role in the *in vivo* clearance of SPIONs. In fact, the plasma proteins did not mask entirely the surfaces of the SPIONs, signifying that the surfaces of SPIONs could be directly known by macrophages (regardless of protein coating).

The circulation time is closely dependent of the size of coated SPIONs; for instance, a diameter size less than 20 nm exhibited prolonged circulation time in comparison to the bigger coated SPIONs (e.g. 50-150 nm) which showed a faster removal from the blood circulation by RES organs such as liver and spleen [230,231] It is notable that dextran-coated SPIONs (Ferridex®) is recognized as the only US Food and Drug Administration (FDA) approved product for use in MRI technique as a contrast agents. Since the type of coating adsorption to the surface of SPIONs is significantly important for biological applications, *Jung et al.* [232] introduced a model for this adsorption where hydrogen bonds, through polymer hydroxyl groups, take place at different segment of the dextran. Via this sophisticated model, stability of the dextran at the surface of SPIONs can be predicted in different mediums (e.g. aqueous, cell culture and biological media). The colloidal forces considered in this model consist of electrostatic, steric and electrosteric repulsion forces.

One of the major problems of the nanoparticles designed for brain drug delivery and imaging, is the presence of the blood brain barrier (BBB) which prohibit the entrance of nanoparticles to the desired sites. The distinguished feature of dextran-coated SPIONs is their ability to pass through BBB and target the brain tumor (demonstrated on the glial tumors in rat) [233].

4.2.7. Poly(D,L-lactide)

Due to its low toxicity together with its contribution to the improvement of functional hybrid nanoparticles for the biomedical applications, poly(D,L-lactide) is employed as a coating of SPIONs. [234, 235] Poly(D,L-lactide) shows the capability of enhancing the colloidal stability of ferrofluids after its polymerization on the SPIONs' surface via ring-opening polymerization (ROP) of cyclic lactone derivative of its monomers. [236, 237] The catalyst (e.g. tin(II) 2-ethylhexanoate (Stannous octoate)) was employed to offer a high number of end-attached polymer chains, which covered the surface of SPIONs at suitable temperature (e.g. 130 °C). [234, 236, 237] The grafting density of this polymer can be enhanced by improving the polymer concentration; however this density is declined via increasing the molar mass of the polymer. [234]

4.2.8. Poly(D,L-lactide-*co*-glycolide)

Poly(D,L-lactide-*co*-glycolide) (PLGA), which is highly accepted in the scientific communities for various biomedical applications, is a major co-polyester of D,L-lactide. [238-250] PLGA has demonstrated very low toxicity, and tunable hydrophilicity and biodegradability according to the fraction of D,L-lactide and glycolide in its structure. [251-258] In addition, the byproducts (e.g. lactic acid and glycolic acid) of PLGA, achieved by its degradation in biological mediums, have shown no toxicity. [259-280] The reason is that these byproducts are consumed by tricarboxylic acid cycle or Krebs cycle as a carbon source; hence can be removed from biological environments by transferring to carbon dioxide and water. [57, 238-241, 281-291] It is also notable that PLGA has been frequently used as scaffold in tissue engineering [251, 276] and drug delivery systems [271, 292], due to its authorization by FDA. [293-298] For instance, the encapsulation of fluorescein isothiocyanate inside PLGA nanoparticles, conjugated with SPIONs, has confirmed its capability for controlled release pattern of the entrapped drug entity [299]. In addition, these engineered nanoparticles demonstrated a high spin spin relaxation time (i.e. $T2$) and that they have also great potential for imaging purposes. [299] Owing to the above mentioned characteristics of PLGA-coated SPIONs, they are currently used for the simultaneous diagnosis and treatment of the breast cancer. [300] It is notable that the crystallinity of PLGA nanoparticles/shells is a key issue for its controllable drug delivery due to its effect on the biodegradability; for instance lower crystallinity leads to a faster degradation of PLGA (i.e. faster drug release). [154, 292, 293, 301-303] The crystallinity of the polymer is highly dependent on the variation of the LA/GA monomer ration.

Another key matter is the yield of drug loading on the polymeric coating of SPIONs. In order to increase the loading efficiency of SPIONs into polymeric sub-micron particles, *Okassa et al.* [59] employed a simple emulsion/evaporation method for preparation of PLGA particles loaded with both magnetite and maghemite nanoparticles. The obtained micro-particles exhibited superparamagnetic property. The emulsification–diffusion method was employed to achieve high magnetically susceptible SPIONs (8-20 nm) encapsulated within PLGA, which has spherical shape (i.e. size of 90–180 nm). [304] Both homogenization strength (in the emulsification step) and agitation speed (in the solvent diffusion stage) have the a diverse relation with the average size of the encapsulated SPIONs; hence, by increasing these parameters, the particle sizes can be decreased to achieve a high magnetic susceptibility. The smaller PLGA-encapsulated SPIONs (with the size of 90 nm) have been prepared using the similar emulsification–diffusion method through optimization of the homogenizer and agitator speed. [305]

4.2.9. Alginate

Alginate is a linear natural polymer, which is composed of both α-L-guluronate (G) and β-D-mannuronate (M) units in irregular proportions and orderly arrangements is an anionic polysaccharide. In addition, alginate is detected in species such as the cell walls of brown algae. [67, 75, 82] Due to the excellent biocompatibility of alginate, this polymer has been extensively employed on the surface of SPIONs. [61, 69, 73, 76, 77, 84, 306-311] The major application of coated SPIONs with alginate is focused on the enhancement of cellular and molecular MR image resolution. [65, 66] For instance, the efficacy of SPIONs as contrast agents was probed for β TC-tet cells which were encapsulated within alginate/poly-L-lysine/alginate (APA) micro-beads. [312] Two various routs were used for labeling; in the first route, the SPIONs were conjugated to the cells, then labeled cells were encapsulated in APA beads. In contrast, SPIONs were attached to alginate (via suspention of nanoparticles in alginate solution before encapsulation) instead of cells for the second route. The major problems of tracking methods are focused on the cell growth or rearrangement; in this case the efficacy of monitoring was reduced significantly.

For the drug delivery purposes, creating of magnetic beads (dispersion of random magnetic nanoparticles inside the polymer matrix) is very important. For the synthesis of the nanobeads, the polymeric gels (e.g. polystyrene-polyacrylate copolymer [313]) should be added during nucleation and growth of SPIONs. [314]

The polymeric gels can control the polydispersity of SPIONs due to their spatial framework for the synthesis of nanoparticles. The SPIONs were synthesized via co-precipitation of iron ions inside semi-interpenetrating polymer networks of alginate and poly-(N-isopropylacrylamide) (Alg-PNiPAAm semi-IPNs). [315] In addition, the influence of the network morphology on the structural organization and SPIONs size distribution inside Alg-PNi-PAAmsemi-IPNs. [309] It has been confirmed that the polydispersity of the SPIONs significantly decreases when the reaction takes place inside the Alg-PNiPAAm semi-IPNs, as compared to the case where the reaction is carried out in an alginate solution.

4.2.10. Chitosan and Polyethylenimine

Due to its precise biological properties (e.g. biocompatibility, biodegradability, antibacterial, wound healing activity and mucoadhesive properties which can cause high affinity for cell membranes), chitosan has attracted much attention as a suitable candidates for the coating of SPIONs. [93, 96, 132] It is well recognized that chitosan have high potential to enhance the interactions between drug and ocular mucosa due to its high mucoadhesive properties. [90, 316-319] In addition to chitosan, polyethylenimine (PEI) has been used as SPIONs' coating for both complex and condense DNA and transfect cell lines. [102, 320, 321] Magnetotransfection and gene therapy by SPIONs with minimal side effects offer the potential of mediating disease through modification of specific cellular functions of target cells. *Kievit et al.* [92] prepared a non-viral copolymer (chitosan-PEI-PEG) coated SPIONs which demonstrated effective gene delivery and transfection both *in vitro* and *in vivo*. Low molecular weight PEI was employed due to its low cytotoxicity compared to high molecular weight PEI. [99, 103] Chitosan was employed to facilitate the efficacy of transfection of low molecular weight PEI via its crosslinking property. [100, 101, 105] *Wang et al.* [88] dispersed mineralized-SPIONs (13 nm) in chitosan hydrogel, homogeneously, for a use biotechnological applications. The chitosan- and starch-coated SPIONs were synthesized as a hyperthermic thermoseed. [91] The chitosan-coated SPIONs generated higher temperature changes (23 °C) under an alternating magnetic field than those of the starch-coated particles. Additionally, the capturing rate of the particles was 96% under an external magnetic field of 0.4 T.

4.3. NON-POLYMERIC COATING

Due to the SPIONs' wrapping in biological media, the toxic Fe^{2+} ions are produced. One of the major problems of the polymeric coatings is their low potential to isolate the SPIONs core from the biological environment; hence, toxic Fe^{2+} ions can penetrate to the in vivo environment through polymer network. [29, 30] In order to overcome this major obstacle of magnetic caries, the biocompatible non-polymeric-rigid coatings such as gold [138, 139, 162, 322-341], silica [106, 117, 119, 327, 332, 337, 342-378] and carbon [334, 339, 368, 379-395] have been used.

One problem of SPIONs for targeting purposes is their low magnetization compared to the pure metallic magnetic materials. More specifically, Fe nanoparticles demonstrate superior magnetic properties with high saturation magnetization (i.e. 218 and 161 emu g^{-1}, respectively). [396] The main problem with the Fe nanoparticles is their simplicity oxidation in the air. Carbon is mainly used on the surface of Fe nanoparticles in order to inhibit this oxidation.

There are several methods for the production of carbon coating on the surface of Fe nanoparticles, including carbon arc, [397] flowing gas plasma, [398] laser-induced pyrolysis, [399] mechano-synthesis, [400] mechanical milling, [401] thermal carbonization, [402] ultrasound, [403] and swelling process [404]. Interestingly, SPIONs are used to produce the carbon coated Fe nanoparticles. For instance, polydivinylbenzene (PDVB)-coated SPIONs were employed to prepare air-stable iron (α-Fe) nanocrystalline nanoparticles by thermal decomposition in an inert atmosphere. [405] It is noteworthy that in order to increase the biocompatibility of SPIONs, pyrocarbon would be a great candidate due to its super compatibility properties. [406, 407] Although this a new polymer and has not been employed as a coating of SPIONs yet, due to its unique biophysiochemical properties, pyrocarbon is likely to be recognized as a very promising candidate in near future. [408-448]

4.3.1. Gold

Gold is recognized as a promising rigid, neutral and biocompatible coating materials for SPIONs; gold can provide a good stability of SPIONs together with significant potential binding capability to various chemical and biological agents. [99, 128-141] The magnetic moments of the magnetite is not changed by employing the gold shell on the particles' surface. [449] The most frequent approaches used in preparing SPIONs@Au core/shell nanoparticles are the chemical reducing processes, where SPIONs are formed and employed as seeds. [449-453] Au-shell coating is performed by reduction of Au^{3+} on the surface of SPIONs. The magnetite nanoparticles are dispersed in $HAuCl_4 \cdot 4H_2O$ solution in a beaker, and slowly mixed in a shaking incubator to allow adsorption of Au^{3+} ions onto the surface of SPIONs. NH_2OH solution is then added to the system while shaking the mixture for 1 h. The above mentioned procedure has been frequently employed for the preparation of gold-coated SPIONs in various biomedical applications. For instance, for antibody immobilization, [141] the core (with the size of 20 nm; spherical morphology, and a narrow size distribution) was obtained via co-precipitation of Fe (II) and Fe (III) ions (with the ratio of 1:2) in alkaline medium. The reduction of Au^{3+} was occurred by hydroxylamine to produce core/shell nanoparticles with an average size of 50 nm. Immunoglobulin G, which is a monomeric immunoglobulin and built of two heavy chains γ and two light chains, was immobilized onto the magnetic carrier by a simple incubation process and employed for the detection of HBV antigen in blood. *Wang et al.* [453] developed a procedure for the preparation of the monodispersed gold coated SPIONs. The magnetic core was obtained via a chemical reaction using iron (III) acetylacetonate, phenyl ether, oleic acid, and hexadecanediol. The obtained colloidal SPIONs were introduced to a mixture of gold acetate, hexadecanediol, oleic acid, and oleylamine at the gold precursor to the iron oxide ratio of approximately 7:1. Consequently, the colloid was heated to 189-190 °C, held for 1.5 h, cooled to the room temperature, and as a final point diluted with ethanol. The prepared Fe_3O_4@Au core/shell nanoparticles were collected via centrifuging of the final colloid. *Tamer et al.* [449] synthesized Fe_3O_4@Au core/shell NPs using a combined co-precipitation and sonochemical method. The magnetic core with an average size of 9.5±3 nm was obtained by co-precipitation of iron salts in the base medium (i.e. NaOH). The sonication process was employed for creating the gold shell by the induction of a $HAuCl_4$ solution containing SPIONs as seeds in the presence of $NaBH_4$. The average hydrodynamic size of the obtained core/shell nanoparticles was 12.5±3 nm whereas the nanoparticles were clustered with non-capped iron nanoparticles. *Pham et al.* [133] used citrate reduction protocol for the preparation of gold-coated SPIONs with a size ranging from 15 to 40 nm. SPIONs were obtained by the co-precipitation of iron salts in a strong alkaline (NaOH) solution. The particles were then oxidized in HNO_3 at 80-90 °C, washed several times, and re-dispersed in TMOH at pH=11. Sodium citrate was introduced to the suspension, in order to exchange absorbed OH^- with citrate ions, and finally $HAuCl_4$ was added when the suspension heated to boiling. Immunoglobulin G separation by the core-shell nanoparticles demonstrated a high yield (i.e. 35%) at an Immunoglobulin G concentration of 0.4 μg/ml. *Jain et al.* [138] synthesized γ-Fe_2O_3@Au nanoparticles employing similar protocol to investigate the surface plasmon resonance in the gold-coated nanoparticles. Similarly, *Xi et al.* [128] used this procedure to synthesize the SPIONs@Au nanoparticles and used them as gene probes to detect HBV DNA. The surface of the nanoparticles was functionalized with oligonucleotide with the maximal percentage of hybridization strands of 14±2. Large network aggregates were formed when the nanoparticles was applied to detect HBV DNA molecules. Interestingly, *Seino et al.* [131] synthesized iron oxide/gold nanoparticles using high-energy electron beam. Maghemite SPIONs with the size of 26 nm were dispersed in an aqueous solution containing HAuCl$_4$, 2-propanil and PVA, and then closed up in a glass vial. The suspension was sonicated and irradiated with high-energy electron beam (10 MeV; 800 kGy/h) for 10-20 s or [60] Co gamma rays (2 KGy/h) for 3 h. Using this method, the surface of SPIONs is almost coated with very fine gold shells. The development of the gold nanoparticles was due to the reducing of Au^{n+} by the alcohol radicals and hydrogen atoms generated from water radiolysis. The hetero-interparticle coalescence strategy has recently been enhanced by *Park et al.* [452] for the preparation of gold-coated SPIONs (maghemite) with controllable size together with very narrow size distribution. Fe(CO)$_5$, olic acid or oleyamine, and phenyl ether were stirred at 100 °C under an argon purge. The solution was heated to 253 °C and refluxed for 1 h. After the solution was cooled to room temperature, (CH$_3$)$_3$NO.2H$_2$O was introduced and the solution was stirred at 130 °C for 2 h. The temperature was then increased to 253 °C. The solution was refluxed for 2 h and stirred overnight. The resulting nanoparticles were precipitated with ethanol, rinsed several times, and finally the particles were dispersed in toluene. The standard two-phase method reported by Brust and Schiffrin [454] was employed for the preparation of gold nanoparticles with the mean diameter of 2 nm, which were encapsulated by decanethiol monolayer shells. A thermally activated processing protocol was used to obtain the core/shell structure. A mixture of the gold nanoparticles encapsulated with the monolayer shell, olic-SPIONs or (oleyamine – SPIONs), and tetraocytylammonim bromide in tolune was prepared and heated at 149 °C for 1 h. The resulting nanoparticles were found to be SPIONs@Gold and can be employed for bio-separation and as bio-functional nanoprobes (for surface enhanced Ramman scattering assay). [136]

4.3.2. Silica

Similar to gold, silica is a biocompatiable, nontoxic, functionalisable and chemically stable material, which is suitable for preventing degradation and agglomeration of SPIONs during their stay in human body and other biological media. Silica has been recognized as one of the most ideal materials for creating very safe shell for SPIONs. [83, 84, 106-127, 254, 306, 342, 343, 354, 455-466] Amorphous silica coating on magnetite nanoparticles was first reported by Philipse *et al* [467] by sol gel approach. As silica is hydrophilic in nature, the silica coated core-shell particles are well dispersed in aqueous suspension. The silica coated SPIONs can be negatively charged above the isoelectic point of silica (pH~2); hence, they have been used for separation of biomolecules through electrostatic interactions. In addition, silica also has hydroxyl groups (silanol group: Si-OH), useful for the attachment of further functionality through covalent bonding with organosilanes.

The encapsulation of SPIONs in silica is usually achieved by Stöber process [116], sol-gel [468] or microemulsion synthesis [119, 459]. In the former, silica shell is formed through hydrolysis and condensation reaction of tetraethoxy silane (TEOS); for instance, this process was employed for the preparation of SPIONs (with the mean size of 9.2 nm) coated with 2 nm silica layer. [457] Similar to the pervious section, primary SPIONs were synthesized by the alkaline co-precipitation technique. SPIONs were re-dispersed in DI water and a solution of TEOS in ethanol was added together with triethylamine (catalyzer). The processing parameters are recognized as a key for controlling the shape and structure of the final product; for example, silica beads (random deispersion of SPIONs in silica matrix) down to silica-coated SPIONs can be obtained.

Sol-gel method is also a common approach for the preparation of silica-coated SPIONs, which is applied based on the hydrolysis of TEOS with conformal and uniform shells. [107] The concentration of the sol-gel precursor controls the thickness of silica shell. Fluorescent dyes could also be integrated into the silica shell via covalent coupling/bonding between the organic dyes and the sol-gel precursor with the purpose of *in situ* characterization of coated nanoparticles.

Microemulsion is another promising way for preparing uniform silica shell on the surface of SPIONs. For instance, SiO_2-coated SPIONs (maghemite) were synthesized in a water-in-oil microemulsion. [117] The primary SPIONs, which was prepared via co-precipitation method, were coated either with PVA or citric acid through simple dispersion. Consequently, the coated magnetic nanoparticles were introduced to sodium-bis(2-ethylhexyl)sulfosuccinate (AOT) in octane phase. Pre-hydrolyzed tetramethoxysilane (TMOS) was added and the mixture was sonicated for a suitable period of time (e.g., 25 min). Condensation of emulsified TMOS was initiated upon addition of $(CH_3)_4OH$ under sonication for 30 min. Newly, Lee et al. [465] have introduced a large-scale method for the preparation of magnetite@silica nanoparticles by the addition of TEOS in reverse micelles through the development of uniformly sized SPIONs. It is notable that the microemulsion was achieved by dissolving sodium dodecylbenzenesulfonate in xylene via sonication. Typically, iron metal salts are dissolved in deionized (DI) water and added to the microemulsion under vigorous stirring. Subsequently, the reverse-micelle solution was slowly heated to 90 °C under continuously flowing argon gas for an hour. Subsequently, hydrazine (34 wt% aqueous solution) was injected into the solution, held for 3h, cooled down to 40 °C, followed by TEOS injection into the mixture, which initially mixed with the organic xylene phase and started to hydrolyze in the water region of the reverse micelles to form amorphous silica shells on the surface of the SPIONs. The core size of the obtained nanoparticles depends on the w-value (i.e. [polar solvent]/[surfactant]) in the reverse-micelle solution while the thickness of the shell could be dependent on the amount of injected TEOS after the synthesis of the magnetic core. The surface of silica-coated nanoparticles and porous silica-encapsulated SPIONs were then functionalized by amine groups and crosslinked by enzymes in order to assay the applicability of the core/shell nanoparticles as high-performance biocatalysts. [112] Reverse microemulsion technique in conjunction with templating strategies were employed for the preparation of homogeneous SiO_2-coated maghemite SPIONs with controlled SiO_2 shell thickness (1.8-30 nm) or with a mesoporous silica shell. [469] In this case, the primary γ-Fe_2O_3 nanoparticles with very narrow size distribution were obtained by the thermal decomposition of iron pentacarbonyl precursor in the presence of both an oleic acid stabilizer and octylether. Silica shell were applied on the surface of SPIONs through the formation of water-in-cyclohexane reverse microemulsion. TEOS was added to the microemulsion and the reaction was continued for 16 h at room temperature. When methanol was added into the reaction solution, SiO_2/SPIONs were precipitated. Mesoporous silica-coated SPIONs were also obtained through stirring TEOS and octadecyltrimethoxysilane (C18TMS) in a mixture of ethanol and aqueous 15% NH_4OH solution with silica-coated SPIONs at room temperature for 6 h. The silica shell thickness and porosity of the silica overlayer could be tuned by controlling the synthesis and processing parameters.

4.4. FUNCTIONALIZATION OF SPIONS

Functionalized-SPIONs offer exciting new opportunities toward developing high yield targeted drug delivery systems, via tailoring their functional properties for drug delivery applications. [153, 470-473] In addition to the magnetic properties of SPIONs, functionalization of their surfaces, through the surface modification with specific functional groups, can enhance the targeting applications via interaction with the target tumor/cancer. [474-476]

It is notable that a simple external magnetic field may be all that is needed to target a drug to a specific site inside the body without the need to functionalize the nanoparticles. However, to increase the targeting efficiency of magnetic nanoparticles, functionalization should be applied. [477]

The various coatings are used not only for creating a protective and biocompatible shell on the surface of SPIONs, but also for further functionalization purposes via targeting species (e.g. aptamers, antibodies and functional groups). The flexibility of the conjugation chemistry presents an smart way to the preparation of a range of biomolecules-nanoparticle conjugates for targeted bio-applications. [478-497] For instance, functionalized SPIONs can be used for the molecular, cellular and tissue imaging, [498-501] targeted drug delivery, [502, 503] and therapeutics applications [504, 505].

The SPIONs with the core size of 4.5 nm was coated with 4-methylcatechol (4-MC). [11] The obtained materials were directly conjugated with a peptide, c(RGDyK), by the Mannich reaction (see Figure 4.5) and the final product (with size of 8.4 nm) was highly stable in physiological environment. These functionalized nanoparticles have capability to target the integrin $\alpha_v\beta_3$-rich tumor cells. The MRI results confirmed the accumulation of engineered nanoparticles in the targeted site (see Figure 4.6).

The immune-targeted SPIONs, which are biocompatible, have also been developed for *in vivo* MR diagnostic and drug delivery purposes for kidney disease. [506] The In these nanoparticle complexes, SPIONs core with a functionalized phospholipid coating is conjugated to antibodies for targeting cells expressing specific target antigens. The life time of functionalized SPIONs in the blood is recognized as key for increasing the yield of targeted delivery and imaging. Radio-conjugated antibodies or radio-drugs are promising methods for the visualization of SPIONs in blood, especially to track the circulation half life of employed nanoparticles. [507, 508]

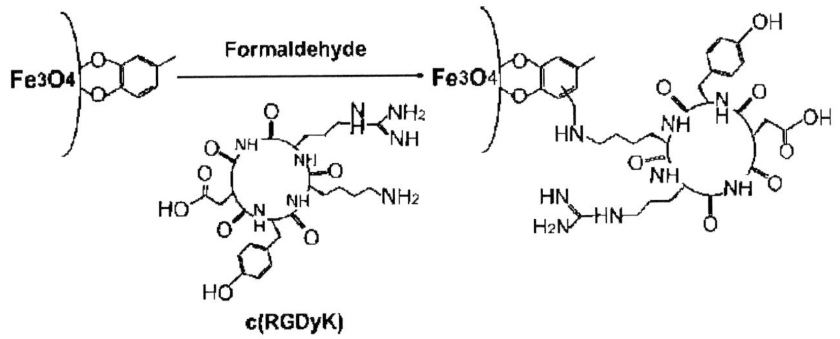

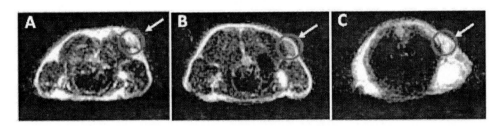

Figure 4.5. Schematic preparation of the engineered SPIONs for targeting purposes. With permission from ref. [11]

Hafaid et al. [481] explored the enhancement of the immune-sensors based on magnetic iron nanoparticles functionalized with streptavidin to which a biotinylated FAB part of the antibody has been bound using a biotin-streptavidin interaction.

The functionalization of the SPIONs are designed according to the desired biomedical applications (such as targeted drug delivery and imaging, hyperthermia, separation, transfection and labeling); [42, 57, 354, 355, 386, 482, 483, 486, 487, 490, 509-539, 334, 492, 494-497, 540-549] Comprehensive explanations of functionalization of SPONs according to their biomedical applications are presented in Chapter 6.

Figure 4.6. MRI images which were taken from the cross section of the U87MG (human glioblastoma) tumors which implanted in mice: (A) without engineered-SPIONs, (B) with the injection of 300 μg of engineered-SPIONs, and (C) with the injection of engineered-SPIONs, and blocking dose of c(RGDyK). With permission from ref. [11].

REFERENCES

[1] Baudry J, Bertrand E, Rouzeau C, Greffier O, Koenig A, Dreyfus R, et al. Colloids for studying molecular recognition. Annales de Chimie: *Science des Materiaux* 2004;29(1):97-106.

[2] Mylon SE, Chen KL, Elimelech M. Influence of natural organic matter and ionic composition on the kinetics and structure of hematite colloid aggregation: Implications to iron depletion in estuaries. *Langmuir* 2004;20(21):9000-6.

[3] Holthoff H, Egelhaaf SU, Borkovec M, Schurtenberger P, Sticher H. Coagulation rate measurements of colloidal particles by simultaneous static and dynamic light scattering. *Langmuir* 1996;12(23):5541-9.

[4] Holthoff H, Schmitt A, Fernández-Barbero A, Borkovec M, Cabrerizo-Vílchez MA, Schurtenberger P, et al. Measurement of absolute coagulation rate constants for colloidal particles: Comparison of single and multiparticle light scattering techniques. *Journal of Colloid and Interface Science 1997*;192(2):463-70.

[5] Lattuada M, Sandkühler P, Wu H, Sefcik J, Morbidelli M. Aggregation kinetics of polymer colloids in reaction limited regime: Experiments and simulations. *Advances in Colloid and Interface Science* 2003;103(1):33-56.

[6] Bloomfield VA. Static and dynamic light scattering from aggregating particles. *Biopolymers* 2000;54(3):168-72.

[7] Bacri JC, Perzynski R, Salin D, Cabuil V, Massart R. Ionic ferrofluids: A crossing of chemistry and physics. *Journal of Magnetism and Magnetic Materials* 1990;85(1-3):27-32.

[8] Fritz G, Schadler V, Willenbacher N, Wagner NJ. Electrosteric stabilization of colloidal dispersions. *Langmuir* 2002 Aug;18(16):6381-90.

[9] Napper DH. Flocculation studies of sterically stabilized dispersions. *Journal of Colloid and Interface Science* 1970;32(1):106-&.

[10] LaVan DA, McGuire T, Langer R. Small-scale systems for in vivo drug delivery. *Nature Biotechnology 2003* Oct;21(10):1184-91.

[11] Xie J, Chen K, Lee HY, Xu CJ, Hsu AR, Peng S, et al. Ultrasmall c(RGDyK)-coated Fe3O4 nanoparticles and their specific targeting to integrin alpha(v)beta(3)-rich tumor cells. *Journal of the American Chemical Society 2008 Jun;*130(24):7542-3.

[12] Gref R. Biodegradable long-circulation polymer nanospheres Science 1994;263:3.

[13] Akerman ME, Chan WCW, Laakkonen P, Ruoslahti E. *Nanocrystal targeting in vivo Proceeding of Natural Academy of Science,* USA 2002;99:5.

[14] Hamley IW. Nanotechnology with soft materials. *Angew Chem-Int Edit* 2003; 42(15):1692-712.

[15] Mahmoudi M, Milani AS, Stroeve P. Surface Architecture of Superparamagnetic Iron Oxide Nanoparticles for Application in Drug Delivery and Their Biological Response: A Review. *International Journal of Biomedical Nanoscience and Nanotechnology* 2010;in press.

[16] Mahmoudi M, Shokrgozar MA, Simchi A, Imani M, Milani AS, Stroeve P, et al. Multiphysics Flow Modeling and in Vitro Toxicity of Iron Oxide Nanoparticles Coated with Poly(vinyl alcohol). *Journal of Physical Chemistry C 2009 Feb;*113(6):2322-31.

[17] Mahmoudi M, Simchi A, Imani M. Cytotoxicity of Uncoated and Polyvinyl Alcohol Coated Superparamagnetic Iron Oxide Nanoparticles. *The Journal of Physical Chemistry C* 2009;113(22):9573-80.

[18] Mahmoudi M, Simchi A, Imani M, Hafeli UO. Superparamagnetic Iron Oxide Nanoparticles with Rigid Cross-linked Polyethylene Glycol Fumarate Coating for Application in Imaging and Drug Delivery. *Journal of Physical Chemistry C* 2009 May;113(19):8124-31.

[19] Mahmoudi M, Simchi A, Imani M, Milani AS, Stroeve P. Optimal Design and Characterization of Superparamagnetic Iron Oxide Nanoparticles Coated with Polyvinyl Alcohol for Targeted Delivery and Imaging. *Journal of Physical Chemistry B* 2008;112(46):14470-81.

[20] Mahmoudi M, Simchi A, Imani M, Milani AS, Stroeve P. An in vitro study of bare and poly(ethylene glycol)-co-fumarate-coated superparamagnetic iron oxide nanoparticles: a new toxicity identification procedure. *Nanotechnology* 2009 Jun;20(22).

[21] Mahmoudi M, Simchi A, Imani M, Shokrgozar MA, Milani AS, Hafeli U, et al. A new approach for the in vitro identification of the cytotoxicity of superparamagnetic iron oxide nanoparticles. *Colloids and Surfaces B: Biointerfaces* 2010;75:300-9.

[22] Mahmoudi M, Simchi A, Imani M, Stroeve P, Sohrabi A. *Template growth of superparamagnetic iron oxide nanoparticles in poly(vinyl alcohol)/water mixture*. Thin Solid Films 2010;in press doi:10.1016/j.tsf.2009.12.112.

[23] Mahmoudi M, Simchi A, Milani AS, Stroeve P. *Proc 2nd Conf Nanostruct (Kish Island, Iran)* 2008:2.

[24] Mahmoudi M, Simchi A, Milani AS, Stroeve P. Cell toxicity of superparamagnetic iron oxide nanoparticles. *Journal of Colloidal and Interface Science* 2009;336:510-8.

[25] Mahmoudi M, Simchi A, Vali H, Imani M, Shokrgozar MA, Azadmanesh K, et al. Cytotoxicity and cell cycle effects of bare and polyvinyl alcohol coated iron oxide nanoparticles in mouse fibroblasts *Advanced Biomaterials 2009;*12:in press.

[26] Wang Y, Wong JF, Teng XW, Lin XZ, Yang H. "Pulling" nanoparticles into water: Phase transfer of oleic acid stabilized monodisperse nanoparticles into aqueous solutions of alpha-cyclodextrin. *Nano Lett* 2003 Nov;3(11):1555-9.

[27] Pellegrino T, Manna L, Kudera S, Liedl T, Koktysh D, Rogach AL, et al. Hydrophobic nanocrystals coated with an amphiphilic polymer shell: A general route to water soluble nanocrystals. *Nano Lett* 2004 Apr;4(4):703-7.

[28] Zhao X, Harris JM. Novel degradable poly(ethylene glycol) hydrogels for controlled release of protein. *J Pharm Sci* 1998 Nov;87(11):1450-8.

[29] Nel A, Xia T, Madler L, Li N. Toxic potential of materials at the nanolevel. *Science* 2006;311:622-7.

[30] Nel AE, Modler L, Velegol D, Xia T, Hoek EMV, Somasundaran P, et al. Understanding biophysicochemical interactions at the nano-bio interface. *Nature Materials* 2009;8(7):543-57.

[31] Mahmoudi M, Simchi A, Milani AS, Stroeve P. Cell toxicity of superparamagnetic iron oxide nanoparticles *Journal of Colloid and Interface Science* 2009.

[32] Merikhi J, Jungk HO, Feldmann C. Sub-micrometer CoAl2O4 pigment particles - synthesis and preparation of coatings. *Journal of Materials Chemistry* 2000;10(6):1311-4.

[33] Miller MM, Prinz GA, Cheng SF, Bounnak S. Detection of a micron-sized magnetic sphere using a ring-shaped anisotropic magnetoresistance-based sensor: A model for a magnetoresistance-based biosensor. *Applied Physics Letters* 2002 Sep;81(12):2211-3.

[34] Working PK. In *Chemistry and Biological Applications of Polyethylene Glycol. American Chemical Society Symposium Series* 1997;680:13.

[35] Zhang J, Rana S, Srivastava RS, Misra RDK. On the chemical synthesis and drug delivery response of folate receptor-activated, polyethylene glycol-functionalized magnetite nanoparticles. *Acta Biomaterialia* 2008 Jan;4(1):40-8.

[36] Kumar RV, Koltypin Y, Cohen YS, Cohen Y, Aurbach D, Palchik O, et al. Preparation of amorphous magnetite nanoparticles embedded in polyvinyl alcohol using ultrasound radiation. *Journal of Materials Chemistry* 2000 May;10(5):1125-9.

[37] Wang X, Chen C, Liu H, Ma J. Preparation and characterization of PAA/PVDF membrane-immobilized Pd/Fe nanoparticles for dechlorination of trichloroacetic acid. *Water Research* 2008;42(18):4656-64.

[38] Ma YH, Wu SY, Wu T, Chang YJ, Hua MY, Chen JP. Magnetically targeted thrombolysis with recombinant tissue plasminogen activator bound to polyacrylic acid-coated nanoparticles. *Biomaterials* 2009;30(19):3343-51.

[39] Lin CL, Lee CF, Chiu WY. Preparation and properties of poly(acrylic acid) oligomer stabilized superparamagnetic ferrofluid. *Journal of Colloid and Interface Science* 2005;291(2):411-20.

[40] Zhang S, Zhang L, He B, Wu Z. Preparation and characterization of thermosensitive PNIPAA-coated iron oxide nanoparticles. *Nanotechnology* 2008;19(32).

[41] Meenach SA, Anderson AA, Suthar M, Anderson KW, Hilt JZ. Biocompatibility analysis of magnetic hydrogel nanocomposites based on poly(N-isopropylacrylamide) and iron oxide. *Journal of Biomedical Materials Research - Part A* 2009;91(3):903-9.

[42] Kim GC, Li YY, Chu YF, Cheng SX, Zhang XZ, Zhuo RX. A nanosized, thermo-sensitive drug carrier: Self-assembled Fe 3O4-OA-g-P(OA-co-NIPAAm) magnetomicelles. Journal of Biomaterials Science, *Polymer Edition 2008*;19(9):1249-59.

[43] Simberg D, Park JH, Karmali PP, Zhang WM, Merkulov S, McCrae K, et al. Differential proteomics analysis of the surface heterogeneity of dextran iron oxide nanoparticles and the implications for their in vivo clearance. *Biomaterials* 2009;30(23-24):3926-33.

[44] Pardoe H, Chua-anusorn W, St. Pierre TG, Dobson J. Structural and magnetic properties of nanoscale iron oxide particles synthesized in the presence of dextran or polyvinyl alcohol. *Journal of Magnetism and Magnetic Materials* 2001;225(1-2):41-6.

[45] Pardoe H, Chua-anusorn W, St Pierre TG, Dobson J. Structural and magnetic properties of nanoscale iron oxide particles synthesized in the presence of dextran or polyvinyl alcohol. 3rd International Conference on Scientific and Clinical Applications of Magnetic Carriers; 2000 May 03-06; Rostock, Germany: *Elsevier Science Bv;* 2000. p. 41-6.

[46] Mornet S, Portier J, Duguet E. A method for synthesis and functionalization of ultrasmall superparamagnetic covalent carriers based on maghemite and dextran. 5th International Conference on Scientific and Clinical Applications of Magnetic Carriers; 2004 May 20-22; Lyon, FRANCE: *Elsevier Science Bv;* 2004. p. 127-34.

[47] Moore A, Weissleder R, Bogdanov Jr A. Uptake of dextran-coated monocrystalline iron oxides in tumor cells and macrophages. *Journal of Magnetic Resonance Imaging* 1997;7(6):1140-5.

[48] Moore A, Marecos E, Bogdanov Jr A, Weissleder R. Tumoral distribution of long-circulating dextran-coated iron oxide nanoparticles in a rodent model. *Radiology* 2000;214(2):568-74.

[49] Massia SP, Stark J, Letbetter DS. Surface-immobilized dextran limits cell adhesion and spreading. *Biomaterials* 2000 Nov;21(22):2253-61.

[50] Hradil J, Pisarev A, Babic M, Horak D. Dextran-modified iron oxide nanoparticles. *China Particuology* 2007 Feb-Apr;5(1-2):162-8.

[51] Heinze T, Liebert T, Heublein B, Hornig S. Functional Polymers Based on Dextran. *Adv Polym Sci* 2006;205:199–291.

[52] Chen T-J, Cheng T-H, Chen C-Y, Hsu SCN, Cheng T-L, Liu G-C, et al. Targeted Herceptin–dextran iron oxide nanoparticles for noninvasive imaging of HER2/neu receptors using MRI. *J Biol Inorg Chem* 2009;14:253-60.

[53] Carmen Bautista M, Bomati-Miguel O, Del Puerto Morales M, Serna CJ, Veintemillas-Verdaguer S. Surface characterisation of dextran-coated iron oxide nanoparticles prepared by laser pyrolysis and coprecipitation. *Journal of Magnetism and Magnetic Materials* 2005;293(1):20-7.

[54] Berry CC, Wells S, Charles S, Curtis ASG. Dextran and albumin derivatised iron oxide nanoparticles: Influence on fibroblasts in vitro. *Biomaterials* 2003;24(25):4551-7.

[55] Bautista MC, Bomati-Miguel O, Morales MD, Serna CJ, Veintemillas-Verdaguer S. Surface characterisation of dextran-coated iron oxide nanoparticles prepared by laser pyrolysis and coprecipitation. *5th International Conference on Scientific and Clinical Applications of Magnetic Carriers;* 2004 May 20-22; Lyon, FRANCE: Elsevier Science Bv; 2004. p. 20-7.

[56] Wassel RA, Grady B, Kopke RD, Dormer KJ. Dispersion of super paramagnetic iron oxide nanoparticles in poly(d,l-lactide-co-glycolide) microparticles. *Colloids and Surfaces A: Physicochemical and Engineering Aspects* 2007;292(2-3):125-30.

[57] Patel D, Moon JY, Chang Y, Kim TJ, Lee GH. Poly(d,l-lactide-co-glycolide) coated superparamagnetic iron oxide nanoparticles: Synthesis, characterization and in vivo study as MRI contrast agent. *Colloids and Surfaces A: Physicochemical and Engineering Aspects* 2008;313-314:91-4.

[58] Astete CE, Kumar CSSR, Sabliov CM. Size control of poly(d,l-lactide-co-glycolide) and poly(d,l-lactide-co-glycolide)-magnetite nanoparticles synthesized by emulsion evaporation technique. *Colloids and Surfaces A: Physicochemical and Engineering Aspects* 2007;299(1-3):209-16.

[59] Okassa LN, Marchais H, Douziech-Eyrolles L, Herve K, Cohen-Jonathan S, Munnier E, et al. Optimization of iron oxide nanoparticles encapsulation within poly(D,L-lactide-co-glycolide) sub-micron particles. *European Journal of Pharmaceutics and Biopharmaceutics* 2007;67:8.

[60] Okassa LN, Marchais H, Douziech-Eyrolles L, Cohen-Jonathan S, Souce M, Dubois P, et al. Development and characterization of sub-micron poly(D,L-lactide-co-glycolide) particles loaded with magnetite/maghemite nanoparticles. *International Journal of Pharmaceutics* 2005 Sep;302(1-2):187-96.

[61] Zouboulis AI, Katsoyiannis IA. Arsenic Removal Using Iron Oxide Loaded Alginate Beads. *Ind Eng Chem Res* 2002;41:7.

[62] Vold IMN, Kristiansen KA, Christensen BE. A study of the chain stiffness and extension of alginates, in vitro epimerized alginates, and periodate-oxidized alginates using size-exclusion chromatography combined with light scattering and viscosity detectors. *Biomacromolecules* 2006;7(7):2136-46.

[63] Sreeram KJ, Yamini Shrivastava H, Nair BU. Studies on the nature of interaction of iron (III) with alginates. BBA-Gen Subj 2004;1670:5.

[64] Siew CK, Williams PA, Young NWG. New insights into the mechanism of gelation of alginate and pectin: Charge annihilation and reversal mechanism. *Biomacromolecules* 2005;6(2):963-9.

[65] Shen F, Poncet-Legrand C, Somers S, Slade A, Yip C, Duft AM, et al. Properties of a novel magnetized alginate for magnetic resonance imaging. *Biotechnology and Bioengineering* 2003;83(3):282-92.

[66] Shen F, Li AA, Gong YK, Somers S, Potter MA, Winnik FM, et al. Encapsulation of recombinant cells with a novel magnetized alginate for magnetic resonance imaging. *Human Gene Therapy* 2005;16(8):971-84.

[67] Robitaille R, Pariseau JF, Leblond FA, Lamoureux M, Lepage Y, Halle JP. Studies on small alginate-poly-l-lysine microcapsules. III. Biocompatibility of smaller versus standard microcapsules. *J Biomed Mater Res* 1999;44:5.

[68] Nishio Y, Yamada A, Ezaki K, Miyashita Y, Furukawa H, Horie K. Preparation and magnetometric characterization of iron oxide-containing alginate/poly(vinyl alcohol) networks. *Polymer* 2004;45:8.

[69] Morales MA, Finotelli PV, Coaquira JAH, Rocha-Leao MHM, Diaz-Aguila C, Baggio-Saitovitch EM, et al. In situ synthesis and magnetic studies of iron oxide nanoparticles in calcium-alginate matrix for biomedical applications. *Materials Science and Engineering C* 2008;28:5.

[70] Ma HL, Xu YF, Qi XR, Maitani Y, Nagai T. Superparamagnetic iron oxide nanoparticles stabilized by alginate: Pharmacokinetics, tissue distribution, and applications in detecting liver cancers. *International Journal of Pharmaceutics* 2008 Apr;354(1-2):217-26.

[71] Ma HL, Qi XT, Maitani Y, Nagai T. Preparation and characterization of superparamagnetic iron oxide nanoparticles stabilized by alginate. *International Journal of Pharmaceutics* 2007 Mar;333(1-2):177-86.

[72] LeRoux MA, Guilak F, Setton LA. Compressive and shear properties of alginate gel: Effects of sodium ions and alginate concentration. *Journal of Biomedical Materials Research* 1999;47(1):46-53.

[73] Kroll E, Winnik FM, Ziolo RF. In situ preparation of nanocrystalline gamma-Fe2O3 in iron(II) cross-linked alginate gels. *Chemistry of Materials* 1996 Aug;8(8):1594-&.

[74] Haug A, Smidsrod O. Strontium-calcium selectivity of alginates. *Nature* 1967;215(102):757.

[75] Gombotz WR, Wee SF. Protein release from alginate matrices. *Adv Drug Deliv Rev* 1998;31(3):267-85.

[76] Finotelli PV, Morales MA, Rocha-Leao MH, Baggio-Saitovitch EM, Rossi AM. Magnetic studies of iron (III) nanoparticles in alginate polymer for drug delivery applications. *Mater Sci Eng* C 2004;24:5.

[77] Finotelli PV, Morales MA, Rocha-Leao MH, Baggio-Saitovitch EM, Rossi AM. Magnetic studies of iron(III) nanoparticles in alginate polymer for drug delivery applications. *2nd Meeting of the Brazilian-Society-for-Materials-Research* (Brazil-MRS); 2003 Oct 26-29; Rio de Janeiro, Brazil: Elsevier Science Bv; 2003. p. 625-9.

[78] De Boisseson MR, Leonard M, Hubert P, Marchal P, Stequert A, Castel C, et al. Physical alginate hydrogels based on hydrophobic or dual hydrophobic/ionic interactions: Bead formation, structure, and stability. *Journal of Colloid and Interface Science* 2004;273(1):131-9.

[79] Chen KL, Mylon SE, Elimelech M. Enhanced aggregation of alginate-coated iron oxide (Hematite) nanoparticles in the presence of calcium, strontium, and barium cations. *Langmuir* 2007;23(11):5920-8.

[80] Chen KL, Mylon SE, Elimelech M. Aggregation kinetics of alginate-coated hematite nanoparticles in monovalent and divalent electrolytes. *Environmental Science and Technology* 2006;40(5):1516-23.

[81] Bu H, Kjįniksen AL, Knudsen KD, Nystroȥ^m B. Rheological and structural properties of aqueous alginate during gelation via the Ugi multicomponent condesation reaction. *Biomacromolecules* 2004;5(4):1470-9.

[82] Bouhadir KH, Lee KY, Alsberg E, Damm KL, Anderson KW, Mooney DJ. Degradation of partially oxidized alginate and its potential application for tissue engineering. *Biotechnol Prog* 2001;17:6.

[83] Boissiere M, Allouche J, Chaneac C, Brayner R, Devoisselle J-M, Livage J, et al. Potentialities of silica/alginate nanoparticles as Hybrid Magnetic Carriers. *International Journal of Pharmaceutics* 2007;344:7.

[84] Boissiere M, Allouche J, Brayner R, Chaneac C, Livage J, Coradin T. Design of iron oxide/silica/alginate HYbird MAgnetic Carriers (HYMAC). *5th International Workshop on Biomedical Applications and Nanotechnology;* 2006 Feb 16-17; Berlin, GERMANY; 2006. p. 4649-54.

[85] Zhu AP, Yuan LH, Liao TQ. Suspension of Fe3O4 nanoparticles stabilized by chitosan and o-carboxymethylchitosan. *International Journal of Pharmaceutics* 2008 Feb;350(1-2):361-8.

[86] Zhi J, Wang YJ, Lu YC, Ma JY, Luo GS. In situ preparation of magnetic chitosan/Fe3O4 composite nanoparticles in tiny pools of water-in-oil microemulsion. *Reactive & Functional Polymers 2006* Dec;66(12):1552-8.

[87] Zheng N, Zhou X, Yang W, Li X, Yuan Z. Direct electrochemistry and electrocatalysis of hemoglobin immobilized in a magnetic nanoparticles-chitosan film. *Talanta* 2009;79(3):780-6.

[88] Wang Y, Li B, Zhou Y, Jia D. In Situ Mineralization of Magnetite Nanoparticles in Chitosan Hydrogel. *Nanoscale Res Lett* 2009;4:1041-6.

[89] Li P, Zhu AM, Liu QL, Zhang QG. Fe3O4/poly(N-Isopropylacrylamide)/Chitosan Composite Microspheres with Multiresponsive Properties. *Ind Eng Chem Res* 2008;47(20):7.

[90] Lehr CM, Bouwstra JA, Schacht EH, Junginger HE. In vitro evaluation of mucoadhesive properties of chitosan and some other natural polymers. *Int J Pharm* 1992;78:43-8.

[91] Kim D-H, Kim K-N, Kim K-M, Lee Y-K. Targeting to carcinoma cells with chitosan- and starch-coated magnetic nanoparticles for magnetic hyperthermia. *Journal of Biomedical Materials Research Part A* 2009;88:1-11.

[92] Kievit FM, Veiseh O, Bhattarai N, Fang C, Gunn JW, Lee D, et al. PEI–PEG–Chitosan-Copolymer-Coated Iron Oxide Nanoparticles for Safe Gene Delivery: Synthesis, Complexation, and Transfection. *Advanced Functional Materials* 2009;19:2244–51.

[93] Khor E, Lim LY. Implantable applications of chitin and chitosan. *Biomaterials* 2003 Jun;24(13):2339-49.

[94] Kang B, Chang SQ, Dai YD, Chen D. Radiation synthesis and magnetic properties of novel Co0.7MFe0.3/chitosan compound nanoparticles for targeted drug carrier. *Radiation Physics and Chemistry* 2007 Jun;76(6):968-73.

[95] Janardhanan SK, Ramasamy I, Nair BU. Synthesis of iron oxide nanoparticles using chitosan and starch templates. *Transition Metal Chemistry* 2008 Feb;33(1):127-31.
[96] Illum L, Farraj NF, Davis SS. Chitosan as a novel nasal delivery. *Pharm Res* 1994;11:1186–9.
[97] Belessi V, Zboril R, Tucek J, Mashlan M, Tzitzios V, Petridis D. Ferrofluids from magnetic-chitosan hybrids. *Chemistry of Materials* 2008 May;20(10):3298-305.
[98] An XN, Su ZX. Characterization and application of high magnetic property chitosan particles. *Journal of Applied Polymer Science* 2001 Aug;81(5):1175-81.
[99] Thomas M, Klibanov AM. Conjugation to gold nanoparticles enhances polyethylenimine's transfer of plasmid dna into mammalian cells. *Proceedings of the National Academy of Sciences of the United States of America* 2003;100(16):9138-43.
[100] Thomas M, Ge Q, Lu JJ, Chen J, Klibanov AM. Cross-linked small polyethylenimines: While still nontoxic, deliver DNA efficiently to mammalian cells in vitro and in vivo. *Pharmaceutical Research* 2005;22(3):373-80.
[101] Tang GP, Guo HY, Alexis F, Wang X, Zeng S, Lim TM, et al. Low molecular weight polyethylenimines linked by ²ζ-cyclodextrin for gene transfer into the nervous system. *Journal of Gene Medicine* 2006;8(6):736-44.
[102] Lungwitz U, Breunig M, Blunk T, G¶pferich A. Polyethylenimine-based non-viral gene delivery systems. *European Journal of Pharmaceutics and Biopharmaceutics* 2005;60(2):247-66.
[103] Fischer D, Bieber T, Li Y, Elsässer HP, Kissel T. A novel non-viral vector for DNA delivery based on low molecular weight, branched polyethylenimine: Effect of molecular weight on transfection efficiency and cytotoxicity. *Pharmaceutical Research* 1999;16(8):1273-9.
[104] Brus C, Petersen H, Aigner A, Czubayko F, Kissel T. Physicochemical and biological characterization of polyethylenimine-graft-poly(ethylene glycol) block copolymers as a delivery system for oligonucleotides and ribozymes. *Bioconjugate Chemistry* 2004 Jul-Aug;15(4):677-84.
[105] Arote R, Kim TH, Kim YK, Hwang SK, Jiang HL, Song HH, et al. A biodegradable poly(ester amine) based on polycaprolactone and polyethylenimine as a gene carrier. *Biomaterials* 2007;28(4):735-44.
[106] Zhao X, Shi Y, Wang T, Cai Y, Jiang G. Preparation of silica-magnetite nanoparticle mixed hemimicelle sorbents for extraction of several typical phenolic compounds from environmental water samples. *Journal of Chromatography A 2008*;1188(2):140-7.
[107] Yu CH, Al-Saadi A, Shih SJ, Qiu L, Tam KY, Tsang SC. Immobilization of BSA on silica-coated magnetic iron oxide nanoparticle. *Journal of Physical Chemistry C* 2009;113(2):537-43.
[108] Yiu HHP, Keane MA, Lethbridge ZAD, Lees MR, El Haj AJ, Dobson J. Synthesis of novel magnetic iron metal-silica (Fe-SBA-15) and magnetite-silica (Fe3O4-SBA-15) nanocomposites with a high iron content using temperature-programed reduction. *Nanotechnology* 2008 Jun;19(25).
[109] Yang PP, Quan ZW, Lu LL, Huang SS, Lin J. Luminescence functionalization of mesoporous silica with different morphologies and applications as drug delivery systems. *Biomaterials* 2008 Feb;29(6):692-702.

[110] Yague C, Moros M, Grazu V, Arruebo M, Santamaria J. Synthesis and stealthing study of bare and PEGylated silica micro- and nanoparticles as potential drug-delivery vectors. *Chemical Engineering Journal* 2008 Mar;137(1):45-53.

[111] Vertegel AA, Siegel RW, Dordick JS. Silica nanoparticle size influences the structure and enzymatic activity of adsorbed lysozyme. Langmuir 2004;20:6800-7.

[112] Tsang SC, Yu CH, Gao X, Tam K. Silica-encapsulated nanomagnetic particle as a new recoverable biocatalyst carrier. *Journal of Physical Chemistry* B 2006;110(34):16914-22.

[113] Thierry B, Zimmer L, McNiven S, Finnie K, Barbe C, Griessert HJ. Electrostatic self-assembly of PEG copolymers onto porous silica nanoparticles. *Langmuir* 2008 Aug;24(15):8143-50.

[114] Tartaj P, Serna CJ. Synthesis of monodisperse superparamagnetic Fe/silica nanospherical composites. *Journal of the American Chemical Society* 2003 Dec;125(51):15754-5.

[115] Tadic M, Markovic D, Spasojevic V, Kusigerski V, Remskar M, Pirnat J, et al. Synthesis and magnetic properties of concentrated alpha-Fe2O3 nanoparticles in a silica matrix. *Journal of Alloys and Compounds* 2007 Aug;441(1-2):291-6.

[116] Stober W, Fink A, Bohn E. Controlled growth of monodisperse silica spheres in the micron size range. *Journal of Colloid and Interface Science* 1968;26(1):62-9.

[117] Steitz B, Krauss F, Rousseau S, Hofmann H, Petri-Fink A. Positional control of superparamagnetic iron oxide nanoparticles in silica beads. *Advanced Engineering Materials* 2007 May;9(5):375-80.

[118] Santra S, Tapec R, Theodoropoulou N, Dobson J, Hebard A, Tan WH. Synthesis and characterization of silica-coated iron oxide nanoparticles in microemulsion: The effect of nonionic surfactants. *Langmuir* 2001 May;17(10):2900-6.

[119] Santra S, Tapec R, Theodoropoulou N, Dobson J, Hebard A, Tan W. Synthesis and characterization of silica-coated iron oxide nanoparticles in microemulsion: The effect of nonionic surfactants. *Langmuir* 2001;17(10):2900-6.

[120] Niznansky D, Rehspringer JL, Drillon M. Preparation of magnetic manoparticles (gamma-fe2o3) in the silica matrix. 5th European Magnetic Materials and Applications Conference (EMMA 93); 1993 Aug 24-27; Kosice, Slovakia: *Ieee-Inst Electrical Electronics Engineers Inc;* 1993. p. 821-3.

[121] Mulvaney P, Liz-Marzan LM, Giersig M, Ung T. Silica encapsulation of quantum dots and metal clusters. *Journal of Materials Chemistry* 2000;10(6):1259-70.

[122] Lukehart CM, Milne SB, Stock SR. Formation of crystalline nanoclusters of Fe2P, RuP, Co2P, Rh2P, Ni2P, Pd5P2, or PtP2 in a silica xerogel matrix from single-source molecular precursors. *Chemistry of Materials* 1998 Mar;10(3):903-8.

[123] Lu ZY, Qin YQ, Fang JY, Sun J, Li J, Liu FQ, et al. Monodisperse magnetizable silica composite particles from heteroaggregate of carboxylic polystyrene latex and Fe3O4 nanoparticles. *Nanotechnology* 2008 Feb;19(5).

[124] Lu ZY, Dai J, Song XN, Wang G, Yang WS. Facile synthesis of Fe3O4/SiO2 composite nanoparticles from primary silica particles. *Colloids and Surfaces a-Physicochemical and Engineering Aspects* 2008 Mar;317(1-3):450-6.

[125] Lu CW, Hung Y, Hsiao JK, Yao M, Chung TH, Lin YS, et al. Bifunctional magnetic silica nanoparticles for highly efficient human stem cell labeling. *Nano Letters* 2007;7(1):149-54.

[126] Lu AH, Li WC, Kiefer A, Schmidt W, Bill E, Fink G, et al. Fabrication of magnetically separable mesostructured silica with an open pore system. *Journal of the American Chemical Society* 2004 Jul;126(28):8616-7.

[127] Lim YT, Kim JK, Noh YW, Cho MY, Chung BH. Multifunctional Silica Nanocapsule with a Single Surface Hole. *Small 2009 Feb*;5(3):324-8.

[128] Xi D, Luo XP, Lu QH, Yao K, Liu ZL, Ning Q. The detection of HBV DNA with gold-coated iron oxide nanoparticle gene probes *J Nanopart Res* 2008;10:393-400.

[129] Wang SG, Lu WT, Tovmachenko O, Rai US, Yu HT, Ray PC. Challenge in understanding size and shape dependent toxicity of gold nanomaterials in human skin keratinocytes. *Chemical Physics Letters* 2008 Sep;463(1-3):145-9.

[130] Tkachenko AG, Xie H, Coleman D, Glomm W, Ryan J, Anderson MF, et al. Multifunctional gold nanoparticle-peptide complexes for nuclear targeting. *Journal of the American Chemical Society* 2003 Apr;125(16):4700-1.

[131] Seino S, Kinoshita T, Nakagawa T, Kojima T, Taniguci R, Okuda S, et al. Radiation induced synthesis of gold/iron-oxide composite nanoparticles using high-energy electron beam. *J Nanopart Res* 2008;10:1071-6.

[132] Robert E, James JS. Selective colorimetric detection of polynucleotides based on the distance-dependent optical properties of gold nanoparticles. *Science* 1977;277:1078-81.

[133] Pham TTH, Cao C, Sim SJ. Application of citrate-stabilized gold-coated ferric oxide composite nanoparticles for biological separations *Journal of Magnetism and Magnetic Materials* 2008;320:2049-55.

[134] Lo CK, Xiao D, Choi MMF. Homocysteine-protected gold-coated magnetic nanoparticles: synthesis and characterisationt. *Journal of Materials Chemistry* 2007 Jun;17(23):2418-27.

[135] Lim YT, Cho MY, Choi BS, Lee JM, Chung BH. Paramagnetic gold nanostructures for dual modal bioimaging and phototherapy of cancer cells. *Chemical Communications* 2008(40):4930-2.

[136] Lim IIS, Njoki PN, Park HY, Wang X, Wang LY, Mott D, et al. Gold and magnetic oxide/gold core/shell nanoparticles as bio-functional nanoprobes. *Nanotechnology* 2008 Jul;19(30).

[137] Kundu S, Peng LH, Liang H. A new route to obtain high-yield multiple-shaped gold nanoparticles in aqueous solution using microwave irradiation. *Inorganic Chemistry* 2008 Jul;47(14):6344-52.

[138] Jain PK, Xiao Y, Walsworth R, Cohen AE. Surface Plasmon Resonance Enhanced Magneto-Optics (SuPREMO): Faraday Rotation Enhancement in Gold-Coated Iron Oxide Nanocrystals. *Nano Letters* 2009;9(4):1644-50.

[139] Hwu JR, Lin YS, Josephrajan T, Hsu MH, Cheng FY, Yeh CS, et al. Targeted Paclitaxel by Conjugation to Iron Oxide and Gold Nanoparticles. *Journal of the American Chemical Society* 2009 Jan;131(1):66-+.

[140] De Palma R, Liu CX, Barbagini F, Reekmans G, Bonroy K, Laureyn W, et al. Magnetic particles as labels in bioassays: Interactions between a biotinylated gold substrate and streptavidin magnetic particles. *Journal of Physical Chemistry* C 2007 Aug;111(33):12227-35.

[141] Cui Y, Wang Y, Hui W, Zhang Z, Xin X, Chen C. The Synthesis of GoldMag Nano-Particles and their Application for Antibody Immobilization. *Biomedical Microdevices* 2005;7(2).

[142] Roberts MJ, Bentley MD, Harris JM. Chemistry for peptide and protein PEGylation. *Advanced Drug Delivery Reviews* 2002;54:459-76.
[143] Lyczak JB, Morrison SL. Biological and pharmacokinetic properties of a novel immunoglobulin-CD4 fusion protein. *Archives of Virology* 1994;139(1-2):189-96.
[144] Syed S, Schuyler PD, Kulczycky M, Sheffield WP. Potent antithrombin activity and delayed clearance from the circulation characterize recombinant hirudin genetically fused to albumin. *Blood* 1997;89(9):3243-52.
[145] Davis FF. *Enzyme Engineering* 1978;4.
[146] Milton Harris J, Chess RB. Effect of pegylation on pharmaceuticals. *Nature Reviews Drug Discovery* 2003;2(3):214-21.
[147] Bailon P, Berthold W. Polyethylene glycol-conjugated pharmaceutical proteins. *Pharmaceutical Science and Technology Today* 1998;1(8):352-6.
[148] Drake WM, Parkinson C, Akker SA, Monson JP, Besser GM, Trainer PJ. Successful treatment of resistant acromegaly with a growth hormone receptor antagonist. *Eur J Endocrinol* 2001 Oct;145(4):451-6.
[149] Harris JM, Chess RB. Effect of PEGylation on Pharmaceuticals. *Nature Reviews, Drug Discovery* 2003;2:8.
[150] Richter AW, Akerblom E. *Int Arch Allergy Appl Immunol 1983*;70:8.
[151] Kim DK, Zhang Y, Kehr J, Klason T, Bjelke B, Muhammed M. Characterization and MRI study of surfactant-coated superparamagnetic nanoparticles administered into the rat brain. 3rd International Conference on Scientific and Clinical Applications of Magnetic Carriers; 2000 May 03-06; Rostock, Germany: *Elsevier Science Bv;* 2000. p. 256-61.
[152] Kohler N, Sun C, Fichtenholtz A, Gunn J, Fang C, Zhang MQ. *Methotrexate-immobilized poly(ethylene glycol) magnetic nanoparticles for MR imaging and drug delivery*. Small 2006 Jun;2(6):785-92.
[153] Zhang Y, Kohler N, Zhang M. Surface modification of superparamagnetic magnetite nanoparticles and their interacellular uptake *Biomaterials* 2002;23(7):9.
[154] Moghimi SM, Hunter AC, Murray JC. Long-circulating and target-specific nanoparticles: Theory to practice. *Pharmacological Reviews* 2001 Jun;53(2):283-318.
[155] Sun SH, Zeng H. Size-controlled synthesis of magnetite nanoparticies. *Journal of the American Chemical Society* 2002 Jul;124(28):8204-5.
[156] Wang BD, Xu CJ, Xie J, Yang ZY, Sun SL. pH Controlled Release of Chromone from Chromone-Fe3O4 Nanoparticles. *Journal of the American Chemical Society* 2008 Nov;130(44):14436-+.
[157] Xie J, Xu C, Kohler N, Hou Y, Sun S. Controlled PEGylation of monodisperse Fe3O4 nanoparticles for reduced non-specific uptake by macrophage cells. *Adv Mater* 2007 Oct;19(20):3163-+.
[158] Shafi K, U-Iman A, Yan XZ, Yang NL, Estournes C, White H, et al. Sonochemical synthesis of functionalized amorphous iron oxide nanoparticles. *Langmuir* 2001 Aug;17(16):5093-7.
[159] Gupta AK, Curtis ASG. Surface modified superparamagnetic nanoparticles for drug delivery: interaction studies with human fibroblasts in culture. *Journal of Materials Science and Materials Medicine* 2004;15(4):493-6.

[160] Gupta AK, Gupta M. Synthesis and surface engineering superparamagnetic iron oxide nanoparticles for drug delivery and cellular targeting. *Handbook of Particulate Drug Delivery* American Scientific Publishers, USA 2007.

[161] Herve K, Douziech-Eyrolles L, Munnier E, Cohen-Jonathan S, Souce M, Marchais H, et al. The development of stable aqueous suspensions of PEGylated SPIONs for biomedical applications. *Nanotechnology* 2008 Nov;19(46).

[162] Gupta AK, Naregalkar RR, Vaidya VD, Gupta M. Recent advances on surface engineering of magnetic iron oxide nanoparticles and their biomedical applications. *Nanomedicine* 2007;2(1):23-39.

[163] Shi N, Boado RJ, Pardridge WM. Receptor-mediated gene targeting to tissues in vivo following intravenous administration of pegylated immunoliposomes. *Pharmaceutical Research* 2001;18(8):1091-5.

[164] Calvo P, Gouritin B, Chacun H, Desmaile D, D'Angelo J, Noel JP, et al. Long-circulating pegylated polycyanoacrylate nanoparticles as new drug carrier for brain delivery. *Pharmaceutical Research* 2001;18(8):1157-66.

[165] He XZ, Ma JY, Mercado AE, Xu WJ, Jabbari E. Cytotoxicity of Paclitaxel in biodegradable self-assembled core-shell poly(lactide-co-glycolide ethylene oxide fumarate) nanoparticles. *Pharmaceutical Research* 2008 Jul;25(7):1552-62.

[166] Shu XZ, Liu YC, Palumbo FS, Lu Y, Prestwich GD. In situ crosslinkable hyaluronan hydrogels for tissue engineering. *Biomaterials* 2004 Mar-Apr;25(7-8):1339-48.

[167] Sanborn TJ, Messersmith PB, Barron AE. In situ crosslinking of a biomimetic peptide-PEG hydrogel via thermally triggered activation of factor XIII. *Biomaterials* 2002 Jul;23(13):2703-10.

[168] Temenoff JS, Shin H, Conway DE, Engel PS, Mikos AG. In vitro cytotoxicity of redox radical initiators for cross-linking of oligo(poly(ethylene glycol) fumarate) macromers. *Biomacromolecules* 2003 Nov-Dec;4(6):1605-13.

[169] Peter SJ, Yaszemski MJ, Suggs LJ, Payne RG, Langer R, Hayes WC, et al. Characterization of partially saturated poly(propylene fumarate) for orthopaedic application. *Journal of Biomaterials Science, Polymer Edition* 1997;8(11):893-904.

[170] Peter SJ, Suggs LJ, Yaszemski MJ, Engel PS, Mikos AG. Synthesis of poly(propylene fumarate) by acylation of propylene glycol in the presence of a proton scavenger. *Journal of Biomaterials Science, Polymer Edition* 1999;10(3):363-73.

[171] Jo S, Shin H, Shung AK, Fisher JP, Mikos AG. Synthesis and characterization of oligo(poly(ethylene glycol) fumarate) macromer. *Macromolecules* 2001 Apr;34(9):2839-44.

[172] Suggs LJ, Kao EY, Palombo LL, Krishnan RS, Widmer MS, Mikos AG. Preparation and characterization of poly(propylene fumarate-co-ethylene glycol) hydrogels. *Journal of Biomaterials Science, Polymer Edition* 1998;9(7):653-66.

[173] Park H, Temenoff JS, Tabata Y, Caplan AI, Mikos AG. Injectable biodegradable hydrogel composites for rabbit marrow mesenchymal stem cell and growth factor delivery for cartilage tissue engineering. *Biomaterials* 2007;28(21):3217-27.

[174] Hashemi Doulabi AS, Mirzadeh H, Imani M, Sharifi S, Atai M, Mehdipour-Ataei S. Synthesis and preparation of biodegradable and visible light crosslinkable unsaturated fumarate-based networks for biomedical applications. *Polymers for Advanced Technologies* 2008;19(9):1199-208.

[175] Shin H, Temenoff JS, Mikos AG. In vitro cytotoxicity of unsaturated oligo[poly(ethylene glycol)fumarate] macromers and their cross-linked hydrogels. *Biomacromolecules* 2003;4(3):552-60.

[176] Temenoff JS, Park H, Jabbari E, Conway DE, Sheffield TL, Ambrose CG, et al. Thermally cross-linked oligo(poly(ethylene glycol) fumarate) hydrogels support osteogenic differentiation of encapsulated marrow stromal cells in vitro. *Biomacromolecules* 2004;5(1):5-10.

[177] Shin H, Quinten Ruh|© P, Mikos AG, Jansen JA. In vivo bone and soft tissue response to injectable, biodegradable oligo(poly(ethylene glycol) fumarate) hydrogels. *Biomaterials* 2003;24(19):3201-11.

[178] Wang S, Lu L, Gruetzmacher JA, Currier BL, Yaszemski MJ. Synthesis and characterizations of biodegradable and crosslinkable poly($\varepsilon\mu$-caprolactone fumarate), poly(ethylene glycol fumarate), and their amphiphilic copolymer. *Biomaterials* 2006;27(6):832-41.

[179] Okino H, Nakayama Y, Tanaka M, Matsuda T. In situ hydrogelation of photocurable gelatin and drug release. *Journal of Biomedical Materials Research* 2002;59(2):233-45.

[180] Timmer MD, Jo SB, Wang CY, Ambrose CG, Mikos AG. Characterization of the cross-linked structure of fumarate-based degradable polymer networks. *Macromolecules* 2002 May;35(11):4373-9.

[181] McGuire WP, Rowinsky EK, Rosenhein NB, Grumbine FC, Ettinger DS, Armstrong DK, et al. Taxol: A unique antineoplastic agent with significant activity in advanced ovarian epithelial neoplasms. *Annals of Internal Medicine* 1989;111(4):273-9.

[182] Gagandeep S, Novikoff PM, Ott M, Gupta S. Paclitaxel shows cytotoxic activity in human hepatocellular carcinoma cell lines. *Cancer Letters* 1999;136(1):109-18.

[183] Alani AWG, Bae Y, Rao DA, Kwon GS. Polymeric micelles for the pH-dependent controlled, continuous low dose release of paclitaxel. *Biomaterials;*31(7):1765-72.

[184] Azad N, Perroy A, Gardner E, Imamura CK, Graves C, Sarosy GA, et al. A phase I study of paclitaxel and continuous daily CAI in patients with refractory solid tumors. *Cancer Biology and Therapy;*8(19):1800-5.

[185] Diehm NA, Hoppe H, Do DD. Drug Eluting Balloons. *Techniques in Vascular and Interventional Radiology;*13(1):59-63.

[186] Zhang SQ, Song YN, He XH, Zhong BH, Zhang ZQ. Liquid chromatography-tandem mass spectrometry for the determination of paclitaxel in rat plasma after intravenous administration of poly(l-glutamic acid)-alanine-paclitaxel conjugate. *Journal of Pharmaceutical and Biomedical Analysis;*51(5):1169-74.

[187] Roff WJ, Scott JR, Pacitti J. *Handbook of Common Polymers CRC Press,* Cleveland 1971.

[188] Pritchard JG. *MacDonald Technical and Scientific,* London 1970.

[189] Zainuddin Z, D.J.T. H, Le TT. *Radiation Physics and Chemistry* 2001;62:9.

[190] Xue B, Sun Y. Protein adsorption equilibria and kinetics to a poly(vinyl alcohol)-based magnetic affinity support. *Journal of Chromatography A* 2001 Jul;921(2):109-19.

[191] Yokoi H, Kantoh T. Thermal-decomposition of the iron(iii) hydroxide and magnetite composites of poly(vinyl alcohol) - preparation of magnetite and metallic iron particles. *Bulletin of the Chemical Society of Japan* 1993 May;66(5):1536-41.

[192] Sairam M, Naidu BVK, Nataraj SK, Sreedhar B, Aminabhavi TM. Poly(vinyl alcohol)-iron oxide nanocomposite membranes for pervaporation dehydration of isopropanol,

1,4-dioxane and tetrahydrofuran. *Journal of Membrane Science* 2006 Oct;283(1-2):65-73.

[193] Schopf B, Neuberger T, Schulze K, Petri A, Chastellain M, Hofmann M, et al. Methodology description for detection of cellular uptake of PVA coated superparamagnetic iron oxide nanoparticles (SPION) in synovial cells of sheep. 5th International Conference on Scientific and Clinical Applications of Magnetic Carriers; 2004 May 20-22; Lyon, FRANCE: *Elsevier Science Bv;* 2004. p. 411-8.

[194] Chastellain A, Petri A, Hofmann H. Particle size investigations of a multistep synthesis of PVA coated superparamagnetic nanoparticles. *Journal of Colloid and Interface Science* 2004 Oct;278(2):353-60.

[195] Finch CA. *Polyvinyl alcohol; properties and applications Interscience*, Div of Wiley London; New York 1973:622.

[196] Maruoka S, Matsuura T, Kawasaki K, Okamoto M, Yoshiaki H, Kodama M, et al. Biocompatibility of polyvinylalcohol gel as a vitreous substitute. *Current Eye Research* 2006 Jul-Aug;31(7-8):599-606.

[197] Osada Y, Gong JP. Soft and wet materials: Polymer gels. *Advanced Materials* 1998 Aug;10(11):827-37.

[198] Petri-Fink A, Chastellain M, Juillerat-Jeanneret L, Ferrari A, Hofmann H. Development of functionalized superparamagnetic iron oxide nanoparticles for interaction with human cancer cells. *Biomaterials* 2005 May;26(15):2685-94.

[199] Nel AE, madler I, Velegol D, Xia T, Hoek EMV, Somasundaran P, et al. understanding biophysicochemical interactions at the nano–bio interface. *Nature Materials* 2009;8:543-57.

[200] Biswas S, Srivastava VK, Ram S, Fecht HJ. Nanorods of silver-coated magnetic CrO2 particles from a polymer template in hot water. *Journal of Physical Chemistry* C 2007;111(21):7593-8.

[201] Ricciardi R, Auriemma F, De Rosa C, Lauprâitre F. X-ray Diffraction Analysis of Poly(vinyl alcohol) Hydrogels, Obtained by Freezing and Thawing Techniques. *Macromolecules* 2004;37(5):1921-7.

[202] Ricciardi R, Auriemma F, Gaillet C, De Rosa C, Lauprâitre F. Investigation of the crystallinity of freeze/thaw poly(vinyl alcohol) hydrogels by different techniques. *Macromolecules* 2004;37(25):9510-6.

[203] Park JH, Von Maltzahn G, Zhang L, Derfus AM, Simberg D, Harris TJ, et al. Systematic surface engineering of magnetic nanoworms for in vivo tumor targeting. *Small* 2009;5(6):694-700.

[204] Park JH, Von Maltzahn G, Zhang L, Schwartz MP, Ruoslahti E, Bhatia SN, et al. Magnetic iron oxide nanoworms for tumor targeting and imaging. *Advanced Materials* 2008;20(9):1630-5.

[205] Burugapalli K, Koul V, Dinda AK. Effect of composition of interpenetrating polymer network hydrogels based on poly(acrylic acid) and gelatin on tissue response: A quantitative in vivo study. *Journal of Biomedical Materials Research Part* A 2004 Feb;68A(2):210-8.

[206] Fahlvik AK, Holtz E, Schroder U, Klaveness J. *J InVest Radiol* 1990;25.

[207] Kellar KE, Fujii DK, Wolfgang H, Gunther WH, Briley-Saebo K, Spiller M, et al. *Magn Reson Mater Phys, Biol Med* 1999;8.

[208] Kim DK, Mikhaylova M, Zhang Y, Tsakalakos T, Muhammed M. *Chem Mater* 2003;15.
[209] Fahlvik AK, Holtz E, Schroder U, Klaveness J. *InVest Radiol* 1990;25.
[210] Wu Y, Guo J, Yang W, Wang C, Fu S. *Polymer* 2006;47:8.
[211] Moffat BA, Reddy GR, McConville P. *MRI Mol Imaging* 2003;2:9.
[212] Iijima M, Yonemochi Y, Tsukada M, Kamiya H. Microstructure control of iron hydroxide nanoparticles using surfactants with different molecular structures. *Journal of Colloid and Interface Science* 2006;298:202-8.
[213] Mak S-Y, Chen D-H. Binding and Sulfonation of Poly(acrylic acid) on Iron Oxide Nanoparticles: a Novel, Magnetic, Strong Acid Cation Nano-Adsorbent. *Macromolecular Rapid Communications* 2005;26:1567-71.
[214] Schild HG. Poly(N-isopropylacrylamide): Experiment, theory and application. Progress in Polymer Science (Oxford) 1992;17(2):163-249.
[215] Chen G, Hofmann AS. Preparation and properties of thermoreversible, phaseseparating enzyme-oligo(N-isopropylacrylamide) conjugates. *Bioconjugate Chemistry* 1993;4(6):6.
[216] Schmaljohann D. *Advanced Drug Delivery Reviews* 2006;58(15).
[217] Saunders BR, Vincent B. *Advanced Colloid and Interface Science* 1999;80(1).
[218] Chung JE, Yokoyama M, Yamato M, Aoyagi T, Sakurai Y, Okano T. Thermoresponsive drug delivery from polymeric micelles constructed using block copolymers of poly(N-isopropylacrylamide) and poly(butylmethacrylate). *Journal of Controlled Release* 1999;62(1-2):115-27.
[219] Yan H, Tsujii K. Potential application of poly(N-isopropylacrylamide) gel containing polymeric micelles to drug delivery systems. *Colloids and Surfaces B: Biointerfaces* 2005;46(3):142-6.
[220] Antunes FE, Gentile L, Tavano L, Rossi CO. Rheological characterization of the thermal gelation of poly(N-isopropylacrylamide) and poly(N-isopropylacrylamide) co-Acrylic acid. *Applied Rheology* 2009;19(4).
[221] Lin CL, Chiu WY. *J Polym Sci, Part A: Polym Chem* 2005;43(23).
[222] Cai J, Guo J, Ji ML, Yang WL, Wang CC, Fu SK. Preparation and characterization of multiresponsive polymer composite microspheres with core-shell structure. *Colloid Polym Sci* 2007;285(14).
[223] Sun YB, Ding XB, Zheng ZH, Cheng X, Hu XH, Peng YX. Chem Commun 2006;26.
[224] Lai JJ, Hoffman JM, Ebara M, Hoffman AS, Estournes C, Wattiaux A, et al. *Langmuir* 2007;23(13).
[225] Carter S, Hunt B, Rimmer S. Highly branched poly(N-isopropylacrylamide)s with imidazole end groups prepared by radical polymerization in the presence of a styryl monomer containing a dithioester group. *Macromolecules* 2005;38(11):4595-603.
[226] Rimmer S, Carter S, Rutkaite R, Haycock JW, Swanson L. Highly branched poly-(N-isopropylacrylamide)s with arginine-glycine- aspartic acid (RGD)- or COOH-chain ends that form sub-micron stimulus-responsive particles above the critical solution temperature. *Soft Matter* 2007;3(8):971-3.
[227] Kamps JA, Scherphof GL. Receptor versus non-receptor mediated clearance of liposomes. *Adv Drug Deliv Rev* 1998;32(1-2):81-97.

[228] Lynch I. Are there generic mechanisms governing interactions between nanoparticles and cells? Epitope mapping the outer layer of the protein–material interface. *Physica A* 2007;373:511-20.

[229] Bertholon I, Ponchel G, Labarre D, Couvreur P, Vauthier C. Bioadhesive properties of poly(alkylcyanoacrylate) nanoparticles coated with polysaccharide. *J Nanosci Nanotechnol* 2006;6(9-10):3102-9.

[230] Bulte JW, Kraitchman DL. Iron oxide MR contrast agents for molecular and cellular imaging. *NMR Biomed* 2004;17(7):484-99.

[231] Weissleder R, Bogdanov JA, Neuwelt EA, Papisov M. Long-circulating iron oxides for MR imaging. *Adv Drug Deliv Rev* 1995;16:321-34.

[232] Jung CW, Jacobs P. Surface properties of superparamagnetic iron oxide MR contrast agents: ferumoxides ferumoxtran ferumoxsil. *Magn Reson Imaging* 1995;13:675-91.

[233] Mykhaylyk O, Cherchenko A, Ilkin A, Dudchenko N, Ruditsa V, Novoseletz M, et al. Glial brain tumor targeting of magnetite nanoparticles in rats. *Journal of Magnetism and Magnetic Materials* 2001;225(1-2):241-7.

[234] Tian J, Feng YK, Xu YS. Ring Oxpening Polymerization of D,L-Lactide on Magnetite Nanoparticles. *Macromolecular Research* 2006;14(2):5.

[235] Ruiz JM, Benoit JP. *Journal of Controlled Release* 1991;86(8):5.

[236] Lahann J, Langer R. *Macromol Rapid Comm* 2001;22.

[237] Choi IS, Langer R. *Macromolecules* 2001;34.

[238] Rowlands AS, Lim SA, Martin D, Cooper-White JJ. Polyurethane/poly(lactic-co-glycolic) acid composite scaffolds fabricated by thermally induced phase separation. *Biomaterials* 2007;28(12):2109-21.

[239] Sahoo S, Ouyang H, James CH, Tay TE, Toh SL. Characterization of a novel polymeric scaffold for potential application in tendon/ligament tissue engineering. *Tissue Engineering* 2006;12(1):91-9.

[240] Salehi R, Nowruzi K, Entezami AA, Asgharzadeh V, Davaran S. Thermosensitive polylactide-glycolide delivery systems for treatment of narcotic addictions. *Polymers for Advanced Technologies* 2009;20(4):416-22.

[241] Sander EA, Alb AM, Nauman EA, Reed WF, Dee KC. Solvent effects on the microstructure and properties of 75/25 poly(D,L-lactide-co-glycolide) tissue scaffolds. *Journal of Biomedical Materials Research - Part A* 2004;70(3):506-13.

[242] Saunders BR, Fraylich M, Alexander C, Wang W, Liang H, Cheikh Al Ghanami R, et al. Biodegradable thermoresponsive microparticle dispersions for injectable cell delivery prepared using a single-step process. *Advanced Materials* 2009;21(18):1809-13.

[243] Schneider OD, Loher S, Brunner TJ, Uebersax L, Simonet M, Grass RN, et al. Cotton wool-like nanocomposite biomaterials prepared by electrospinning: In vitro bioactivity and osteogenic differentiation of human mesenchymal stem cells. *Journal of Biomedical Materials Research - Part B Applied Biomaterials* 2008;84(2):350-62.

[244] Shum AWT, Li J, Mak AFT. Fabrication and structural characterization of porous biodegradable poly(DL-lactic-co-glycolic acid) scaffolds with controlled range of pore sizes. *Polymer Degradation and Stability* 2005;87(3):487-93.

[245] Singh L, Kumar V, Ratner BD. Generation of Porous Microcellular 85/15 poly (DL-Lactide-co-Glycolide) Foams Using Supercritical CO2 for Biomedical Applications.

American Society of Mechanical Engineers, Materials Division (Publication) MD 2000;91:29-34.

[246] Singh L, Kumar V, Ratner BD. Generation of porous microcellular 85/15 poly (DL-lactide-co-glycolide) foams for biomedical applications. *Biomaterials* 2004;25(13):2611-7.

[247] Spiller KL, Laurencin SJ, Lowman AM. Characterization of the behavior of porous hydrogels in model osmotically-conditioned articular cartilage systems. *Journal of Biomedical Materials Research - Part B Applied Biomaterials* 2009;90(2):752-9.

[248] Suh KY, Khademhosseini A, Yang JM, Eng G, Langer R. Direct immobilization and patterning of hyaluronic acid on hydrophilic substrates. 2005 3rd IEEE/EMBS Special *Topic Conference on Microtechnology in Medicine and Biology*; 2005; 2005. p. 331-2.

[249] Terai H, Hannouche D, Ochoa E, Yamano Y, Vacanti JP. In vitro engineering of bone using a rotational oxygen-permeable bioreactor system. *Materials Science and Engineering C* 2002;20(1-2):3-8.

[250] Toh SL, Teh TKH, Vallaya S, Goh JCH. Novel silk scaffolds for ligament tissue engineering applications. *Key Engineering Materials* 2006;326-328 I:727-30.

[251] Vozzi G, Flaim CJ, Bianchi F, Ahluwalia A, Bhatia S. Microfabricated PLGA scaffolds: A comparative study for application to tissue engineering. *Materials Science and Engineering* C 2002;20(1-2):43-7.

[252] Yang YI, Seol DL, Kim HI, Cho MH, Lee SJ. Continuous perfusion culture for generation of functional tissue-engineered soft tissues. *Current Applied Physics* 2007;7(SUPPL.1):e80-e4.

[253] Yoon OJ, Kim HW, Kim DJ, Lee HJ, Yun JY, Noh YH, et al. Nanocomposites of electrospun poly[(D,L-lactic)-co-(glycolic acid)] and plasma-functionalized single-walled carbon nanotubes for biomedical applications. *Plasma Processes and Polymers* 2009;6(2):101-9.

[254] Yu J, Bai X, Suh J, Sang BL, Sang JS. Mechanical capping of silica nanotubes for encapsulation of molecules. *Journal of the American Chemical Society* 2009;131(43):15574-5.

[255] Yu L, Zhang Z, Zhang H, Ding J. Mixing a sol and a precipitate of block copolymers with different block ratios leads to an injectable hydrogel. *Biomacromolecules* 2009;10(6):1547-53.

[256] Zhang J, Liu L, Gao Z, Li L, Feng X, Wu W, et al. Novel Approach to Engineer Implantable Nasal Alar Cartilage Employing Marrow Precursor Cell Sheet and Biodegradable Scaffold. *Journal of Oral and Maxillofacial Surgery* 2009;67(2):257-64.

[257] Zhang N, Nichols HL, Tylor S, Wen X. Fabrication of nanocrystalline hydroxyapatite doped degradable composite hollow fiber for guided and biomimetic bone tissue engineering. *Materials Science and Engineering* C 2007;27(3):599-606.

[258] Zhang Z, Feng SS. In vitro investigation on poly(lactide)-tween 80 copolymer nanoparticles fabricated by dialysis method for chemotherapy. *Biomacromolecules* 2006;7(4):1139-46.

[259] Alexis F. Factors affecting the degradation and drug-release mechanism of poly(lactic acid) and poly[(lactic acid)-co-(glycolic acid)]. *Polymer International* 2005;54(1):36-46.

[260] Alexis F, Venkatraman S, Rath SK, Gan LH. Some insight into hydrolytic scission mechanisms in bioerodible polyesters. *Journal of Applied Polymer Science* 2006;102(4):3111-7.

[261] Bertram JP, Jay SM, Hynes SR, Robinson R, Criscione JM, Lavik EB. Functionalized poly(lactic-co-glycolic acid) enhances drug delivery and provides chemical moieties for surface engineering while preserving biocompatibility. *Acta Biomaterialia* 2009;5(8):2860-71.

[262] Bini TB, Gao S, Xu X, Wang S, Ramakrishna S, Leong KW. Peripheral nerve regeneration by microbraided poly(L-lactide-co-glycolide) biodegradable polymer fibers. *Journal of Biomedical Materials Research - Part A 2004*;68(2):286-95.

[263] Brochhausen C, Zehbe R, Watzer B, Halstenberg S, Gabler F, Schubert H, et al. Immobilization and controlled release of prostaglandin E2 from poly-L-lactide-co-glycolide microspheres. *Journal of Biomedical Materials Research - Part A* 2009;91(2):454-62.

[264] Cai Q, Bei J, Wang S. Synthesis and characterization of glycolide/lactide/caprolactone tri-component copolymer. *Acta Polymerica Sinica* 1999(6):764.

[265] Casper CL, Yamaguchi N, Kiick KL, Rabolt JF. Functionalizing electrospun fibers with biologically relevant macromolecules. Biomacromolecules 2005;6(4):1998-2007.

[266] Chiellini E, Errico C, Bartoli C, Chiellini F. Poly(hydroxyalkanoates)-based polymeric nanoparticles for drug delivery. *Journal of Biomedicine and* Biotechnology 2009;2009.

[267] Chiu J, Kim K, Zhong S, Hsiao B, Chu B, Hadjiargyrou M. Development of a cell-delivery vehicle derived from electrospun non-woven nanostructured membranes. *Annual International Conference of the IEEE Engineering in Medicine and Biology - Proceedings;* 2002; 2002. p. 759-60.

[268] Cima LG, Ingber DE, Vacanti JP, Langer R. Hepatocyte culture on biodegradable polymeric substrates. *Biotechnology and Bioengineering* 1991;38(2):145-58.

[269] Deng L, Shi K, Zhang Y, Wang H, Zeng J, Guo X, et al. Synthesis of well-defined poly(N-isopropylacrylamide)-b-poly(L-glutamic acid) by a versatile approach and micellization. *Journal of Colloid and Interface Science* 2008;323(1):169-75.

[270] Doneva TA, Yin HB, Stephens P, Bowen WR, Thomas DW. Development and AFM study of porous scaffolds for wound healing applications. *Spectroscopy* 2004;18(4):587-96.

[271] Galeska I, Kim TK, Patil SD, Bhardwaj U, Chattopadhyay D, Papadimitrakopoulos F, et al. Controlled release of dexamethasone from PLGA microspheres embedded within polyacid-containing PVA hydrogels. *AAPS Journal* 2005;7(1):E231-E40.

[272] Gelain F, Lomander A, Vescovi AL, Zhang S. Systematic studies of a self-assembling peptide nanofiber scaffold with other scaffolds. *Journal of Nanoscience and Nanotechnology* 2007;7(2):424-34.

[273] Harris L, Kim B-S, Mooney DJ. Open pore matrices formed with gas foaming. American Society of Mechanical Engineers, *Bioengineering Division (Publication) BED;* 1997; 1997. p. 351.

[274] Hasrc V, Lewandrowski K, Gresser JD, Wise DL, Trantolo DJ. Versatility of biodegradable biopolymers: Degradability and an in vivo application. *Journal of Biotechnology* 2001;86(2):135-50.

[275] He H, Lee LJ. Poly(lactic-co-glycolic acid) and functional hydrogels for drug delivery applications. *Annual Technical Conference - ANTEC, Conference Proceedings;* 2004; 2004. p. 3356-60.
[276] Holy CE, Cheng C, Davies JE, Shoichet MS. Optimizing the sterilization of PLGA scaffolds for use in tissue engineering. *Biomaterials* 2001;22(1):25-31.
[277] Holy CE, Dang SM, Davies JE, Shoichet MS. In vitro degradation of a novel poly(lactide-co-glycolide) 75/25 foam. *Biomaterials* 1999;20(13):1177-85.
[278] Holy CE, Yakubovich R. Processing cell-seeded polyester scaffolds for histology. *Journal of Biomedical Materials Research* 2000;50(2):276-9.
[279] Hua FJ, Park TG, Lee DS. A facile preparation of highly interconnected macroporous poly(D,L-lactic acid-co-glycolic acid) (PLGA) scaffolds by liquid-liquid phase separation of a PLGA-dioxane-water ternary system. *Polymer* 2003;44(6):1911-20.
[280] Hyun H, Cho JS, Kim BS, Lee JW, Kim MS, Khang G, et al. Comparison of micelles formed by amphiphilic star block copolymers prepared in the presence of a nonmetallic monomer activator. *Journal of Polymer Science, Part A: Polymer Chemistry* 2008;46(6):2084-96.
[281] Maspero FA, Ruffieux K, Müller B, Wintermantel E. Resorbable defect analog PLGA scaffolds using CO2 as solvent: Structural characterization. *Journal of Biomedical Materials Research* 2002;62(1):89-98.
[282] McDonald PF, Lyons JG, Geever LM, Higginbotham CL. In vitro degradation and drug release from polymer blends based on poly(dl-lactide), poly(l-lactide-glycolide) and poly(ζμ-caprolactone). *Journal of Materials Science* 2009:1-9.
[283] McNally-Heintzelman KM, Riley JN, Heintzelman DL. Scaffold-enhanced albumin and n-butyl-cyanoacrylate adhesives for tissue repair: Ex vivo evaluation in a porcine model. *Biomedical Sciences Instrumentation* 2003;39:312-7.
[284] Miller DC, Haberstroh KM, Webster TJ. Mechanisms controlling increased vascular cell adhesion to nano-structured polymer films. Bioengineering, *Proceedings of the Northeast Conference;* 2004; 2004. p. 120-1.
[285] Miller DC, Thapa A, Haberstroh KM, Webster TJ. Enhanced functions of cells on polymers with nanostructured surfaces. *Annual International Conference of the IEEE Engineering in Medicine and Biology - Proceedings;* 2002; 2002. p. 755-6.
[286] Oh SH, Cho SH, Lee JH. Preparation and characterization of hydrophilic PLGA/TWEEN 80 films and porous scaffolds. *Molecular Crystals and Liquid Crystals* 2004;418:229/[957]-241/[969].
[287] Patrick Jr CW, Chauvin PB, Hobley J, Reece GP. Preadipocyte seeded PLGA scaffolds for adipose tissue engineering. *Tissue Engineering* 1999;5(2):139-51.
[288] Penco M, Ranucci E, Ferruti P. A New Chain Extension Reaction on Poly(lactic-glycolic acid) (PLGA) Thermal Oligomers Leading to High Molecular Weight PLGA-Based Polymeric Products. *Polymer International* 1998;46(3):203-16.
[289] Perets A, Baruch Y, Weisbuch F, Shoshany G, Neufeld G, Cohen S. Enhancing the vascularization of three-dimensional porous alginate scaffolds by incorporating controlled release basic fibroblast growth factor microspheres. *Journal of Biomedical Materials Research - Part A* 2003;65(4):489-97.
[290] Price RL, Ellison K, Haberstroh KM, Webster TJ. Nanometer surface roughness increases select osteoblast adhesion on carbon nanofiber compacts. *Journal of Biomedical Materials Research - Part A 2004;*70(1):129-38.

[291] Qiu LY, Zhu KJ. Novel blends of poly[bis(glycine ethyl ester) phosphazene] and polyesters or polyanhydrides: compatibility and degradation characteristics in vitro. *Polymer International* 2000;49(11):1283-8.
[292] Bala I, Haribaran S, Kumar R. PLGA nanoparticles in drug delivery: The state of the art. *Critical Reviews in therapeutic Drug Carrier Systems* 2004;21:36.
[293] Anderson JM, Shive MS. *Advanced Drug Delivery Reviews* 1997;28:20.
[294] S.K. S, A.K. P, V. L. *Biomacromolecules* 2005;6:8.
[295] Jain RA. *Biomaterials* 2000;21:16.
[296] Prabha S, Labhasetwar V. *Molecular Pharmaceutics* 2004;1:9.
[297] Gvili K, Benny O, Danino D, Machluf M, Labhasetwar V. *Biopolymers* 2007;85:13.
[298] Na K, Kim S, Park K, Kim K, Woo DG, Kwon IC, et al. *Journal of American Chemical Society* 2007;129:2.
[299] Cheng FY, Wang SPH, Su CH, Tsai TL, Wu PC, Shieh DB, et al. Stabilizer-free poly(lactide-co-glycolide) nanoparticles for multimodal biomedical probes. *Biomaterials* 2008;29:9.
[300] Yang J, Lee CH, Park J, Seo S, Lim EK, Song YJ, et al. Antibody conjugated magnetic PLGA nanoparticles for diagnosis and treatment of breast cancer. *Journal of Materials Chemistry* 2007;17(26):2695-705.
[301] Ghosh S. Recent research and development in synthetic polymer-based drug delivery systems *Journal of Chemical Reaserach* 2004:6.
[302] Gombotz W, Pettit D. Biodegradable polymers for protein and peptide delivery, review. *Bioconjugate Chemistry* 1995;6:20.
[303] Ugo B, Eric A, Eric D. Development of a nanoprecipitation method intended for the entrapment of hydrophilic drugs into nanoparticles. *Eur J Pharm Sci* 2005;24:9.
[304] Lee SJ, Jeong JR, Shin SC, Kim JC, Chang YH, Lee KH, et al. Magnetic enhancement of iron oxide nanoparticles encapsulated with poly(D,L-latide-co-glycolide). *Colloids and Surfaces a-Physicochemical and Engineering Aspects* 2005 Mar;255(1-3):19-25.
[305] Jeong JR, Lee SJ, Kim JD, Shin SC. *IEEE Transactions on Magnetics* 2004;40(4):3.
[306] Boissiï"re M, Allouche J, Brayner R, Chanï©ac C, Livage J, Coradin T. Design of iron oxide/silica/alginate Hybrid MAgnetic Carriers (HYMAC). *Journal of Nanoscience and Nanotechnology* 2007;7(12):4649-54.
[307] Finotelli PV, Morales MA, Rocha-Leï£o MH, Baggio-Saitovitch EM, Rossi AM. Magnetic studies of iron(III) nanoparticles in alginate polymer for drug delivery applications. *Materials Science and Engineering C* 2004;24(5):625-9.
[308] Finotelli PV, Sampaio DA, Morales MA, Rossi AM, Rocha-Lï$J½ao MH. Ca alginate as scaffold for iron oxide nanoparticles synthesis. *Brazilian Journal of Chemical Engineering* 2008;25(4):759-64.
[309] Hernandez R, Sacristan J, Nogales A, Ezquerra TA, Mijangos C. Structural organization of iron oxide nanoparticles synthesized inside hybrid polymer gels derived from alginate studied with small-angle X-ray scattering. *Langmuir* 2009;25(22):13212-8.
[310] Morales MA, Finotelli PV, Coaquira JAH, Rocha-Leï£o MHM, Diaz-Aguila C, Baggio-Saitovitch EM, et al. In situ synthesis and magnetic studies of iron oxide nanoparticles in calcium-alginate matrix for biomedical applications. *Materials Science and Engineering C 2008*;28(2):253-7.

[311] Naik R, Senaratne U, Powell N, Buc EC, Tsoi GM, Naik VM, et al. Magnetic properties of nanosized iron oxide particles precipitated in alginate hydrogels. *Journal of Applied Physics* 2005;97(10):1-3.

[312] Constantinidis I, Grant SC, Simpson NE, Oca-Cossio JA, Sweeney CA, Mao H, et al. Use of magnetic nanoparticles to monitor alginate-encapsulated ²ζTC-tet cells. *Magnetic Resonance in Medicine* 2009;61(2):282-90.

[313] Breulmann M, Colfen H, Hentze H-P, Antonietti M, Walsh D, Mann S. Elastic magnets: Template-controlled mineralization of iron oxide colloids in a sponge-like gel matrix. *Advanced Materials* 1998;10:237-41.

[314] Ziolo RF, Giannelis EP, Weinstein BA, O'Horo MP, Ganguly BN, Mehrotra V, et al. Matrix-Mediated Synthesis of Nanocrystalline ggr-Fe2O3: A New Optically Transparent Magnetic Material. *Science* 1992;257:219-23.

[315] Hernandez R, Mijangos C. In situ synthesis of magnetic iron oxide nanoparticles in thermally responsive alginate-poly(N-isopropylacrylamide) semi-interpenetrating polymer networks. *Macromolecular Rapid Communications* 2009;30(3):176-81.

[316] Gurny R, Ibrahim H, Aebi A, Buri P, Wilson CG, Washington N, et al. Design and evaluation of controlled release systems for the eye. *J Control Release* 1987;6:367–73.

[317] Saettone MF, Chetoni P, Torracca MT, Burgalassi S, Giannaccini B. Evaluation of muco-adhesive properties and in vivo activity of ophthalmic vehicles based on hyaluronic acid. *Int J Pharm* 1989;51:203–12.

[318] Durrani, A.M., Farr SJ, Kellaway IW. Influence of molecular weight and formulation pH on the precorneal clearance rate of hyaluronic acid in the rabbit eye. Int J Pharm 1995;118:243-50.

[319] Zimmer A, Chetoni P, Saettone MF, Zerbe H, Kreuter j. Evaluation of pilocarpine-loaded albumin particles as controlled drug delivery systems for the eye. II. Co-administration with bioadhesive and viscous polymers. *J Control Release* 1995;33:31-46.

[320] Scherer F, Anton M, Schillinger U, Henke J, Bergemann C, Krüger A, et al. Magnetofection: Enhancing and targeting gene delivery by magnetic force in vitro and in vivo. *Gene Therapy* 2002;9(2):102-9.

[321] Huth S, Lausier J, Gersting SW, Rudolph C, Plank C, Welsch U, et al. Insights into the mechanism of magnetofection using PEI-based magnetofectins for gene transfer. *Journal of Gene Medicine* 2004;6(8):923-36.

[322] Caseri W. Nanocomposites of polymers and metals or semiconductors: Historical background and optical properties. *Macromolecular Rapid Communications* 2000;21(11):705-22.

[323] Gole A, Stone JW, Gemmill WR, Loye HCZ, Murphy CJ. Iron oxide coated gold nanorods: Synthesis, characterization, and magnetic manipulation. *Langmuir* 2008;24(12):6232-7.

[324] Goon IY, Lai LMH, Lim M, Munroe P, Gooding JJ, Amal R. Fabrication and dispersion of gold-shell-protected magnetite nanoparticles: Systematic control using polyethyleneimine. *Chemistry of Materials* 2009;21(4):673-81.

[325] Kim DK, Mikhailova M, Toprak M, Zhang Y, Bjelke B, Kehr J, et al. In-situ gold coating of superparamagnetic nanoparticles by microemulsion method. *Materials Research Society Symposium - Proceedings; 2002*; 2002. p. 137-42.

[326] Liu ZL, Peng L, Yao KL, Lu QH, Wang HB. Preparation and character of ultrasmall Fe3O4/Au nanoparticles. *Gongneng Cailiao/Journal of Functional Materials* 2005;36(2):196-9.

[327] Lu Z, Dai J, Song X, Wang G, Yang W. Facile synthesis of Fe3O4/SiO2 composite nanoparticles from primary silica particles. *Colloids and Surfaces A: Physicochemical and Engineering Aspects* 2008;317(1-3):450-6.

[328] Ma LL, Feldman MD, Tam JM, Paranjape AS, Cheruku KK, Larson TA, et al. Small multifunctional nanoclusters (Nanoroses) for targeted cellular imaging and therapy. *ACS Nano* 2009;3(9):2686-96.

[329] Ma Z, Han H, Tu S, Xue J. Fabrication of shape-controlled hematite particles and growth of gold nanoshells. *Colloids and Surfaces A: Physicochemical and Engineering Aspects* 2009;334(1-3):142-6.

[330] Mandal M, Kundu S, Ghosh SK, Panigrahi S, Sau TK, Yusuf SM, et al. Magnetite nanoparticles with tunable gold or silver shell. *Journal of Colloid and Interface Science* 2005;286(1):187-94.

[331] Mikhaylova M, Kim DK, Bobrysheva N, Osmolowsky M, Semenov V, Tsakalakos T, et al. Superparamagnetism of Magnetite Nanoparticles: *Dependence on Surface Modification. Langmuir 2004*;20(6):2472-7.

[332] Pacifico J, Van Leeuwen YM, Spuch-Calvar M, ِJnchez-Iglesias A, Rodrí-guez-Lorenzo L, Pí©rez-Juste J, et al. Field gradient imaging of nanoparticle systems: Analysis of geometry and surface coating effects. *Nanotechnology* 2009;20(9).

[333] Pal S, Morales M, Mukherjee P, Srikanth H. Synthesis and magnetic properties of gold coated iron oxide nanoparticles. *Journal of Applied Physics* 2009;105(7).

[334] Shubayev VI, Pisanic Ii TR, Jin S. Magnetic nanoparticles for theragnostics. *Advanced Drug Delivery Reviews* 2009;61(6):467-77.

[335] Wang L, Luo J, Maye MM, Fan Q, Rendeng Q, Engelhard MH, et al. Iron oxide-gold core-shell nanoparticles and thin film assembly. *Journal of Materials Chemistry* 2005;15(18):1821-32.

[336] Wang L, Shi X, Mahs S, Choi J, Sarup K, Wang GR, et al. Iron oxide composite nanoparticles and sensing properties. *Materials Research Society Symposium Proceedings;* 2005; 2005. p. 170-5.

[337] Winkenwerder WA, Ekerdt JG. Interaction of germanium with silicon dioxide. *Surface Science* 2008;602(16):2796-800.

[338] Xi D, Luo X, Lu Q, Yao K, Liu Z, Ning Q. The detection of HBV DNA with gold-coated iron oxide nanoparticle gene probes. *Journal of Nanoparticle Research* 2008;10(3):393-400.

[339] Xu P, Ji X, Yang H, Qi J, Zheng W, Abetz V, et al. Controllable fabrication of carbon nanotubes on catalysts derived from PS-b-P2VP block copolymer template and in situ synthesis of carbon nanotubes/Au nanoparticles composite materials. *Materials Chemistry and Physics;*119(1-2):249-53.

[340] Zeleِíٕkov ِíA, Zeleِíٕk V, Degmov ِíJ, Kovچڑِí J, Sedlچڑِíkov ِíK, Kusíٍ½ M, et al. The iron-gold magnetic nanoparticles: Preparation, characterization and magnetic properties. *Reviews on Advanced Materials Science* 2008;18(6):501-4.

[341] Zhou YF, Gulaka P, Zhou J, Xiao M, Xu D, Hsieh JT, et al. Preparation and evaluation of a radioisotope-incorporated iron oxide core/Au shell nanoplatform for dual modality imaging. *Journal of Biomedical Nanotechnology* 2008;4(4):474-81.

[342] Giri S, Trewyn BG, Stellmaker MP, Lin VSY. Stimuli-responsive controlled-release delivery system based on mesoporous silica nanorods capped with magnetic nanoparticles. *Angewandte Chemie-International Edition* 2005;44(32):5038-44.

[343] Liang Y, Gong JL, Huang Y, Zheng Y, Jiang JH, Shen GL, et al. Biocompatible core-shell nanoparticle-based surface-enhanced Raman scattering probes for detection of DNA related to HIV gene using silica-coated magnetic nanoparticles as separation tools. *Talanta* 2007 Apr;72(2):443-9.

[344] Ashtari P, He X, Wang K, Gong P. An efficient method for recovery of target ssDNA based on amino-modified silica-coated magnetic nanoparticles. *Talanta* 2005;67(3):548-54.

[345] Balakrishnan S, Launikonis A, Osvath P, Swiegers GF, Douvalis AP, Wilson GJ. Synthesis and characterisation of optically tuneable, magnetic phosphors. *Materials Chemistry and Physics*.

[346] Bele M, Hribar G, ȝŒampelj S, Makovec D, Gaberc-Porekar V, Zorko M, et al. Zinc-decorated silica-coated magnetic nanoparticles for protein binding and controlled release. *Journal of Chromatography B: Analytical Technologies in the Biomedical and Life Sciences* 2008;867(1):160-4.

[347] Chakrabarti S, Mandal SK, Nath BK, Das D, Ganguli D, Chaudhuri S. Synthesis of $^3\zeta$-Fe$_2$O$_3$ nanoparticles coated on silica spheres: Structural and magnetic properties. *European Physical Journal* B 2003;34(2):163-71.

[348] El-Shall SS, Li S. Synthesis of nanoparticles by a laser vaporization - Controlled condensation technique. *Proceedings of SPIE - The International Society for Optical Engineering; 1997*; 1997. p. 98-109.

[349] Froȝˆba M, Koȝˆhn R, Bouffaud G, Richard O, Van Tendeloo G. Fe$_2$O$_3$ nanoparticles within mesoporous MCM-48 silica: In situ formation and characterization. *Chemistry of Materials* 1999;11(10):2858-65.

[350] Guo X, Deng Y, Gu D, Che R, Zhao D. Synthesis and microwave absorption of uniform hematite nanoparticles and their core-shell mesoporous silica nanocomposites. *Journal of Materials Chemistry* 2009;19(37):6706-12.

[351] Haddad PS, Duarte EL, Baptista MS, Goya GF, Leite CAP, Itri R. Synthesis and characterization of silica-coated magnetic nanoparticles. *Progress in Colloid and Polymer Science* 2004:232-8.

[352] Ii, Yun SH, Lee CW, Lee JS, Seo CW, Lee EK. Fabrication of SiO$_2$-coated magnetic nanoparticles for applications to protein separation and purification. *Materials Science Forum* 2004:1033-6.

[353] Iijima M, Yonemochi Y, Kimata M, Hasegawa M, Tsukada M, Kamiya H. Preparation of agglomeration-free hematite particles coated with silica and their reduction behavior in hydrogen. *Journal of Colloid and Interface Science* 2005;287(2):526-33.

[354] Kang K, Choi J, Nam JH, Lee SC, Kim KJ, Lee SW, et al. Preparation and characterization of chemically functionalized silica-coated magnetic nanoparticles as a DNA separator. *Journal of Physical Chemistry B* 2009;113(2):536-43.

[355] Kang KH, Chang JH. High throughput magnetic separation for human DNA by aminosilanized iron oxide nanoparticles. *Journal of the Korean Ceramic Society* 2008;45(10):605-9.

[356] Kobayashi Y, Yoshida M, Nagao D, Ando Y, Miyazaki T, Konno M. Synthesis of SiO2-coated magnetite nanoparticles and immobilization of proteins on them. *Ceramic Transactions* 2007:135-41.

[357] Koc̆hn R, Bouffaud G, Richard O, Van Tendeloo G, Froc̆ba M. Iron (III) oxide within mesoporous MCM-48 silica phases: synthesis and characterization. *Materials Research Society Symposium - Proceedings* 1999;547:81-6.

[358] Koc̆hn R, Brieler F, Froc̆ba M. Ternary transition metal oxides within mesoporous MCM-48 silica phases: Synthesis and characterization. *Studies in Surface Science and Catalysis* 2000:341-8.

[359] Koc̆hn R, Froc̆ba M. Nanoparticles of 3d transition metal oxides in mesoporous MCM-48 silica host structures: Synthesis and characterization. *Catalysis Today* 2001;68(1-3):227-36.

[360] Lee J, Chang JH. Magnetic DNA separation process with functionalized magnetic silica nanoparticles. *Progress in Biomedical Optics and Imaging - Proceedings of SPIE;* 2009; 2009.

[361] Lu Y, Yin Y, Mayers BT, Xia Y. Modifying the Surface Properties of Superparamagnetic Iron Oxide Nanoparticles through a Sol-Gel Approach. *Nano Letters* 2002;2(3):183-6.

[362] Ma D, Veres T, Clime L, Normandin F, Guan J, Kingston D, et al. Superparamagnetic FexOy@SiO2 core-shell nanostructures: Controlled synthesis and magnetic characterization. *Journal of Physical Chemistry C* 2007;111(5):1999-2007.

[363] Ma H, Tarr J, Decoster MA, McNamara J, Caruntu D, Chen JF, et al. Synthesis of magnetic porous hollow silica nanotubes for drug delivery. *Journal of Applied Physics* 2009;105(7).

[364] Maver U, Bele M, Makovec D, C̆ampelj S, Jamnik J, Gaberšc̆ek M. Incorporation and release of drug into/from superparamagnetic iron oxide nanoparticles. *Journal of Magnetism and Magnetic Materials* 2009;321(19):3187-92.

[365] Morel AL, Nikitenko SI, Gionnet K, Wattiaux A, Lai-Kee-Him J, Labrugere C, et al. Sonochemical approach to the synthesis of Fe3O4 @SiO2 core - Shell nanoparticles with tunable properties. *ACS Nano* 2008;2(5):847-56.

[366] Nagao D, Yokoyama M, Saeki S, Kobayashi Y, Konno M. Preparation of composite particles with magnetic silica core and fluorescent polymer shell. *Colloid and Polymer Science* 2008;286(8-9):959-64.

[367] Narita A, Naka K, Chujo Y. Facile control of silica shell layer thickness on hydrophilic iron oxide nanoparticles via reverse micelle method. *Colloids and Surfaces A: Physicochemical and Engineering Aspects* 2009;336(1-3):46-56.

[368] Shen R, Camargo PHC, Xia Y, Yang H. Silane-based poly(ethylene glycol) as a primer for surface modification of nonhydrolytically synthesized nanoparticles using the Stöber method. *Langmuir* 2008;24(19):11189-95.

[369] Wang G, Fang Y, Kim P, Hayek A, Weatherspoon MR, Perry JW, et al. Layer-by-layer dendritic growth of hyperbranched thin films for surface sol-gel syntheses of conformal, functional, nanocrystalline oxide coatings on complex 3D (bio)silica templates. *Advanced Functional Materials* 2009;19(17):2768-76.

[370] Wang HH, Wang YXJ, Leung KCF, Sheng H, Zhang G, Lee SKM, et al. Mesenchymal stern cell intracellular labeling using silica-coated superparamagnetic iron oxide nanoparticles with amine functional peripheries. 5th Int Conference on Information

Technology and Applications in Biomedicine, ITAB 2008 in conjunction with 2nd Int Symposium and Summer School on Biomedical and Health Engineering, *IS3BHE* 2008; 2008; 2008. p. 187-9.

[371] Wang X, Wang L, He X, Zhang Y, Chen L. A molecularly imprinted polymer-coated nanocomposite of magnetic nanoparticles for estrone recognition. *Talanta* 2009;78(2):327-32.

[372] Wang Z, Guo Y, Li S, Sun Y, He N. Synthesis and characterization of SiO2/(PMMA/Fe 3O4) magnetic nanocomposites. *Journal of Nanoscience and Nanotechnology* 2008;8(4):1797-802.

[373] Yang D, Hu J, Fu S. Controlled synthesis of Magnetite-silica nanocomposites via a seeded sol-gel approach. *Journal of Physical Chemistry C* 2009;113(18):7646-51.

[374] Yiu HHP, McBain SC, Lethbridge ZAD, Lees MR, Dobson J. Preparation and characterization of polyethylenimine-coated Fe 3O4-MCM-48 nanocomposite particles as a novel agent for magnet-assisted transfection. *Journal of Biomedical Materials Research - Part A;*92(1):386-92.

[375] Yu CH, Lo CCH, Yeung CMY, Tam K, Tsang SC. Nano-engineering of magnetic particles for bio catalysis and bioseparation. *2006 NSTI Nanotechnology Conference and Trade Show - NSTI Nanotech 2006 Technical Proceedings*; 2006; 2006. p. 513-6.

[376] Zhang F, Wang CC. Fabrication of one-dimensional iron oxide/silica nanostructures with high magnetic sensitivity by dipole-directed self-assembly. *Journal of Physical Chemistry C* 2008;112(39):15151-6.

[377] Zhang G, Liu Y, Zhang C, Hu W, Xu W, Li Z, et al. Aqueous immune magnetite nanoparticles for immunoassay. *Journal of Nanoparticle Research* 2009;11(2):441-8.

[378] Œampelj S, Makovec D, Drofenik M. Functionalization of magnetic nanoparticles with 3-aminopropyl silane. *Journal of Magnetism and Magnetic Materials* 2009;321(10):1346-50.

[379] Fernandez MP, Schmool DS, Silva AS, Sevilla M, Fuertes AB, Gorria P, et al. Exchange-bias and superparamagnetic behaviour of Fe nanoparticles embedded in a porous carbon matrix. *9th International Workshop on Non Crystalline Solids;* 2008 Apr 27-30; Porto, PORTUGAL; 2008. p. 5219-21.

[380] Kastner JR, Ganagavaram R, Kolar P, Teja A, Xu CB. Catalytic ozonation of propanal using wood fly ash and metal oxide nanoparticle impregnated carbon. *Environmental Science & Technology* 2008 Jan;42(2):556-62.

[381] Ayyappan S, Gnanaprakash G, Panneerselvam G, Antony MP, Philip J. Effect of surfactant monolayer on reduction of Fe3O4 nanoparticles under vacuum. *Journal of Physical Chemistry* C 2008;112(47):18376-83.

[382] Boguslavsky Y, Margel S. Synthesis and characterization of poly(divinylbenzene)-coated magnetic iron oxide nanoparticles as precursor for the formation of air-stable carbon-coated iron crystalline nanoparticles. *Journal of Colloid and Interface Science* 2008;317(1):101-14.

[383] Chen YH, Franzreb M, Lin RH, Chen LL, Chang CY, Yu YH, et al. Platinum-doped tio-/magnetic poly(methyl methacrylate) microspheres as a novel photocatalyst. Industrial and Engineering Chemistry Research 2009;48(16):7616-23.

[384] Decker S, Lagadic I, Klabunde KJ, Moscovici J, Michalowicz A. EXAFS Observation of the Sr and Fe Site Structural Environment in SrO and Fe2O3-Coated SrO

Nanoparticles Used as Carbon Tetrachloride Destructive Adsorbents. *Chemistry of Materials* 1998;10(2):674-8.

[385] Fulton JL, Matson DW, Pecher KH, Amonette JE, Linehan JC. Metal-based nanoparticle synthesis from the rapid expansion of carbon dioxide solutions. *Journal of Nanoscience and Nanotechnology* 2006;6(2):562-7.

[386] Herrmann IK, Grass RN, Mazunin D, Stark WJ. Synthesis and covalent surface functionalization of nonoxidic iron core-shell nanomagnets. *Chemistry of Materials* 2009;21(14):3275-81.

[387] Huwe H, Froẑba M. Synthesis and characterization of transition metal and metal oxide nanoparticles inside mesoporous carbon CMK-3. *Carbon* 2007;45(2):304-14.

[388] Karimi A, Denizot B, Hindr|© F, Filmon R, Greneche JM, Laurent S, et al. Effect of chain length and electrical charge on properties of ammonium-bearing bisphosphonate-coated superparamagnetic iron oxide nanoparticles: formulation and physicochemical studies. *Journal of Nanoparticle Research* 2009:1-10.

[389] Khanfekr A, Arzani K, Nemati A, Hosseini M. Production of perovskite catalysts on ceramic monoliths with nanoparticles for dual fuel system automobiles. *International Journal of Environmental Science and Technology* 2009;6(1):105-12.

[390] Llarena I, Romero G, Ziolo RF, Moya SE. Carbon nanotube surface modification with polyelectrolyte brushes endowed with quantum dots and metal oxide nanoparticles through in situ synthesis. *Nanotechnology;*21(5).

[391] Moscovici J, Benzakour M, Decker S, Carnes C, Klabunde K, Michalowicz A. Unexpected Fe local order in iron oxide-coated nanocrystalline magnesium oxides with exceptional reactivities against environmental toxins. *Journal of Synchrotron Radiation* 2001;8(2):925-7.

[392] Nurmi JT, Tratnyek PG, Sarathy V, Baer DR, Amonette JE, Pecher K, et al. Characterization and properties of metallic iron nanoparticles: Spectroscopy, electrochemistry, and kinetics. *Environmental Science and Technology* 2005;39(5):1221-30.

[393] Rabias I, Fardis M, Devlin E, Boukos N, Tsitrouli D, Papavassiliou G. No aging phenomena in ferrofluids: The influence of coating on interparticle interactions of maghemite nanoparticles. *ACS Nano* 2008;2(5):977-83.

[394] Roonasi P, Holmgren A. A Fourier transform infrared (FTIR) and thermogravimetric analysis (TGA) study of oleate adsorbed on magnetite nano-particle surface. *Applied Surface Science* 2009;255(11):5891-5.

[395] Tan F, Fan X, Zhang G, Zhang F. Coating and filling of carbon nanotubes with homogeneous magnetic nanoparticles. *Materials Letters* 2007;61(8-9):1805-8.

[396] Cullitty BD, Graham CD. *Introduction to Magnetic Materials.* John Wiley & Sons 2009.

[397] McHenry ME, Majetich SA, Kirkpatrick EM. Synthesis, structure, propoerties and magnetic applications of carbon-coated nanocrystals produced by a carbon arc. *Materials science and Engineering A* 1995;204(1-2):19-24.

[398] Teunissen W, Geus JW. in: H Hattori, K Otsuka (Eds), *Science and Technology in Catalysis,* Elsevier, Amsterdam 1998.

[399] Bi XX, Ganguly B, Huffman GP, Huggins FE, Endo M, Eklund PC. Nanocrystalline a–Fe, Fe3C, and Fe7C3 produced by CO2 laser pyrolysis. *Journal of Materials Research Society* 1993;8(7):1666-74.

[400] Goodwin TJ, Yoo SH, Matteazzi, P.Groza JR. *Cementite-iron nanocomposite Nanostructured Materials* 1997;8(5):559-66.

[401] Yelsukov EP, Lomayeva SF, Konygin GN, Dorofeev GA, Povstugar VI, Mikhailova SS, et al. Structure, phase composition and magnetic characteristics of the nanocrystalline iron obtained by mechanical milling in heptane *Nanostructured Materials 1999*;12(1-4):483-6.

[402] Hirano SI, Tajima S. Synthesis and magnetic properties of Fe5C2 by reaction of iron oxide and carbon monoxide *Journal of Materials Science* 1990;25(10):4457-61.

[403] Nikitenko SI, Koltypin Y, Felner I, Yeshurun I, Shames AI, Jiang JZ, et al. Tailoring the Properties of Feâ˜'Fe3C Nanocrystalline Particles Prepared by Sonochemistry. *The Journal of Physical Chemistry B* 2004;108(23):7620-6.

[404] Shpaisman N, Margel S. Synthesis and Characterization of Air-Stable Iron Nanocrystalline Particles Based on a Single-Step Swelling Process of Uniform Polystyrene Template Microspheres. *Chemistry of Materials* 2005;18(2):396-402.

[405] Boguslavsky Y, Margel S. Synthesis and characterization of poly(divinylbenzene)-coated magnetic iron oxide nanoparticles as precursor for the formation of air-stable carbon-coated iron crystalline nanoparticles. *Journal of Colloid and Interface Science* 2008;317:101-14.

[406] Ely JL. Process for depositing pyrocarbon coatings in a fluidized bed. *European Patent* EP0779939 1999.

[407] Blitz JP, Gunko VM. Surface Chemistry in Biomedical and Environmental Science. *Proceedings of the NATO Advanced Research Workshop on Pure and Applied Surface Chemistry and Nanomaterials for Human Life and Environmental Protection*, Kyiv, Ukraine, 14-17 September 2005.

[408] Calderon NR, Voytovych R, Narciso J, Eustathopoulos N. Wetting dynamics versus interfacial reactivity of AlSi alloys on carbon. *Journal of Materials Science* 2009:1-7.

[409] Chang X, Huang QZ, Wang XF, Xiao Y, Yang X, Xie ZY, et al. Carbon paper fabricated by molding and CVD process for PEMFC application. *Fenmo Yejin Cailiao Kexue yu Gongcheng/Materials Science and Engineering of Powder Metallurgy* 2009;14(3):184-8.

[410] Chen Jh, Chen Gl, Geng Hr, Wang Y. Microstructure and properties of SiC gradiently coated Cf/C composites prepared by a RCLD method. *International Journal of Minerals, Metallurgy and Materials* 2009;16(3):334-8.

[411] Chen Z, Xiong X, Huang B, Li G, Xiao P, Zhang H, et al. Ablation behaviors of C/C composites with pyrocarbon (PyC)-TaC-PyC multi-interlayers in oxyacetylene flame. *Fuhe Cailiao Xuebao/Acta Materiae Compositae Sinica* 2009;26(3):155-61.

[412] Chung KC, Ram AN, Shauver MJ. Outcomes of pyrolytic carbon arthroplasty for the proximal interphalangeal joint. *Plastic and Reconstructive Surgery* 2009;123(5):1521-32.

[413] de Aragon JSM, Moran SL, Rizzo M, Reggin KB, Beckenbaugh RD. Early Outcomes of Pyrolytic Carbon Hemiarthroplasty for the Treatment of Trapezial-Metacarpal Arthritis. *Journal of Hand Surgery* 2009;34(2):205-12.

[414] Feldscher SB. Postoperative Management for PIP Joint Pyrocarbon Arthroplasty. *Journal of Hand Therapy* 2009.

[415] Guo L, Zhang D, Li K, Li H. Fabrication of isotropic pyrocarbon at 1400 i°c by thermal gradient chemical vapor deposition apparatus. *Journal Wuhan University of Technology, Materials Science Edition* 2009;24(5):728-31.

[416] He Y, Li K, Li H, Guo L, Zhou B. Preparation for 2D-C/C composites with pure RL textures and its characteristic. Cailiao Yanjiu Xuebao/*Chinese Journal of Materials Research* 2009;23(2):138-42.

[417] He YG, Li KZ, Li HJ, Wei JF, Fu QG, Zhang DS. Effect of interface structures on the fracture behavior of two-dimensional carbon/carbon composites by isothermal chemical vapor infiltration. *Journal of Materials Science* 2009:1-6.

[418] He YG, Li KZ, Wei JF, Guo LJ, Li HJ, Zhang LL. Microstructure and fabrication of C/C composites at super-short gas residence time. Cailiao Gongcheng/*Journal of Materials Engineering* 2009(9):33-7.

[419] Hu Y, Luo R, Zhang Y, Zhang J, Li J. Effect of preform density on densification rate and mechanical properties of carbon/carbon composites. *Materials science and Engineering A*;527(3):797-801.

[420] Jang KS, Lee E, Kim TW, Han IS, Woo SK, Lee KS. Mechanical behavior of SiCf reinforced SiC composites with fiber coating - Stress alleviation in SiC fiber by soft coating layer. Nippon Seramikkusu Kyokai Gakujutsu Ronbunshi/*Journal of the Ceramic Society of Japan* 2009;117(1365):582-7.

[421] Kasem H, Bonnamy S, Berthier Y, Dufrḯ©noy P, Jacquemard P. Tribological, physicochemical and thermal study of the abrupt friction transition during carbon/carbon composite friction. *Wear* 2009;267(5-8):846-52.

[422] Kashyap YS, Yadav PS, Sarkar PS, Agrawal A, Roy T, Sinha A, et al. Application of X-ray phase-contrast imaging technique in the study of pyrocarbon-coated zirconia kernels. *NDT and E International* 2009;42(5):384-8.

[423] Kurbakov SD. The carbon phases formed during pyrolysis of gaseous hydrocarbons in reaction volume of a fluidized bed apparatus. *Inorganic Materials* 2009;45(1):23-34.

[424] Lacroix R, Fournet R, Ziegler-Devin I, Marquaire PM. Kinetic modeling of surface reactions involved in CVI of pyrocarbon obtained by propane pyrolysis. *Carbon;*48(1):132-44.

[425] Li H, Xu G, Li K, Wang C, Li W, Li M. The infiltration process and texture transition of 2D C/C composites. *Journal of Materials Science and Technology* 2009;25(1):109-14.

[426] Li M, Qi L, Li H, Xu G. Fractal characterization of pore microstructure evolution in carbon/carbon composites. Science in China, Series E: *Technological Sciences* 2009;52(4):871-7.

[427] Liu H, Cheng H, Wang J, Che R, Tang G, Ma Q. Microstructural investigations of the pyrocarbon interphase in SiC fiber-reinforced SiC matrix composites. *Materials Letters* 2009;63(23):2029-31.

[428] Liu H, Cheng H, Wang J, Tang G, Che R, Ma Q. Effects of the fiber surface characteristics on the interfacial microstructure and mechanical properties of the KD SiC fiber reinforced SiC matrix composites. *Materials science and Engineering* A 2009;525(1-2):121-7.

[429] Makunin AV, Agafonov KN. Power-technological cyclic procedure with a high-temperature pyrolysis-gasification unit. *Theoretical Foundations of Chemical Engineering* 2009;43(4):575-82.

[430] Manocha LM, Patel H, Manocha S, Roy AK, Sngh JR. Development of carbon/carbon composite.s with carbon nanotubes as reinforcement and chemical vapor infiltration carbon as matrix. *Journal of Nanoscience and Nanotechnology* 2009;9(5):3119-24.

[431] Ohzawa Y, Suzuki T, Achiha T, Nakajima T. Surface-modification of anode carbon for lithium-ion battery using chemical vapor infiltration technique. *Journal of Physics and Chemistry of Solids* 2009.

[432] Ozcan S, Tezcan J, Filip P. Microstructure and elastic properties of individual components of C/C composites. *Carbon* 2009;47(15):3403-14.

[433] Ozcan S, Tezcan J, Howe JY, Filip P. Study on elasto-plastic behavior of different carbon types in carbon/carbon composites. *Ceramic Engineering and Science Proceedings;* 2009; 2009. p. 141-9.

[434] Perrone V. Preliminary experience with pyrolitic carbon implant in hand artroplasty and spacer. Gli impianti protesici e gli spaziatori in pirocarbonio nella chirurgia della mano: *Risultati preliminari* 2009;60(1):9-19.

[435] Pu Q, Qi L, Li M, Li H. Image registration algorithm in measuring the extinction angle of pyrocarbons in C/C composites. *Fuhe Cailiao Xuebao/Acta Materiae Compositae Sinica 2009*;26(4):141-5.

[436] Qi L, Li M, Li H, Xu G, Wang C. Research on precision-calibration techniques for selected area electron diffraction patterns of pyrocarbon. *Microscopy Research and Technique* 2009;72(4):338-42.

[437] Szalay G, Meyer C, Jürgensen I, Stigler B, Schnettler R. The operative treatment of rhizarthrosis for patients with extreme exposure of the hand by athletic activies isthe trapezektomy with interposition of an pyrocarbon spacer a option for treatment ? Die therapierefraktäre rhizarthrose des sportlers mit belastung der oberen extremität *Ist die trapezektomie mit interposition eines pyrocarbonspacer eine behandlungsoption?* 2009;23(3):161-4.

[438] Szalay G, Meyer C, Kraus R, Heiss C, Schnettler R. The operative treatment of rhizarthrosis with pyrocarbon spacer as replacement of the trapezium. *Die operative Versorgung der Rhizarthrose mittels Pyrocarbonspacer als Trapeziumersatz* 2009;41(5):300-5.

[439] Wijk U, Wollmark M, Kopylov P, Tägil M. Outcomes of Proximal Interphalangeal Joint Pyrocarbon Implants. *Journal of Hand Surgery;*35(1):38-43.

[440] Wijk U, Wollmark M, Kopylov P, Tägil M. Outcomes of Proximal Interphalangeal Joint Pyrocarbon Implants. *Journal of Hand Surgery* 2009.

[441] Wu X, Luo R, Ni Y, Xiang Q. Microstructure and mechanical properties of carbon foams and fibers reinforced carbon composites densified by CLVI and pitch impregnation. *Composites Part A: Applied Science and Manufacturing* 2009;40(2):225-31.

[442] Wu X, Luo R, Zhang J, Xiang Q, Ni Y. Deposition mechanism and microstructure of pyrocarbon prepared by chemical vapor infiltration with kerosene as precursor. *Carbon* 2009;47(6):1429-35.

[443] Xia Y, Qiao SR, Wang QQ, Zhang CY, Hou JT. Preparation, microstructure and oxidation resistance of SiCN ceramic matrix composites with glass-like carbon interface. Carbon - *Science and Technology* 2009;2(1):78-81.

[444] Yu XM, Zhou WC, Luo F, Zheng WJ. Mechanical properties of SiC/SiC composites. Hangkong Cailiao Xuebao/*Journal of Aeronautical Materials* 2009;29(3):93-7.

[445] Zeng Fh, Xiong X, Li Gd, Huang By, Luo J. Microstructure and mechanical properties of 3D fine-woven punctured C/C composites with PyC/SiC/TaC interphases. *Transactions of Nonferrous Metals Society of* China (English Edition) 2009;19(6):1428-35.

[446] Zhang JC, Luo RY, Wu XW, Li Q. Fabrication and characteristics of carbon nanofiber-reinforced carbon/carbon composites by fast catalytic infiltration processes. *Chemical Vapor Deposition* 2009;15(1-3):33-8.

[447] Zhang Y, Xiao Z, Wang J, Yang J, Jin Z. Effect of pyrocarbon content in C/C preforms on microstructure and mechanical properties of the C/C-SiC composites. *Materials science and Engineering A* 2009;502(1-2):64-9.

[448] Zou JZ, Zeng XR. Preparation of carbon/carbon composites by microwave pyrolysis chemical vapour infiltration. *Materials Research Innovations* 2009;13(4):421-4.

[449] Tamer U, Gundogdu Y, Boyaci IH, Pekmez K. Synthesis of magnetic core-shell Fe3O4-Au nanoparticle for biomolecule immobilization and detection. *J Nanopart Res* 2009;DOI 10.1007/s11051-009-9749-0.

[450] Lyon JL, Fleming DA, Stone MB, Schiffer P, Williams ME. Synthesis of Fe oxide core/Au shell nanoparticles by iterative hydroxylamine seeding. *Nano Letters* 2004 Apr;4(4):719-23.

[451] Pana O, Teodorescu CM, Chauvet O, Payen C, Macovei D, Turcu R, et al. Structure, morphology and magnetic properties of Fe-Au core-shell nanoparticles. *24th European Conference on Surface Science (ECOSS-24);* 2006 Sep 04-08; Paris, FRANCE; 2006. p. 4352-7.

[452] Park HY, Schadt MJ, Wang L, Lim IIS, Njoki PN, Kim SH, et al. Fabrication of magnetic core @ shell Fe oxide @ Au nanoparticles for interfacial bioactivity and bio-separation. *Langmuir* 2007 Aug;23(17):9050-6.

[453] Wang L, Luo J, Fan Q, Suzuki M, Suzuki IS, Engelhard MH, et al. Monodispersed Core-Shell Fe3O4@Au Nanoparticles. *J Phys Chem B* 2005;109:21593-601.

[454] Brust M, Walker M, Bethell D, Schiffrin DJ, Whyman R. J Chem Soc, Chem Commun 1994:801-2.

[455] Boissiìre M, Allouche J, Chanìac C, Brayner R, Devoisselle JM, Livage J, et al. Potentialities of silica/alginate nanoparticles as Hybrid Magnetic Carriers. *International Journal of Pharmaceutics* 2007;344(1-2):128-34.

[456] Borak B, Arkowski J, Skrzypiec M, Ziolkowski P, Krajewska B, Wawrzynska M, et al. Behavior of silica particles introduced into an isolated rat heart as potential drug carriers. *Biomedical Materials* 2007 Dec;2(4):220-3.

[457] Bumb A, Brechbiel MW, Choyke PL, Fugger L, Eggeman A, Prabhakaran D, et al. Synthesis and characterization of ultra-small superparamagnetic iron oxide nanoparticles thinly coated with silica *Nanotechnology* 2008;19:335601 (6pp).

[458] Carniato F, Tei L, Dastru W, Marchese L, Botta M. Relaxivity modulation in Gd-functionalised mesoporous silicas. *Chemical Communications* 2009(10):1246-8.

[459] Chang SY, Liu L, Asher SA. Creation of templated complex topological morphologies in colloidal silica. *Journal of the American Chemical Society* 1994;116(15):6745-7.

[460] Gong JL, Liang Y, Huang Y, Chen JW, Jiang JH, Shen GL, et al. Ag/SiO2 core-shell nanoparticle-based surface-enhanced Raman probes for immunoassay of cancer marker using silica-coated magnetic nanoparticles as separation tools. *Biosensors & Bioelectronics 2007* Feb;22(7):1501-7.

[461] Graf C, Vossen DLJ, Imhof A, van Blaaderen A. A general method to coat colloidal particles with silica. *Langmuir* 2003 Aug;19(17):6693-700.
[462] Hu SH, Liu DM, Tung WL, Liao CF, Chen SY. Surfactant-Free, Self-Assembled PVA-Iron Oxide/Silica Core-Shell Nanocarriers for Highly Sensitive, Magnetically Controlled Drug Release and Ultrahigh Cancer Cell Uptake Efficiency. *Advanced Functional Materials* 2008 Oct;18(19):2946-55.
[463] Jia GW, Cao ZQ, Xue H, Xu YS, Jiang SY. Novel Zwitterionic-Polymer-Coated Silica Nanoparticles. *Langmuir* 2009 Mar;25(5):3196-9.
[464] Kant KM, Sethupathi K, Rao MSR. Tuning the magnetization dynamics of silica-coated Fe3O4 core-shell nanoparticles by shell thickness control. *52nd Annual Conference on Magnetism and Magnetic Materials;* 2007 Nov 05-09; Tampa, FL; 2007.
[465] Lee J, Lee Y, Youn JK, Na HB, Yu T, Kim H, et al. Simple synthesis of functionalized superparamagnetic magnetite/silica core/shell nanoparticles and their application as magnetically separable high-performance biocatalysts. *Small* 2008;4(1):143-52.
[466] Lien YH, Wu TM. Preparation and characterization of thermosensitive polymers grafted onto silica-coated iron oxide nanoparticles. *Journal of Colloid and Interface Science* 2008;326(2):517-21.
[467] Philipse AP, Vanbruggen MPB, Pathmamanoharan C. Magnetic Silica Dispersions - Preparation and Stability of Surface-Modified Silica Particles with a Magnetic Core. *Langmuir* 1994 Jan;10(1):92-9.
[468] Lu Y, Yin YD, Mayers BT, Xia YN. Modifying the surface properties of superparamagnetic iron oxide nanoparticles through a sol-gel approach. *Nano Letters* 2002 Mar;2(3):183-6.
[469] Yi DK, Lee SS, Papaefthymiou GC, Ying JY. Nanoparticle architectures templated by SiO2/Fe 2O3 nanocomposites. *Chemistry of Materials* 2006;18(3):614-9.
[470] Gupta AK, Berry C, Gupta M, Curtis A. Receptor-mediated targeting of magnetic nanoparticles using insulin as a surface ligand to prevent endocytosis. *IEEE Trans Nanobiosci* 2003;2:255-61.
[471] Gupta AK, Wells S. Surface-modified superparamagnetic nanoparticles for drug delivery: preparation, characterization, and cytotoxicity studies. *IEEE Trans Nanobiosci* 2004;3:66-73.
[472] Berry CC, Charles S, Wells S, Dalby MJ, Curtis AS. The influence of transferrin stabilised magnetic nanoparticles on human dermal fibroblasts in culture. *International Journal of Pharmaceutics* 2002;269:211-25.
[473] Tiefenauer LX, Kuhne G, Andres RY. Antibody-magnetite nanoparticles: in vitro characterization of a potential tumorspecific contrast agent for magnetic resonance imaging. *Bioconjugate Chemistry* 1993;4:347-52.
[474] Gautherie M. *Biological Basis of Oncologic Thermotherapy*. Springer: Berlin 1990.
[475] Hand J. *Physical Techniques in Clinical Hyperthermia*. Wiley: New York 1986.
[476] Vrba JL, M. *Microwave Applicators for Medical Applications*. CVUT Press: Prague 1997.
[477] Kumar A, Jena PK, Behera S, Lockey RF, Mohapatra S. Multifunctional magnetic nanoparticles for targeted delivery. *Nanomedicine: Nanotechnology, Biology, and Medicine;*6(1).

[478] Amstad E, Gillich T, Bilecka I, Textor M, Reimhult E. Ultrastable iron oxide nanoparticle colloidal suspensions using dispersants with catechol-derived anchor groups. *Nano Letters* 2009;9(12):4042-8.

[479] Cheng CM, Kou G, Wang XL, Wang SH, Gu HC, Guo YJ. Synthesis of carboxyl superparamagnetic ultrasmall iron oxide (USPIO) nanoparticles by a novel flocculation-redispersion process. *Journal of Magnetism and Magnetic Materials* 2009;321(17):2663-9.

[480] Earhart CM, Nguyen EM, Wilson RJ, Wang YA, Wang SX. Designs for a microfabricated magnetic sifter. *IEEE Transactions on Magnetics* 2009;45(10):4884-7.

[481] Hafaid I, Gallouz A, Mohamed Hassen W, Abdelghani A, Sassi Z, Bessueille F, et al. Sensitivity improvement of an impedimetric immunosensor using functionalized iron oxide nanoparticles. *Journal of Sensors* 2009;2009.

[482] Huang FK, Chen WC, Lai SF, Liu CJ, Wang CL, Wang CH, et al. Enhancement of irradiation effects on cancer cells by cross-linked dextran-coated iron oxide (CLIO) nanoparticles. *Physics in Medicine and Biology;*55(2):469-82.

[483] *Jin S, Leach JC, Ye K. Nanoparticle-mediated gene delivery. Methods in molecular biology (Clifton, NJ)* 2009;544:547-57.

[484] Kolasinska M, Gutberlet T, Krastev R. Ordering of Fe3O4 nanoparticles in polyelectrolyte multilayer films. *Langmuir* 2009;25(17):10292-7.

[485] Kumar A, Jena PK, Behera S, Lockey RF, Mohapatra S. Multifunctional magnetic nanoparticles for targeted delivery. *Nanomedicine: Nanotechnology, Biology, and Medicine* 2009.

[486] Lee CM, Jeong HJ, Kim EM, Kim DW, Lim ST, Kim HT, et al. Superparamagnetic iron oxide nanoparticles as a dual imaging probe for targeting hepatocytes in vivo. *Magnetic Resonance in Medicine 2009*;62(6):1440-6.

[487] Llusar M, Royo V, Badenes JA, Tena MA, Monr³¡s G. Nanocomposite Fe2O3-SiO2 inclusion pigments from post-functionalized mesoporous silicas. *Journal of the European Ceramic Society* 2009;29(16):3319-32.

[488] Mohapatra S, Panda N, Pramanik P. Boronic acid functionalized superparamagnetic iron oxide nanoparticle as a novel tool for adsorption of sugar. *Materials Science and Engineering C* 2009;29(7):2254-60.

[489] Pal S, Chandra S, Phan MH, Mukherjee P, Srikanth H. Carbon nanostraws: Nanotubes filled with superparamagnetic nanoparticles. *Nanotechnology* 2009;20(48).

[490] Panella B, Vargas A, Ferri D, Baiker A. Chemical availability and reactivity of functional groups grafted to magnetic nanoparticles monitored in situ by ATR-IR spectroscopy. Chemistry of Materials 2009;21(18):4316-22.

[491] Perlstein B, Lublin-Tennenbaum T, Marom I, Margel S. Synthesis and characterization of functionalized magnetic maghemite nanoparticles with fluorescent probe capabilities for biological applications. Journal of Biomedical Materials Research - *Part B Applied Biomaterials;92*(2):353-60.

[492] Roque ACA, Bicho A, Batalha IL, Cardoso AS, Hussain A. Biocompatible and bioactive gum Arabic coated iron oxide magnetic nanoparticles. *Journal of Biotechnology* 2009;144(4):313-20.

[493] Taylor A, Lipert K, Krämer K, Hampel S, Füssel S, Meye A, et al. Biocompatibility of iron filled carbon nanotubes in vitro. *Journal of Nanoscience and Nanotechnology* 2009;9(10):5709-16.

[494] Wang G, Wang C, Dou W, Ma Q, Yuan P, Su X. The synthesis of magnetic and fluorescent bi-functional silica composite nanoparticles via reverse microemulsion method. *Journal of Fluorescence* 2009;19(6):939-46.

[495] Wang X, Liu LH, Ramström O, Yan M. Engineering nanomaterial surfaces for biomedical applications. *Experimental Biology and Medicine* 2009;234(10):1128-39.

[496] Yuan P, Wu Q, Ding Y, Wu H, Yang X. One-step synthesis of iron-oxide-loaded functionalized carbon spheres. *Carbon* 2009;47(11):2648-54.

[497] Zhang Q, Wang C, Qiao L, Yan H, Liu K. Superparamagnetic iron oxide nanoparticles coated with a folate-conjugated polymer. *Journal of Materials Chemistry* 2009;19(44):8393-402.

[498] Isobe M, Narula J, Southern JF, Strauss HW, Khaw BA, Haber E. Imaging the rejecting heart: In vivo detection of major histocompatibility complex class II antigen induction. *Circulation* 1992;85(2):738-46.

[499] Wu YL, Ye Q, Foley LM, Hitchens TK, Sato K, Williams JB, et al. In situ labeling of immune cells with iron oxide particles: An approach to detect organ rejection by cellular MRI. *Proceedings of the National Academy of Sciences of the United States of America* 2006;103(6):1852-7.

[500] Corot C, Robert P, Idée JM, Port M. Recent advances in iron oxide nanocrystal technology for medical imaging. *Advanced Drug Delivery Reviews* 2006;58(14):1471-504.

[501] Jendelov P, Herynek V, Urdzí-kov L, Glogarov K, Kroupov J, Andersson B, et al. Magnetic Resonance Tracking of Transplanted Bone Marrow and Embryonic Stem Cells Labeled by Iron Oxide Nanoparticles in Rat Brain and Spinal Cord. *Journal of Neuroscience Research* 2004;76(2):232-43.

[502] Weissleder R, Elizondo G, Wittenberg J, Lee AS, Josephson L, Brady TJ. Ultrasmall superparamagnetic iron oxide: An intravenous contrast agent for assessing lymph nodes with MR imaging. *Radiology* 1990;175(2):494-8.

[503] Jain TK, Morales MA, Sahoo SK, Leslie-Pelecky DL, Labhasetwar V. Iron oxide nanoparticles for sustained delivery of anticancer agents. *Molecular Pharmaceutics* 2005;2(3):194-205.

[504] Ito A, Kuga Y, Honda H, Kikkawa H, Horiuchi A, Watanabe Y, et al. Magnetite nanoparticle-loaded anti-HER2 immunoliposomes for combination of antibody therapy with hyperthermia. *Cancer Letters* 2004;212(2):167-75.

[505] Simberg D, Duza T, Park JH, Essler M, Pilch J, Zhang L, et al. Biomimetic amplification of nanoparticle homing to tumors. *Proceedings of the National Academy of Sciences of the United States of America* 2007;104(3):932-6.

[506] Hultman KL, Raffo AJ, Grzenda AL, Harris PE, Brown TR, O'Brien S. Magnetic resonance imaging of major histocompatibility class II expression in the renal medulla using immunotargeted superparamagnetic iron oxide nanoparticles. *ACS Nano* 2008;2(3):477-84.

[507] Jalilian AR, Panahifar A, Mahmoudi M, Akhlaghi M, Simchi A. Preparation and biological evaluation of [67Ga]-labeled- superparamagnetic nanoparticles in normal rats. *Radiochimica Acta* 2009;97(1):51-6.

[508] Goldenberg DM, Sharkey RM, Paganelli G, Barbet J, Chatal JF. Antibody Pretargeting Advances Cancer Radioimmunodetection and Radioimmunotherapy. *Journal of clinical oncology* 2006;24:823-34.

[509] Akhtari M, Bragin A, Cohen M, Moats R, Brenker F, Lynch MD, et al. Functionalized magnetonanoparticles for MRI diagnosis and localization in epilepsy. *Epilepsia* 2008;49(8):1419-30.

[510] Barrera C, Herrera AP, Rinaldi C. Colloidal dispersions of monodisperse magnetite nanoparticles modified with poly(ethylene glycol). *Journal of Colloid and Interface Science* 2009;329(1):107-13.

[511] Boyer C, Bulmus V, Priyanto P, Teoh WY, Amal R, Davis TP. The stabilization and bio-functionalization of iron oxide nanoparticles using heterotelechelic polymers. *Journal of Materials Chemistry* 2009;19(1):111-23.

[512] Cengelli F, Grzyb JA, Montoro A, Hofmann H, Hanessian S, Juillerat-Jeanneret L. Surface-functionalized ultrasmall superparamagnetic nanoparticles as magnetic delivery vectors for camptothecin. *ChemMedChem* 2009;4(6):988-97.

[513] Chiu YC, Chen YC. Carboxylate-functionalized iron oxide nanoparticles in surface-assisted laser desorption/ionization mass spectrometry for the analysis of small biomolecules. *Analytical Letters* 2008;41(2):260-7.

[514] Daou TJ, Gren|¨che JM, Pourroy G, Buathong S, Derory A, Ulhaq-Bouillet C, et al. Coupling agent effect on magnetic properties of functionalized magnetite-based nanoparticles. *Chemistry of Materials* 2008;20(18):5869-75.

[515] Earhart C, Jana NR, Erathodiyil N, Ying JY. Synthesis of carbohydrate-conjugated nanoparticles and quantum dots. *Langmuir* 2008;24(12):6215-9.

[516] Feng B, Hong RY, Wang LS, Guo L, Li HZ, Ding J, et al. Synthesis of Fe3O4/APTES/PEG diacid functionalized magnetic nanoparticles for MR imaging. *Colloids and Surfaces A: Physicochemical and Engineering Aspects* 2008;328(1-3):52-9.

[517] Guan N, Liu C, Sun D, Xu J. A facile method to synthesize carboxyl-functionalized magnetic polystyrene nanospheres. Colloids and Surfaces A: *Physicochemical and Engineering Aspects 2009*;335(1-3):174-80.

[518] Guan N, Xu J, Wang L, Sun D. One-step synthesis of amine-functionalized thermo-responsive magnetite nanoparticles and single-crystal hollow structures. *Colloids and Surfaces A: Physicochemical and Engineering Aspects* 2009;346(1-3):221-8.

[519] Hanessian S, Grzyb JA, Cengelli F, Juillerat-Jeanneret L. Synthesis of chemically functionalized superparamagnetic nanoparticles as delivery vectors for chemotherapeutic drugs. B*ioorganic and Medicinal Chemistry* 2008;16(6):2921-31.

[520] Hayashi K, Moriya M, Sakamoto W, Yogo T. Chemoselective synthesis of folic acid-functionalized magnetite nanoparticles via click chemistry for magnetic hyperthermia. *Chemistry of Materials* 2009;21(7):1318-25.

[521] Herrera AP, Barrera C, Rinaldi C. Synthesis and functionalization of magnetite nanoparticles with aminopropylsilane and carboxymethyldextran. *Journal of Materials Chemistry* 2008;18(31):3650-4.

[522] Hess DM, Naik RR, Rinaldi C, Tomczak MM, Watkins JJ. Fabrication of ordered mesoporous silica films with encapsulated iron oxide nanoparticles using ferritin-doped block copolymer templates. *Chemistry of Materials* 2009;21(10):2125-9.

[523] Hnaiein M, Hassen WM, Abdelghani A, Fournier-Wirth C, Coste J, Bessueille F, et al. A conductometric immunosensor based on functionalized magnetite nanoparticles for E. coli detection. *Electrochemistry Communications* 2008;10(8):1152-4.

[524] Huang G, Diakur J, Xu Z, Wiebe LI. Asialoglycoprotein receptor-targeted superparamagnetic iron oxide nanoparticles. *International Journal of Pharmaceutics* 2008;360(1-2):197-203.

[525] Jolivet JP, Cassaignon S, Chanéac C, Chiche D, Tronc E. Design of oxide nanoparticles by aqueous chemistry. *Journal of Sol-Gel Science and Technology* 2008;46(3):299-305.

[526] Kontos AI, Likodimos V, Stergiopoulos T, Tsoukleris DS, Falaras P, Rabias I, et al. Self-organized anodic TiO2 nanotube arrays functionalized by iron oxide nanoparticles. *Chemistry of Materials* 2009;21(4):662-72.

[527] Latham AH, Williams ME. Controlling transport and chemical functionality of magnetic nanoparticles. *Accounts of Chemical Research* 2008;41(3):411-20.

[528] Lin TW, Salzmann CG, Shao LD, Yu CH, Green MLH, Tsang SC. Polyethylene glycol grafting and attachment of encapsulated magnetic iron oxide silica nanoparticles onto chlorosilanized single-wall carbon nanotubes. *Carbon* 2009;47(6):1415-20.

[529] Liu H, Guo J, Jin L, Yang W, Wang C. Fabrication and functionalization of dendritic poly(amidoamine)-immobilized magnetic polymer composite microspheres. *Journal of Physical Chemistry B* 2008;112(11):3315-21.

[530] Luo SC, Jiang J, Liour SS, Gao S, Ying JY, Yu HH. Magnetic PEDOT hollow capsules with single holes. *Chemical Communications* 2009(19):2664-6.

[531] Martin AL, Li B, Gillies ER. Surface functionalization of nanomaterials with dendritic groups: Toward enhanced binding to biological targets. *Journal of the American Chemical Society* 2009;131(2):734-41.

[532] Masotti A, Pitta A, Ortaggi G, Corti M, Innocenti C, Lascialfari A, et al. Synthesis and characterization of polyethylenimine-based iron oxide composites as novel contrast agents for MRI. *Magnetic Resonance Materials in Physics, Biology and Medicine* 2009;22(2):77-87.

[533] Mohapatra S, Pramanik P. Synthesis and stability of functionalized iron oxide nanoparticles using organophosphorus coupling agents. *Colloids and Surfaces A: Physicochemical and Engineering Aspects* 2009;339(1-3):35-42.

[534] Naka K, Narita A, Tanaka H, Chujo Y, Morita M, Inubushi T, et al. Biomedical applications of imidazolium cation-modified iron oxide nanoparticles. *Polymers for Advanced Technologies* 2008;19(10):1421-9.

[535] Natarajan A, Xiong CY, Gruettner C, DeNardo GL, DeNardo SJ. Development of multivalent radioimmunonanoparticles for cancer imaging and therapy. *Cancer Biotherapy and* Radiopharmaceuticals 2008;23(1):82-91.

[536] Oh C, Lee YG, Jon CU, Oh SG. Synthesis and characterization of hollow silica microspheres functionalized with magnetic particles using W/O emulsion method. *Colloids and Surfaces A: Physicochemical and Engineering Aspects* 2009;337(1-3):208-12.

[537] Palani A, Lee JS, Huh J, Kim M, Lee YJ, Chang JH, et al. Selective enrichment of cysteine-containing peptides using SPDP-functionalized superparamagnetic Fe3O4@SiO3 nanoparticles: Application to comprehensive proteomic profiling. *Journal of Proteome Research* 2008;7(8):3591-6.

[538] Park BH, Chang Y, Lee YJ, Park JA, Kim IS, Bae SJ, et al. Targeting of membrane type1-matrix metalloproteinase (MT1-MMP) using superparamagnetic nanoparticles in

human liver cancer cells. *Colloids and Surfaces A: Physicochemical and Engineering Aspects* 2008;313-314:647-50.

[539] Park JA, Lee JJ, Kim IS, Park BH, Lee GH, Kim TJ, et al. Magnetic and MR relaxation properties of avidin-biotin conjugated superparamagnetic nanoparticles. *Colloids and Surfaces A: Physicochemical and Engineering Aspects* 2008;313-314:288-91.

[540] Ragheb RT, Riffle JS. Synthesis and characterization of poly(lactide-b-siloxane-b-lactide) copolymers as magnetite nanoparticle dispersants. *Polymer* 2008;49(25):5397-404.

[541] van Tilborg GAF, Geelen T, Duimel H, Bomans PHH, Frederik PM, Sanders HMHF, et al. Internalization of annexin A5-functionalized iron oxide particles by apoptotic Jurkat cells. *Contrast Media and Molecular Imaging* 2009;4(1):24-32.

[542] Wang L, Gan XX. Biomolecule-functionalized magnetic nanoparticles for flow-through quartz crystal microbalance immunoassay of aflatoxin B1. *Bioprocess and Biosystems Engineering* 2009;32(1):109-16.

[543] White BR, Stackhouse BT, Holcombe JA. Magnetic ^{3}ح-Fe2O3 nanoparticles coated with poly-l-cysteine for chelation of As(III), Cu(II), Cd(II), Ni(II), Pb(II) and Zn(II). *Journal of Hazardous Materials* 2009;161(2-3):848-53.

[544] Wu W, He Q, Chen H. Surface functionalization and application for magnetic iron oxide nanoparticles. *Progress in Chemistry* 2008;20(2-3):265-72.

[545] Yang L, Mao H, Cao Z, Wang YA, Peng X, Wang X, et al. Molecular Imaging of Pancreatic Cancer in an Animal Model Using Targeted Multifunctional Nanoparticles. *Gastroenterology* 2009;136(5):1514-25.e2.

[546] Yang X, Chen Y, Yuan R, Chen G, Blanco E, Gao J, et al. Folate-encoded and Fe3O4-loaded polymeric micelles for dual targeting of cancer cells. *Polymer* 2008;49(16):3477-85.

[547] Yantasee W, Hongsirikarn K, Warner CL, Choi D, Sangvanich T, Toloczko MB, et al. Direct detection of Pb in urine and Cd, Pb, Cu, and Ag in natural waters using electrochemical sensors immobilized with DMSA functionalized magnetic nanoparticles. *Analyst* 2008;133(3):348-55.

[548] Yigit MV, Mazumdar D, Lu Y. MRI detection of thrombin with aptamer functionalized superparamagnetic iron oxide nanoparticles. *Bioconjugate Chemistry* 2008;19(2):412-7.

[549] Zhan J, Zheng T, Piringer G, Day C, McPherson GL, Lu Y, et al. Transport characteristics of nanoscale functional zerovalent iron/silica composites for in situ remediation of trichloroethylene. *Environmental Science and Technology 2008;42(23):*8871-6.

Chapter 5

CYTOTOXICITY OF SPIONs

One of the promising fields, which offer extraordinary opportunities for the use of nanoparticles, is in the medical area. Some examples for these opportunities are: drug delivery, cell/protein separation, biological labeling, targeting, imaging, hyperthermia and early diagnosis. Magnetic nanomaterials in the form of nanoparticles are of great interest in pharmaceutical and medical applications. It is noteworthy that the first successful application of magnetic particles were on the technological side, with ferrofluids, for example, the use to enhance the power of high end loudspeakers and seal high vacuum pumps.[1] On the biomedical side, the first successful Food and Drug Administration (FDA) approved applications include the diagnosis and the use as contrast agents for magnetic resonance imaging (MRI)[2] and the application in extracting tumor cells from bone marrow transplants before injection into a patient.[3] Due to the ultrafine size and biocompatibility, Superparamagnetic iron oxide nanoparticles (SPION) are emerging as promising candidates for biomedical applications, such as in enhanced-resolution magnetic resonance imaging, drug delivery and cellular targeting. The super-paramagnetism of magnetic nanoparticles is particularly useful in applications such as externally guided drug delivery since removal of the external magnetic field prevents agglomeration and subsequent embolism.[4-9] All these benefits justify the exponential growth in the number of publications dealing with nanoparticles for all types of applications and specifically, iron oxide nanoparticles for biomedical applications. The physical properties of SPIONs (i. e. synthesis methods, magnetic behavior, capability of heat production, etc.) have been extensively studied.[10-12] However, the science of toxicity (i.e. toxicology) of magnetic nanoparticles is in the primary stages. Although information about the toxicity of nanoparticles and specifically SPION continues to increase, a significant knowledge gap exists on a complete toxicological profile of these promising nanoparticles proposed for safe future use in many aspects of biomedical engineering. Without the data, risk assessment or regulation for safety of the materials, the technological use shall suffer immeasurably. The main objective of this review paper is to fill the information gap by a comprehensive insight on the different *in vitro* toxicity tests as well as the *in vivo* response to SPION.

5.1. BIOCOMPATIBILITY VERSUS TOXICITY

5.1.1. Biocompatibility

The most important difference between a biomaterial and any other material is its ability to exist in contact with tissues of the human body without causing damage to the body. The co-existence of materials and tissues is usually investigated in the context of biocompatibility studies and has been of interest to scientists and practitioners for many years. The term "biocompatibility" is used extensively in the literature, but there exists a great deal of uncertainty about its actual meaning.[13, 14] According to Williams, it is defined as "the ability of a material to perform with an appropriate host response in a specific application".[15-17] In general, there are three terms in which a biomaterial may be described in or classified into representing tissue responses. These are bioinert, bioresorbableand bioactive and these terms are extensively described in the literature.[18-20] The term bioinert applies to any material that has minimal interaction with its surrounding tissue once placed in the human body. Generally a fibrous capsule might form around bioinert materials. Materials that upon placement within the human body start to dissolve to be slowly replaced by advancing tissue (such as bone) are generally called bioresorbable.[19] Bioactive refers to a material, which upon being placed within the human body interacts with the surrounding bone and in some cases, even soft tissue. This occurs through a time dependent kinetic modification of the material surface, triggered by implantation within the living tissue.

5.1.2. Cytotoxicity

Toxicology is the study of the adverse effects of chemicals on living systems, whether the living systems are human, animal, plant or microbe. Toxicology investigates the relationship between dose and effect on the exposed living system, which is of uttermost importance, since "the dose makes the poison".[21] Toxicity can manifest itself in a wide array of forms, from mild biochemical malfunctions to serious organ damage and death. These events, any of which may be reversible or irreversible, include absorption, transport, and metabolism to more or less toxic metabolites, reaction, excretion, interaction with cellular macromolecules and other modes of toxic action. Toxicology integrates the study of all these events, at all levels of biological organization, from molecules to complex ecosystems. The broad scope of toxicology, from the study of fundamental mechanisms to the measurement of exposure, including toxicity testing and risk analysis, requires an extensively interdisciplinary approach. This approach utilizes the principles and methods of other disciplines such as including molecular biology, chemical physiology, medicine, computer science and informatics.[21] Typical for biological toxicity is a ''threshold dose'' which might be due to the organism's ability to fight the toxin, such as is the case where the toxin is excreted, metabolized, or isolated, for example by encapsulation into a vesicle. A threshold dose might also be due to the fact that the effect is too small to see, or the observation time scale is too short to notice an effect. To analyze biological toxicity, it is very important to choose an adequate (model) system. The release of toxic particle coatings inside a cell, which then interact with chromosomal DNA and transform a cell into a cancer cell, is such a biological effect. To prove that such effects exist, especially when they stem from mixtures of compounds, may be difficult and complex. Theoretically, one virus or bacterium can reproduce to cause a serious infection. However, in a host with an intact immune system the inherent toxicity of the organism is balanced by the host's ability to fight back; the effective toxicity is then a combination of both parts of the relationship. Cytotoxicity is the quality of being toxic to cells and is investigated *in vitro* (the technique of performing a given procedure in a controlled environment outside of a human body) and *in vivo* (materials which takes place inside an organism). *In vivo* testing is routinely carried out in preclinical trials of a drug or compound before it is released for human trials. This enables researchers to investigate the activity of their products *in vivo*, as well as their side effects. Compared to *in vivo* studies, *in vitro* examinations are preferred to obtain preliminary toxicity data due to the ease of use, reproducibility, accessibility, cost effectiveness and lack of ethical restrictions. However, *in vivo* tests are essential for toxicity measurements to probe the overall biological system, while *in vitro* tests allow the determination of the mechanism of action.

Nanoscience is often referred to as "horizontal", "key" or "enabling" since it can pervade virtually all technological sectors. It often brings together different areas of science and benefits from an interdisciplinary or "converging" approach and is expected to lead to innovations that can contribute towards addressing many of the problems facing today's society. Several nanotechnology-based products have been marketed including electronic components, scratch-free paint, sports equipment, wrinkle- and stain-resistant fabrics, sun creams, and medical products. Accordingly, potential occupational and public exposure to manufactured nanoparticles will increase dramatically in the near future. However, while the number of studies on nanoparticle synthesis, characterization and applications continue to increase, information describing the relative health and environmental risk assessment of manufactured nanoparticles or nanomaterials is scarce, and more importantly, understanding of particle-cell interactions on a fundamental level which would allow toxicity predictions is severely lacking.[22] Obviously, understanding the relationship between nanoparticles and/or nanomaterials and their effect on the body is crucial for any potential application as well as health and environmental risk assessment.[23]

Chemical toxicity of (bulk) iron oxide has been investigated in many different studies. Briefly, it was shown that the cytokine response in BEAS-2B cells due to interaction with iron oxide nanoparticles have a lower potency than an equal mass of micron-sized particles of the same nominal composition.[24] Since the isoelectric point of colloidal iron oxide (Fe_2O_3) nanoparticles is around 7, pure electrostatic stabilization at physiological pH is impossible, which makes the development of adequate coatings (for steric stabilization) necessary. Therefore, the particles are, unless in powder form, never considered as bare, uncoated, single iron oxides and the chemical toxicity of the coating material as well as the impact of the coating and possible subsequent derivatization of the particle surfaces on the colloidal behavior and particle-cell interaction has to be considered.

5.2. *IN VITRO* TOXICITY ASSAYS

To obtain reliable and reproducible data from *in vitro* tests, it is crucial to control the conditions of cells during all stages of the experiment to make sure that the measured parameters really monitor an impact of incubated nanoparticles on cells in culture. It is important to note that cell cultures are extraordinary sensitive to changes in their environment such as variation in protein conformation, nutrient, temperature, pH, and waste production. Therefore, it is of uttermost importance to establish adequate and reproducible analytical environments in terms of the choice of cells, growing conditions and sample preparation assay procedure. The toxicity of SPIONs is usually initially determined using *in vitro* cytotoxicity tests. These tests look either at proliferation or morbidity, i.e., cell viability or death. The *in vitro* assays are divided into five major groups including assays measuring metabolic activity of cells (a), assays involving membrane integrity of cells (b), assays to detect apoptosis (c) and dye and cell proliferation assays (d).[23] The currently used *in vitro* assays to explore the cytotoxicity of SPIONs are summarized in Figure 5.1 and will be discussed in detail in the following chapters.[23]

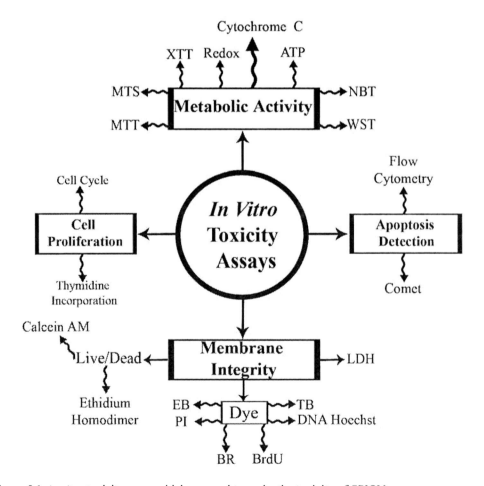

Figure 5.1. *in vitro* toxicity assay which are used to probe the toxicity of SPIONs.

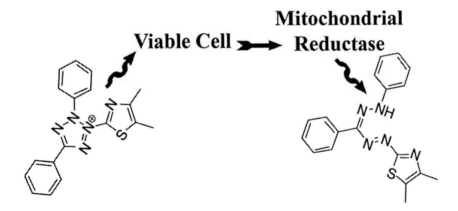

Figure 5.2. Schematic molecular formulation and reduction Methyl thiazole tetrazolium (MTT: (3-(4,5-dimethylthiazol-2-yl)-2,5-diphenyltetrazolium bromide)) composition.

5.2.1. Metabolic Activity

Cytotoxic materials often impact the metabolic activity of cells, which is mainly expressed as mitochondrial function. The most prominent assays to determine metabolic damages are described in this chapter.

5.2.1.1. MTT Assay

Cytotoxicity of SPIONs was assessed by determining mitochondrial damage using MTT (3-(4,5-dimethylthiazol-2-yl)-2,5-diphenyltetrazolium bromide), MTS (3-(4,5-dimet-hylthia zol-2-yl)-5-(3-carboxymethoxyphenyl)-2-(4-sulfophenyl)-2H-tetrazolium), and XTT (sodium(2,3-bis(2-methoxy-4-nitro-5-sulphophenyl)-2H-tetrazolium-5-carboxanilide) assays, which are nonradioactive, colorimetric assays.[25, 26] MTT is used to assess cell viability as a function of redox potential. Actively respiring cells convert the water-soluble MTT to an insoluble purple formazan. The formazan is then solubilized and its concentration determined by optical density. MTT reduction was used to metabolically quantify viable cells after exposure to nanoparticles. Figure 5.2 shows the corresponding chemical equations. The MTT assay is widely used to probe the toxicity of SPIONs since it is fast, easy to use, and in most cases reproducible. Whenever cell viability is below 40% the material is considered to be seriously toxic and investigated with other complementary methods.[27] Table 5.1 summarizes the cytotoxicity data of different SPIONs. However, the data are difficult to compare since the parameters (e.g. incubation time), particles (e.g. particle size and/or coating material), and biological environment (e.g. cell line) vary substantially. From this it becomes clear that there is no universal "nanoparticle" to fit all the cases, and each system must be examined individually.

Gupta et al. [28, 29] showed that uncoated iron oxide nanoparticles were internalized by fibroblasts by endocytosis, which resulted in disruption of the cell membrane. On the other hand, insulin-coated and poly-ethylene-coated nanoparticles attached to the cell membrane, most likely to cell-expressed surface receptors, and were not endocytosed. The presence of insulin caused an apparent increase in human fibroblast cell proliferation and viability measured by MTT assay. Insulin coated SPION did not shown cytotoxicity up to 1 mg/ml, whereas the viability of uncoated nanoparticles decreased significantly from 100% to about 60%. The viability of poly-L-lysine (PLL)-coated SPION treated human cervical carcinoma cells (HeLa) and mesenchymal stem cells (MSC) cells were measured by *Arbab et al.*[30]. Here, no significant decrease in cell viability was observed upon incubation of particles compared to the corresponding control cells at different time intervals. However, transient increases in the activity of the formazen products (P value < 0.01) were observed in particle treated HeLa cells at days 1 and 3 after incubation. Cytotoxicity of superparamagnetic nanoparticles with specific shape and size which were coupled to various proteins (i.e., lactoferrin or ceruloplasmin) has been reported by *Gupta et al.* [31]. The influence on human dermal fibroblasts was probed. The results showed that each nanoparticle type with different surface characteristics caused a distinctly different cell response. The underivatized magnetic particles were internalized by the fibroblasts probably due to endocytosis, which resulted in disruption of the cell membrane and disorganized cell cytoskeleton. In contradiction, lactoferrin or ceruloplasmin coated nanoparticles attached to the cell membrane, most likely to the cell expressed receptors and were not endocytosed. All nanoparticles affected the metabolic activity in a concentration dependent manner in the concentration range of 0–1000 mg/ml. Lactoferrin or ceruloplasmin derivatized nanoparticles showed no cytotoxic effects up to 1 mg/ml and even increased cell viability to about 50–250% in a concentration dependent manner. The increased cell viability was explained by nutrient effects.[32] When incubated with underivatized magnetic particles, the fibroblasts showed significant loss in viability of about 25–50% observed at concentrations ≥ 250 µg/ml. Below this concentration cellular metabolic activity did not change much in comparison with control cells. *Petri-Fink et al.* [33] observed no cytotoxicity in melanoma cell lines after 2 hours of exposure to amino-poly (vinyl alcohol) SPION for all polymer/iron ratios examined (i.e. polymer/iron oxide ratios of 0, 5, 10, 15 and 20). After 24 h, cytotoxicity became apparent at high polymer concentration. A correspondingly coated SPION probed by *Cengelli et al.* [34] was found to be nontoxic as N11 microglial cells only took up amino-PVA-coated SPIONs, and no nitric oxide was produced. One of the main problems, that coated-SPION present, is that once the surface-derivatized nanoparticles are inside the cells, the coating would likely be digested, leaving the uncoated particles exposed to other cellular components and organelles thereby potentially influencing the overall integrity of the cells.[20, 35] As a result it became essential to investigate the cytotoxicity of uncoated SPIONs as well. To achieve this purpose, some groups evaluated the cytotoxicity of uncoated iron oxide nanoparticles. *Hussain et al.*[36] found that magnetite nanoparticles with the size of 30 and 47 nm displayed less or no toxicity effect on rat liver cells (BRL 3A) at the dosage tested (i.e. 10, 50, 100 and 250 µg/ml). Higher concentrations were tested by Cheng *et al.*[37]; at 231 mM of nanoparticles the group found no significant difference between the exposed cells (Cos-7 monkey kidney cells) and the control during 4 h exposure. Because published reports confirm that use of the MTT assay for measuring the toxicity of magnetite nanoparticles has high variability and nonspecificity[38], *Mahmoudi et al.*[39] applied the outlier detection method to minimize variability. They found that cell detachment upon exposure to SPION also necessitated the development of customized protocols for the MTT assay; detachment increased with both increasing SPION concentration and contact time. Following their report, during the assay, cells exposed to SPION detached from the wells after 4 h. In reducing the adhesive properties of L929 cells (primary mouse connective tissue cells), SPION exposure may have increased the error in the MTT assay through the elimination of crystals during removal of the supernatant. To accommodate for cell detachment, cells were examined by optical microscopy to ascertain the density of violet spots prior to detachment. The supernatant was then carefully removed to quantify precipitated formazan. Uncoated SPION showed less or no toxicity at applied dosage (i.e. 0.2, 1, 5 and 20 mM) on both L929 (adhesive) and K562 (human leukemia cells; suspended).[39] In addition, the highest concentration of uncoated SPION (i.e. 1.6 M) was tested by *Mahmoudi et al.*[40] where the toxicity of poly(ethylene glycol)-co-fumarate (PEGF) coated magnetite nanoparticles was clearly observed. Treatment of cell cultures (i.e. rodent or human cells (mesothelioma MSTO-211H and rodent 3T3 fibroblast cells)) with uncoated, nanoparticulate iron oxide revealed a cell-type specific response.[41] Slower proliferating 3T3 cells were only slightly affected by addition of up to 30 ppm iron oxide, while the cell parameters MTT-conversion and DNA content of faster growing MSTO cells were drastically reduced upon exposure to as little as 3.75 ppm iron oxide. The observed toxicity was attributed to iron-induced, free-radical production via the Fenton or Haber–Weiss reactions in addition to internalization of the iron oxide particles.[41]

In order to analyze the potential toxicity of the poly(TMSMA-r-PEGMA) coated SPION, *Lee et al.*[42] used the MTT assay on the control and treated Lewis lung carcinoma (LLC) cell line. Their results clearly indicated that both in situ and stepwise SPION show no toxicity at applied concentrations of the SPION (1, 10, 20, 50, and 100 µg/ cells). *Räty et al.*[43] exposed the HepG2 cell lines to uncoated and coated (conjugated avidin-coated baculoviral vectors (Baavi) with biotinylated superparamagnetic iron oxide particles) SPION, and found no indication of increased cytotoxicity with non-coated or bUSPIO-coated Baavi. In addition, *Kim et al.*[44] reported no toxic effect of silica coated SPION (concentration of 4 mg/ml) on human lung cells (A549). *Muller et al.*[45] found a 20–30% decrease in neutral red uptake in monocyte-macrophages with 10 mg/ml Ferumoxtran-10 across various incubation times.[46] A summary of the experimental setup and results on superparamagnetic iron oxide nanoparticles [45-55] is provided in Table 1.

5.2.1.1.1. Modification of MTT Method

The study of the interactions of surfaces with biological media is important for both uncoated and coated SPION. Studies of surface interactions can give knowledge about toxicity effects. *Mahmoudi et al.*[40, 57] were focused to some extent on SPION surface-passivated with both PEGF and PVA, while also considering the implications for uncoated SPION. For example, they had shown that the medium conditions play an important role in toxicity and this applies to uncoated SPION as well.[57] Previously reported studies on cytotoxicity amount of magnetite nanoparticles have shown that the common *in vitro* examination method can yield erroneous viability values.[58] This is due to the fact that SPION can cause significant changes in the composition cell medium (i.e. due to the interaction with proteins as well as ions), which in turn can cause toxicity. Based on the results of their research, it was shown that a more reliable way of identifying cytotoxicity for *in vitro* assessments is to use particles with saturated surfaces via interactions with Dulbecco's Modified Eagle Medium (DMEM) before usage. The interaction between nanoparticles and proteins in the medium were also confirmed by other researchers.[49-52] By using nanoparticles whose surface have been passivated, even the toxicity of high quantities of uncoated SPION was decreased significantly. It is believed that the existence of Cl compounds on the active surface of particles may be the main source of error in the conventional *in vitro* assessment methods. PEGF and PVA coated nanoparticles illustrate lower toxicity not only due to the presence of biocompatible coating but also due to their lower potency sites for proteins, ions and other components in the cell medium.[40, 57] The effect of surface active and surface passivated SPION on cell medium is summarized on Figure 5.3.

Table 5.1. Cytotoxicity of SPIONs via MTT assay

Cell Line	Average Size of Nanoparticle (nm)	Coating Material	Concentration of SPION	Exposure Conditions	Exposure Duration	Toxicity	Author
human fibroblasts	--------------	PEG and Insulin Coated	0-1 mg/ml	96-well plates at cells/well	24 h	25-50% decrease in viability for bare (250µg/ml); 99% viable for PEG-Coated (1mg/ml)	Gupta [28, 29]
MSC and HeLa cells	--------------	PLL (poly-L-lysine)	--------------	MSC: 96-well plates at 5× Cells/well Hela: 24-well plates at 2×104 Cells/well	1-43 days	The long-term viability, growth rate, and apoptotic indexes of the labeled cells were unaffected by the endosomal incorporation of SPION	Arbab [30]
human dermal fibroblasts	13.6	sodium oleate coated (underivatized and derivatized (lactoferrin or ceruloplasmin))	0-1000 µg/ml	96-well plates at 10000 cells/well	24 h	The underivatized SPION disrupted of the cell membrane and disorganized cell cytoskeleton. Derivatized SPION attached to the cell membrane.	Gupta [31]
human melanoma cells	9 (Core), 14-55 (hydrodynamic)	PVA and amino-PVA	12, 61 and 123 µg/ml	46-well plates at 100000 cells/well	2 and 24 h	High polymer concentration showed toxicity after	Petri-Fink [33]

Cell	Size	Coating	Concentration	Plates/cells	Time	Result	Ref
BRL 3A	30 and 47 nm	--------------	0, 10, 50, 100, 250 µg/ml	96-well plates	24 h	24h 50% decrease in viability (250µg/ml)	Hussain [36]
Cos-7	9 nm	--------------	0.92-23.05 mM	24-well plates at 3× cells/well	4 h	No toxicity detected during applied dosage (0.92-23.05 mM)	Cheng [37]
L929 And K562	17 samples with various shapes and sizes [12]	PVA	0.2, 1, 5 and 20 mM	96-well plates at 10000 cells/well	3,24 and 48 h	Toxicity amounts were Dependent to nanoparticles shape and size	Mahmoudi [39]
L929	82	PEGF and PVA	0.4, 0.8 and 1.6 M	96-well plates at 10000 cells/well	24,48 and 72 h	Toxicity amounts were dependent to nanoparticles shape and size	Mahmoudi [40, 56]
MSTO-211H and rodent 3T3	12	--------------	0-30 ppm	---------	72 h	Viability of MSTO cells was decreased at 3.75 ppm; 3T3 cells were a lived up to 30 ppm	Brunner [41]
LLC	10	poly(TMSMA-r-PEGMA)	1, 10, 20, 50, and 100 µg/cells	96-well plates at cells/well	12 h	No toxicity at applied dosage	Lee [42]
HepG2	42	Uncoated and bUSPIO-coated Baavi	--------------	--------------	--------	no indication of increased cytotoxicity with noncoated or bUSPIO-coated Baavi	Räty [43]
A549	50 nm	Silica	4 mg/ml	--------------	--------	IC50 (50% inhibiting concentration) value was 4 mg/ml	Kim [44]
HMMs	5 (Core), 30 (Hydro-dynamic)	Dextran	0-10 mg/ml	24-well plates at 1-2 × cells/well 48-well plates at 0.5-1× cells/well	24, 48, 72 h	Not (1mg/ml) and mildly toxic (1mg/ml) after 72 h	Müller [45]

Table 5.1. Continued

Cell Line	Average Size of Nanoparticle (nm)	Coating Material	Concentration of SPION	Exposure Conditions	Exposure Duration	Toxicity	Author
H441	9 (core), 63±36 (hydrodynamic)	PEI, PL, LS, Ls and PEI	90 µg/ml	96-well plates at 5000-10000 cells/well	24-48 h	Toxicity of tested complexes is acceptable (cell viability >80%)	Mykhaylyk [46]
HS68	8.7-12	Oleic acid and ethylene glycol	1 mg/ml	12-well plates at 6 x cells/well	24 h	No significant difference in the viability of cells compared to the control group	Lee [47]
GL261	10-100 nm (hydrodynamic)	Dextran	1-200 µg/ml	96-well plates at 10000 cells/well	24 h	The nanoparticles functionalized with dendritic guanidines exhibited somewhat greater toxicity than those functionalized with dendrons having hydroxyl or amine peripheries	Martin [48]
K562	30	Tetraheptylammonium	2.5 µg/ml	96-well plates at 10000 cells/well	72 h	The inhibition rate was 46% for the cell system cultured with Fe3O4-PLA.	Lv [49]
HepG2	61 and 127	amino-functionalized	3 mg/ml, 300, 30, 3, 0.3 and 0.03 µg/ml	96-well plates at 8× Cells/well	5 days	The LD_{50} of Gal-ASPIO-278 was 1500 µg/mL to HepG2 cells	Huang [50]
K562 and K562/A02	----------------	ADM Conjugated	20 µg/ml and 5 mg/ml	96-well plates at 2× cells/well	48 h	Nano-Fe_3O_4 combining with DNR could inhibit two cell lines proliferation significantly ($P < 0.001$)	Chen [51]

LNCaP and PC3	60.8±1.9	TCL-SPION	0.1 mg/ml	48-well plates at 4× cells/	48 h	the cytotoxicity of Dox-loaded TCL-SPION–Apt bioconjugates was nearly as potent as free Dox	Wang [52]
HaCaT and A549	20-40	--------------	0.01-100 mg/ml	96-well plates at cells/ml	24 h	The cell viability decreased in a dose-dependent manner by Fe2O3	Horie [53]
HeLa	4.5-10 nm	dextran, aminodextran, heparin, and dimercapto-succinic acid	0.05, 0.1 and 0.5 mg/ml	24-well plates at 3500 cells/well	24 h	Viability of cell culture is not significantly affected or modified by the presence of the nanoparticles after 24 h of treatment (90–100% viability in relation to the control sample)	Villanueva [54]
SMMC-7721	13.8±5.3	Chitosan	0–123.52 µg/ml	96-well plates at 10000 cells/well	12 h	FITC-CS@MNPs caused a minor reduction (about 10% of control) in cell viability even at high dose (123.52 lg). Naked MNPs caused a significant reduction (90% of control) in cell viability even when tested at the lowest concentration (0.01 mg/ml)	Ge [55]

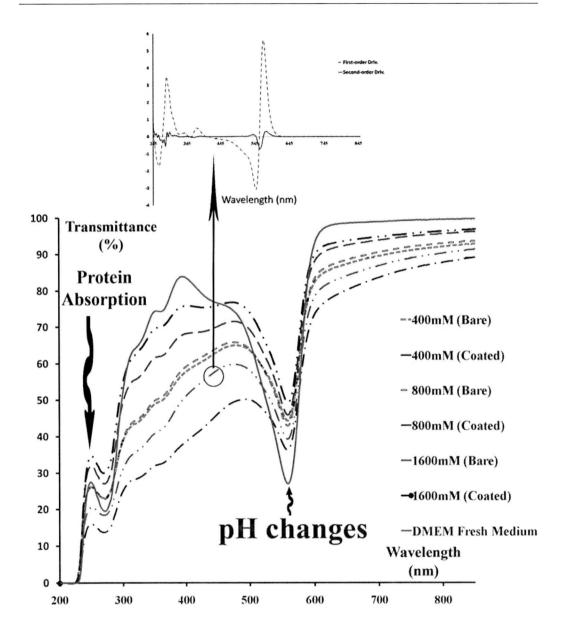

Figure 5.3. (Upper) UV/Vis spectrum of pure cell medium, the extract of coated and uncoated SPION with de-ionized (DI) water as the reference. (Lower) UV/Vis spectrum of pure cell medium and the extract of surface saturated SPION [40].

5.2.1.2. MTS and XTT Assays

The MTS and XTT assays are newer versions of the MTT assay and the final solubilization step, which many studies have shown to produce errors in results.[52] The solubilization step required for the MTT assay destroys the cells under investigation and therefore only allows for a single time-point measurement. MTS and XTT are water soluble formazans and therefore allow continuous monitoring of the cells without having to destroy them. However, unlike the MTT assay they both require the addition of an electron acceptor such as phenazine methosulphate (PMS) for reduction and formazan formation. In a comparison between MTS and XTT assays, the XTT/PMS reagent mixture was found to be unstable, resulting in PMS depletion over time and reduction of formazan formation.[52] The resulting response errors and poor precision were not encountered with the use of MTS/PMS mixtures. For the analysis of higher concentrations of strongly colored (dark) magnetic nanoparticle suspensions, however, it is sometimes necessary to add washing steps to the cell viability measurements. The MTT assay is in this case the preferred method.

5.2.1.2.1. MTS Assay

Huang et al.[59] examined the biocompatibility of SPION (sizes of 5.4 and 7.6 nm) in the concentrations range of 0.1, 1, 10, and 100 μM with the normal breast epithelial cells (H184B5F5/M10) and the three types of breast cancer cells (SKBR3, MB157, and T47D) for 72 h. MTS assay studies revealed that there are no obvious changes in cell viability in the studied concentration range of magnetic nanoparticles. According to *Delcroix et al.*[60], MTS assays showed that rMSC (rat Mesenchymal stem cells) viability was not affected one day post-incubation when using 25 or 50 μg iron/mL (1-hydroxyethylidene-1.1-bisphosphonic acid (HEDP)-coated SPION), whereas it decreased to 70% of control with 100 μg iron/mL. Consequently, 50 μg iron/mL incubated for 48 h was chosen as the best trade-off in terms of iron content, percentage of cell viability.

Pawelczyk et al.[61] investigated the frequency of SPION uptake by activated macrophages using the Boyden chamber model of inflammation. The MTS assay of activated macrophages showed that after 7 days in culture, only 20% of macrophages were viable. On the basis of their results, the BC (Boyden chamber) experiments were designed to include time points at 24 and 96 hours of incubation. The MTS data were validated with the trypan blue exclusion test, which showed a similar survival rate of activated macrophages *in vitro*. A tetrazolium-based cytotoxicity (MTS) assay to detect the activity of mitochondrial enzymes of SPION labeled neural stem cells reported by *Walczak et al.*[62]. Their results showed no detrimental cellular effects following MEP (magnetoelectroporation, MEP treatment (130 V, 17 ms, 1 pulse, 2 mg Fe/mL), as the metabolic rate was similar to that of unelectroporated cells. The decrease in the metabolic rate after MEP was less than 5% compared with that in control cells. MTS assay results by *Au et al.*[63] showed that treatment of astrocytes with magnetic nanoparticles significantly ($p<0.01$) increases MTS production (100%±3.65 vs 112.8%±3.23, for control and treated astrocytes, respectively), indicative of alteration in mitochondrial functioning.

Wan et al.[64] tested the effect of the MPEG(Methoxypoly(ethylene glycol)–oligo(aspartic acid))-Asp3-NH2-, MPEG–PAA- and PAA-coated magnetic nanoparticles on the viability of OCTY mouse cells. They reported that MPEG–Asp3-NH2-coated iron oxide nanoparticles had almost no effect on the cell viability compared with the control. On the contrary, MPEG–PAA- and PAA-coated iron oxide nanoparticles markedly reduced the cell viability, only 16% of the cell viability remained at iron concentration of 400 mg/mL. Following their report, MTS assay data for uncoated iron oxide nanoparticles were not obtained because bare iron oxide nanoparticles were adsorbed on the cells (the cells became black in color after incubation). The adsorbed iron oxide nanoparticles on the cells could not be removed by washing and it interfered with MTS assay. However, the cell number counted after the incubation with uncoated iron oxide nanoparticles was much less than that of the control, indicating that uncoated iron oxide nanoparticles showed significant reduction in the cell viability.[65, 66]

5.2.1.2.2. XTT Assay

SCLN (SPION-loaded WSC(water-soluble chitosan)–LA(linoleic acid) nanoparticles cytotoxicity was investigated *in vitro* using an XTT assay.[67] Cell (Primary hepatocytes separated from mice) viability results after 24 and 48 h of incubation exhibited little cytotoxicity. Cells were approximately 73.4% viable with 100 mg/ml of SCLN after incubation for 48 h. The WSC–LA conjugates encapsulating the SPION effectively reduced the cytotoxicity of iron particles. *Shi et al.*[67] treated KB cells (a human epithelial carcinoma cell line) with dendrimer-stabilized (carboxyl-functionalized poly(amidoamine) (PAMAM) dendrimers of generation 3 (G3)) Fe_3O_4 nanoparticles for 24 h, and the nanoparticles were removed after treating time. The cells were allowed to grow for another 4 days, and their proliferation was determined by an XTT assay. The results indicated that dendrimer-stabilized SPION did not display cytotoxicity to KB cells in the concentration range of 0–80 mg/mL. The cytotoxicity of the CMC(Carboxymethyl Curdlan)-coated SPION (with concentrations of 1, 2.5 and 5 mM) was performed by *Lee et al.*[65] using a B16F10 cell line (breast cancer cell line). According to their results, after incubation for 24 h, the CMC-coated SPION showed no toxicity at high concentration of the SPION. On the other hand, after 48 h, when the concentration of the CMC-coated SPION was higher than 1 mM, the cell viability was reduced about 81%.

5.2.1.3. Redox Assay

A special redox assay is the Alamar blue™ assay in which the soluble blue dye resazurin is converted, by cells, to the pink fluorescent dye resorufin. The distinguished features of this assay include the water solubility of the dye, easy handling, and high sensitivity.[68] Incubation times are normally slightly longer in comparison with the MTT assay, however, its low toxicity makes this assay useful not only for final, but also for serial kinetic measurements. The excitation wavelength is 544 nm and the emission wavelength is 590 nm. Chondrocytes cells are labeled with Endorem (i.e. Endorem™ is a black to reddish-brown aqueous colloid of superparamagnetic iron oxide associated with dextran coating for intravenous administration as a MRI contrast medium for the detection of liver lesions that are associated with an alteration in the RES) combined with Lipofectamine, and their cytotoxicity is measured by Alamar blue assay.[68] Briefly, 70,000 cells are added per well of a 12 well plate and left overnight. They are then labeled overnight with SPION-Lipofectamine. This is followed by two PBS washes and medium is added for 4 h. After this period Alamar Blue is added at a 1:10 concentration to the medium present. 4 h later 100 ml of this medium-Alamar mix is placed in a 96 well plate, read at 570 and 600 nm and analyzed according to the provided calculations. Labeling of chondrocytes at a low doses (<54 mg) had no effect on cell viability as assessed by an Alamar Blue cytotoxicity assay until a slight effect was observed at a dose of 54 mg SPION per well in the combined N/Lipofectamine condition and in the Lipofectamine only condition at 168 mg. This is likely due to the large amount of Lipofectamine present which is known to be cytotoxic at high doses.[69]

Polyak et al.[70] examined the toxicity of SPION-loaded bovine aortic endothelial cells (BAECs). Results showed that cell viability was not adversely affected by internalized magnetic nanoparticles, as assessed by Alamar blue assay. Cell survival of 83±3% relative to untreated cells was observed at the highest applied SPION dose (9 µg per well, corresponding to a SPION loading of 0.3 ng per cell) and at the maximal incubation time of 24 h. Polylactide SPION were formulated using a modified emulsification-solvent evaporation methodology with both the incorporation of oleate-coated iron oxide and a polyethylenimine (PEI) oleate ion pair surface modification for DNA binding by *Chorny et al.*[71] SPION treated (concentration of 1-5 µg/well) rat aortic smooth muscle cells (A10) for 72h showed good viability. *Samanta et al.*[72] developed non-toxic iron oxide nanoparticles that employ bovine serum albumin (BSA) as a biocompatible passivating agent. The particles were stable in a variety of media, and showed no toxicity during interactions with Hela cells.

5.2.1.4. Adenosine Triphosphate Assay

ATP is a biomolecule and has a crucial role in cell metabolism (energy source) as well as neurotransmitter. A depression of the ATP metabolism often occurs upon cell damage or toxicity. The ATP assay can then assess cell toxicity, or cell proliferation as well as cell number.[73] The metabolic activity is determined by bioluminescence measurements in which the enzyme luciferase catalyses the formation of light from ATP and luciferin. The light intensity at 560 nm correlates linearly with the ATP concentration.[74] Some researchers reported usefulness of the ATP assay in comparison to other cytotoxicity assays. These studies have shown the ATP assay to be more sensitive than MTT, MTS, Neutral Red (NR), and the lactate dehydrogenase (LDH) assay.[74, 75] On the other hand, the ATP assay also has some disadvantages. The luminescence readout can be affected by sample quenching, which can be a significant problem when magnetic particles are present. In addition, luminescence intensity is time dependent. Despite these disadvantages, the ATP assay is one of the quickest and most efficient methods, requiring only 10 min of incubation time before readout.

Cheng et al.[76] attached DNS hapten covalently to cross-linked iron oxide (CLIO) to form a 39±0.5nm DNS-CLIO nanoparticle for imaging probe. DNS-CLIO (100 mM Fe) was non-toxic (measured with ATP) to both B16/DNS (DNS receptor positive) and B16/phOx (control receptor positive) cells.

5.12.1.5. Cytochrome C Assay

Cytochrome C assay is considered as a different type of assay measuring metabolic activity which detects extracellular release of superoxide radical anions.[77] Extracellular radicals reduce 100 µM ferricytochrome (cytochrome^{3+}) to cytochrome^{2+} that can be measured at 550 nm. Although *Simko et al.*[77] showed, with this technique, radical formation only in the presence of ultrafine carbon black particles of 12–14 nm diameter, it may also be useful for magnetic particle investigations. *Huang et al.*[50], examined cytotoxicity of amine-functionalized superparamagnetic iron oxide nanoparticles (ASPION), which were derivatized with lactose to form galactose-terminal (Gal-ASPION) nanoparticles, with HepG2 cells. Their results confirmed low cytotoxicity (LD$_{50}$ ~1500 µg/mL) of ASPION. Cytochrome C results have been used by some researchers as a complementary assay in order to confirm their MTT results.[44, 50] In order to assess influences of magnetic iron oxide nanoparticles on O$_2^-$ concentration, *Krötz et al.*[78], examined both cytochrome c assay and L-012 ((8-amino-5-chloro-7-phenylpyrridole-(3,4-d)pyridazine-1,4-(2H,3H)-dione) chemiluminescence. They found that Neodymium–iron–boron magnetic plates and magnetic particles (CombiMAG or PolyMAG), which are composed of Fe$_3$O$_4$, did not decrease O$_2^-$ production. At a concentration of 5 µg/ml (corresponding to 2 µg/400µl, the volume assayed in 1 well of a 24-well dish), they rather caused a small but significant increase in the O$_2^-$ signal.

5.2.1.6. Nitroblue Tetrazolium (NBT) Assay

The formation of superoxide anions during the oxidative burst was assessed using the Nitro blue tetrazolium (NBT) assay, which measures the conversion of NBT to blue formazan.[79] Formazan was solubilized by KOH/dimethyl-sulphoxide (DMSO) per well (mix 1 part of 2 M KOH with 1.167 parts of DMSO just before use) and absorbance was measured at 620 nm.[80] In parallel to MTT assay *Müller et al.*[80], used NBT in order to measure the oxidative burst. Their results showed that there were not significantly atorvastatin interfere with the oxidative burst of human monocyte macrophage, neither with baseline NBT reduction nor after phorbol myristate acetate (PMA) stimulation, with the exception of a reduction of baseline NBT reduction at 5 mM atorvastatin of about 20%. In the absence of statins, PMA stimulated the oxidative burst about 1.65-fold above the medium control. It is noteworthy that in clinical trials, the NBT assay is currently applied to monitor the effect of atorvastatin therapy on macrophage activity in human carotid plaques.[45] The effect of atorvastatin on human monocyte–macrophage uptake of Ferumoxtran-10 *in vitro* using biochemical assays, magnetic resonance imaging and transmission electron microscopy was examined. In addition, the effect of Ferumoxtran-10 on superoxide anion production of HMMs during the oxidative burst was measured. Results confirmed that Ferumoxtran-10 on its own did not modify superoxide anion production significantly compared to controls.[45]

5.2.1.7. WST Assay

WST Cell Proliferation Assay is a sensitive and accurate assay for cell proliferation and cytotoxicity. The assay is highly convenient as it is performed in a single cell tissue culture well and requires no washing, harvesting or solubilization of cells. Adherent or suspension cells are cultured in a micro-plate and then incubated with WST and the assay is monitored with a spectrophotometer. The assay principle is based upon the reduction of the tetrazolium salt WST to formazan by cellular dehydrogenases. The generation of the dark yellow colored formazan is measured at 420-480 nm (optimal at 440 nm) and is directly correlated to cell number. In order to examine the acute toxicity of poly(N,N-dimethylacrylamide) (PDMAAm)-coated maghemite nanoparticles, *Babic et al.*[81] used the WST assay. Both rat and human mesenchymal stem cells were incubated for 72 h with coated and uncoated nanoparticles, together with Endorem (control) at concentrations of 15 μg γ-Fe$_2$O$_3$/mL. The viability of both rat and human MSCs labeled with PDMA nanoparticles did not markedly decrease compared to that of unlabeled MSCs (control). When mesenchymal stem cells were labeled with Endorem, their viability was decreased by 32%; similarly, the viability of uncoated nanoparticles decreased by 15%. Gene therapy mediated by nonviral vectors provides great advantages over conventional drug therapy in inducing immune-suppression after organ transplantation, yet it was rarely reported because T cells are normally difficult to transfect.

Guihua et al. [82] established a new nonviral gene delivery agent bearing CD3 single chain antibody (scAbCD3) as a targeting ligand to T lymphocytes and MRI T_2 agent SPION. Molecular tailoring of these novel gene carriers to effectively transfer both the reporter and therapeutic genes to a rat T lymphocyte cell line was successfully demonstrated. Polyplexes based on these targeted delivery agents exhibited not only high efficacy of gene transfection in HB8521 cells but also low cytotoxicity (probed by WST assay). Similar cytotoxicity examination of SPION using WST reported by other groups.[24, 83]

5.2.2. Membrane Integrity

5.2.2.1. LDH Assay

Due to the cell contents leakage of the damaged membrane, the analysis of cell membrane integrity would be an alternative marker for toxicity.[74] Lactate dehydrogenases are enzymes present in the cytosol of the cell and can only be measured extracellularly when membrane damage occurs. This method is known as the LDH assay. The assay has evolved from its original method, which was simple and cheap, but extremely time consuming, to a coupled assay whereby LDH catalyses the conversion of 2-*p*-(iodophenyl)-3-(*p*-nitrophenyl)-5-phenyltetrazolium chloride (INT), a tetrazolium salt, to a red formazan that can be quantified by optical absorbance, much like the concept of MTT. If toxic agents damage the cell membrane, the LDH assay is the best method for toxicity measurements., In contrast, LDH would be considered as a less sensitive method and may contain error whenever the toxic agent affects the cell intracellularly.[30] The LDH assay is an indicator of irreversible cell death due to membrane damage and is reliable, but is the least sensitive when compared to the tetrazolium assays.[84] The assay is not suitable for the detection of a threshold concentration where the toxicity starts to show but will reliably determine the concentration at which cells can no longer survive. Incubation times of the assay vary between less than an hour and up to more than 4 hours.[85] *Weissleder et al.* [86] used the LDH assay to determine the viability of lymphocytes after magnetic labeling for trafficking studies.

Using LDH release and trypan blue exclusion, *Jang et al.* [87] reported that iron oxides have no toxicity and little biological effect. In another study, the results for LDH leakage for Fe_3O_4 nanoparticles exposure did not produce cytotoxicity up to the concentration of 100 μg/ml, but it produced a significant effect at 250 μg/ml.[36] There was no significant difference when comparing the toxic effects between different sizes of Fe_3O_4 nanoparticles (30 nm versus 47 nm).[36] In addition, the lethality of various SPION concentrations was evaluated with an LDH release assay is reported by *Boutry et al.* [88]. Some other researchers used LDH in order to investigate the toxic effect of SPION on the membrane integrity.[89-93]

5.2.2.2. The Calcein AM, Live/Dead and Ethidium Homodimer Assay

The calcein AM is membrane-permeant and can be introduced into cells via incubation. Once inside the cells, Calcein AM (a nonfluorescent molecule) is hydrolyzed by endogenous esterase into the highly negatively charged green fluorescent calcein. The fluorescent calcein is retained in the cytoplasm in live cells. Calcein AM is as an excellent tool for the studies of cell membrane integrity and for long-term cell tracing due to its lack of cellular toxicity.[94-96] It has also been used to quantifying cell number.[97] It is a true end-point assay for cell viability. The fluorescent signal is monitored using 485 nm excitation wavelength and 530 nm emission wavelength. The fluorescent signal generated from the assay is proportional to the number of living cells in the sample. The calcein AM assay is often combined with other fluorescence-based assays that label dead cells instead of viable cells, such as the Live/Dead® assay by Invitrogen. It incorporates the calcein AM assay with an ethidium homodimer assay.[98] Ethidium homodimer is a hydrophilic dye that is able to enter cells with damaged membranes and bind to nucleic acids, producing bright red fluorescence. After a cell sample has been stained with ethidium homodimer, the dead cells may be viewed and counted under a UV-light microscope. When cells die, the plasma membranes become disrupted. Because of this, ethidium homodimer may enter those cells and bind to DNA within the cells. Due to the absence of compromised membrane in live cells, the ethidium homodimer is not be able to stain them. The use of the Live/Dead® assay has been successfully used to investigate the impact of pancreatic islet labeling with Resovist® superparamagnetic particles and Dynabeads® *in vitro*.[99] The Live/Dead® assay confirmed that islet cells cultivated for 48 hours with Resovist or Dynabeads did not diVer from control cells (90.8 ± 1.9% for Resovist, 89.4 ± 2.0% for Dynabeads, and 91.1 ± 2.5% for control cells). Such labeled islet cells were then used to image the success of pancreatic cell transplantation in diabetes patients using MRI.

5.2.2.3. Dye Assays

Dyes, which are able to cross the cell membranes and stain the cell, are considered as cytotoxicity assays. Depending on the nature of the stain, it can either be taken up or excluded by viable cells. If normally excluded, it will cross the membrane of dead or dying cells. The popular dye assays are discussed in this section.

5.2.2.3.1. Erythrosine B (EB) Assay

Erythrosine B is an iodo-derivative of fluorescein with distinctly bluish shade where as eosin is a bromine derivatives of fluorescing. EB would be considered as exclusion assay, since it only stained dead cells after crossing their damaged cell membranes.[25] Toxicity of some metal nanoparticles have been investigated by this dye.[100]

5.2.2.3.2. Propidium Iodide (PI) Assay

The propidium iodide dye assay, stains the nucleic acids of cells. It may be used simultaneously with the neutral red (NR) assay, since each assay works via different mechanisms and stains different cells. The resulting two-colored cell populations can easily be counted microscopically or by fluorescence activated cell sorting (FACS). In general, fluorescence-based assays are popular methods of assessing cytotoxicity. They are thought to be more sensitive than colorimetric assays and can be analyzed, in plate reader format, by fluorescence microscopy and confocal microscopy and also in FACS.

Toso et al. [101] examined the viability of human islet labeled SPION (280 μg/ml) by propidium iodide and fluorescein diacetate staining. Their results showed that the viability of labeled islets were similar to those of control islets. Regarding the report by *Wang et al.* [102], the cytotoxicity of functionalized Fe_3O_4 nanoparticles was evaluated by fluorescein diacetate (FDA) and propidium iodide (PI) staining and by observing cell morphology changes after incubation with the SPION for 96 h. Their cell viability data showed that the KB cells (a human epithelial carcinoma cell line) treated by functionalized Fe_3O_4 nanoparticles, with or without FA conjugation, display similar percentage of FDA positive cells to the KB cells treated by unmodified magnetite nanoparticles at Fe concentration of 0–150 μg/mL. Dead treated (by Feridex (4–5 nm γ-Fe2O3 core)) cell (mesenchymal stem) percentage were defined by the propidium iodide (PI) staining method measured via flow cytometry.[103]

Song et al.[104] covalently coupled avidin onto the surfaces of fluorescent-magnetic bifunctional nanospheres to construct fluorescent-magnetic-biotargeting trifunctional nanospheres and analyzed the functionality and specificity of these trifunctional nanospheres for their ability to recognize and isolate apoptotic cells labeled with biotinylated annexin V, which recognizes phosphatidylserine exposed on the surfaces of apoptotic cells. Following their results, the multifunctional nanospheres can be used in combination with propidium iodide staining of nuclear DNA to identify cells at different phases of the apoptotic process. Furthermore, they demonstrate that apoptotic cells induced by exposure to ultraviolet light could be isolated simply with a magnet from living cells at an efficiency of at least 80%; these cells can then be easily visualized with a fluorescence microscope. To further examine the effect of SPION labeling on cell viability, *Dunning et al.*[105] performed a propidium iodide exclusion study, which revealed no significant difference in cell viability in Schwann cell (SC) cultures incubated with SPION for up to 48 hr using concentrations up to and including 4 mg/ml of Fe when compared with control SC cultures (control SCs, 96.4±3.2%; 1mg/ml, 96.3±2.1%; 2mg/ml, 96.5±3.2%; 4mg/ml, 96.4±2.8%). Similarly, there was no effect on cell viability in olfactory ensheathing cell (OEC) cultures incubated with 2 mg/ml SPION for 48 hr when compared with control OEC cultures, or SC cultures using an equivalent time point and concentration (control OEC, 97.04±6.62%; SPION-labeled OEC, 99.6±0.8%). PI staining has also been used by many other groups in order to define the cytotoxicity of SPIONs.[31, 106-110]

5.2.2.3.3. Neutral Red (NR) Assay

One of the dyes that is able to quantify the number of viable cells is Neutral Red.[74] This vital dye is taken up via active transport into the cell, accumulates in the lysosomes and can then be spectrophotometrically quantified after releasing the dye with acidified ethanol. Incubation times range from 2–4 hours. The NR assay is sensitive to a wide range of materials, is reliable compared to the MTT and calcein fluorescence assay, and is not influenced by occasional microbial contamination which would usually lead to inaccurately higher cell viability measurements. A disadvantage of the NR assay involves the fixation step that is required before measurements can be read. Cells that do not completely adhere or are loosely attached have a questionable viability status, but are still fixed and included in the final reading, possibly leading to inaccurate cell viabilities.[25] *Müller et al.* [45, 80] used both NR and MTT in order to probe the biocompatibility of Ferumoxtran-10. In addition, in order to investigate the morphology of cells after treatment with PEGF-SPION, the L929 cells were dyed with neutral red and analyzed by fluorescence microscopy and SEM.[40] Cells damaged by unsaturated SPION had spherical shapes; in contrast, saturated SPION with the same concentration (i.e. 800 mM) did not change the cell shapes notably and cells appeared not to be damaged.

5.2.2.3.4. Bromodeoxyuridine (BrdU) Assay

Bromodeoxyuridine (5-bromo-2-deoxyuridine, BrdU) is a synthetic nucleoside that is an analogue of thymidine. BrdU is commonly used in the detection of proliferating cells in living tissues. BrdU can be incorporated into newly synthesized DNA of replicating cells (during the S phase of the cell cycle), substituting for thymidine during DNA replication. Antibodies specific for BrdU can then be used to detect the incorporated chemical, thus indicating cells that were actively replicating their DNA. Binding of the antibody requires denaturation of the DNA, usually by exposing the cells to acid or heat. Because BrdU can replace thymidine during DNA replication, it can cause mutations, and its use is therefore potentially a health hazard. BrdU is a fluorescent labeling system extracted from Lions Mane jellyfish which can be inserted into various multicellular organisms.[111] *Berry et al.*[111] examined the proliferation of fibroblast cells which were treated with dextran derivatised (DD), albumin derivatised (AD) and underivatised plain (P) nanoparticles (8–15nm). Viability staining (Coomassie blue in a methanol/acetic acid aqueous solution) showed that the P and the DD particles caused some cell death. Cells exposed to the AD particles were viable and more densely populated. This result was confirmed by a cell proliferation study (BrdU), where the AD particles induced cell proliferation, whereas the P and DD particles actually inhibited proliferation. In order to evaluate cell (the C17.2 LacZ-transfected mouse NSC line) proliferation after magnetoelectroporation, *Walczak et al.* [62]tested the Bromodeoxyuridine assay. Labeled stem cells exhibited an unaltered viability and proliferation confirmed by trypan blue staining and BrdU, respectively. Some researchers used BrdU as a complementary method in order to confirm their achieved data by perviousprevious described methods.[53, 54]

5.2.2.3.5. Trypan Blue (TB) Assay

The trypan blue exclusion assay can be used to measure cell viability and is the most popular method for cell counting. Limitations of this technique, where dead cells are stained blue, include the method being slow, problems with incomplete re-suspension of adhering cells, and an inability to stain cells undergoing autolysis. In addition, the counting needs to be done within 5 minutes of adding trypan blue in order to avoid getting too high a yield of stained cells. In general, the trypan blue assay should only be used to check overall cell viability or confirm the results of other toxicity assays.[112] The cytotoxicity of bare SPION-treated human lung epithelial cell line A549 was analyzed by *Karlsson et al.* [113] using trypan blue staining. For iron oxide particles (Fe_3O_4, Fe_2O_3), no or low toxicity was observed with the nanoparticle concentration of (80 μg/mL) and treated with cells for 18 h. TB usually is used as complementary *in vitro* method.[55, 114-116]

5.2.2.3.6. DNA Hoechst Assay

The total DNA content of cells is a measure for cell proliferation and can be quantified spectroscopically after converting DNA with an intercalating dye into a highly fluorescent complex (i.e. DNA Hoechst assay). *Brunner et al.* [41] reported cell cultures exposed to uncoated-SPION showed a cell-type specific response.

5.2.2.3.7. Other Dye Assays

Oxidative lesions are determined using the intracellular production of reactive oxygen species (ROS). Intracellular ROS production is measured using the oxidation-sensitive fluoroprobe 2′,7′-dichlorofluorescin diacetate (DCFH-DA). DCFHDA is a nonfluorescent compound that is freely taken up by cells and hydrolyzed by esterases that remove the DA group. DCFH is then oxidized to fluorescent dichlorofluorescein (DCF) by cellular oxidants, thereby indicating the level of intracellular ROS. Magnetite (Fe_3O_4) induced low DNA damage (oxidative lesions) on A549 cells, in contrast maghemite (Fe_2O_3) did not affect DNA in the applied dosage (i.e. 40 and 80 μg/ml). Some researchers used ROS in order to see the toxic effect of SPION on DNA oxidative lesions.[30, 89] Crystal violet is a dye which should be able to determine gas vesicle inside the cells. *Mahmoudi et al.* [57, 117]showed that gas vesicles were formed inside the l929 cells due to the presence of SPION using crystal violet dye.

5.2.3. Cell Proliferation Assay

5.2.3.1. Cell Cycle Assay

The cell cycle assay allows determinations of G_0/G_1 (DNA damaged cells will accumulate in gap1), S (DNA synthesis) and G_2/M (gap2/mitosis) cell cycle phases and distinguishes without interference commonly occurring aggregates. *Mahmoudi et al.* [118]carried out cell cycle assay by staining of the DNA of L929 cells with propidium iodide (PI) followed by flow cytometric measurement of the fluorescence. The stained cells were then analyzed by flow cytometry (FACScan) at an excitation wavelength of 488 nm and emission wavelength of 610 nm. Following their results, PVA-coated SPION treated cells did not showed evident necrosis, apoptosis (via propidium iodide staining) or cell cycle arrest in moderate concentration of nanoparticles (i.e. 200 mM). However, the coated nanoparticles at the highest concentration (400 mM) caused both apoptosis and cell cycle arrest in G_1 phase, possibly due to the irreversible DNA damage and repair of oxidative DNA lesions. Bare nanoparticles showed significant apoptosis amount at the highest concentration. The damaged occurred because of proteins attachment to the surface of nanoparticles, leading to the formation of protein ''corona'' on the shell of magnetic particles.

5.2.3.2. [3H] Thymidine Incorporation Assay

The [^{3}H] thymidine incorporation method was used by some investigators for cytotoxicity of SPION because it is recognized as a simple and rapid determination of cell proliferation.[119, 120] This is not a direct measure of cell division, as DNA synthesis is not always indicative of mitosis. It is important not to label the cells for longer than 24 hr so that one is not measuring DNA repair due to strand breaks within the nucleus itself. *Kamei et al.* [121] probed the cytotoxicity of Gold/iron-oxide Magnetic Nanoparticles (GoldMAN) with Mouse melanoma cells (B16BL6). Cell viability was maintained at more than 90% for 24 h with GoldMAN concentrations less than 30 µg/ml.

5.2.4. Apoptosis Detection

5.2.4.1. Flow Cytometry

An ubiquitous feature of apoptosis is the breakup of chromatin, resulting in the exposure of numerous 3' OH DNA ends. When the DNA of cells undergoing apoptosis is analyzed by gel electrophoresis, a distinctive ladderlike appearance of DNA pieces with discrete molecular weights is observed. A quick way to assess apoptosis is to compare the mobility of DNA extracted from nonapoptotic cells to that of cells which have been induced to undergo apoptosis, such as comparing DNA mobility of untreated Jurkat cells to the mobility of DNA of camptothecininduced Jurkat cells.[122, 123] To determine cellular apoptosis due to the exposure of SPION, there are many of kits, such as the Apoptosis APO-BRDUTM kit, that are used with dual color flow cytometry and microscopy (Invitrogen Corp., Carlsbad, CA).

This kit provides an easy method of assessing apoptosis; however, the use of the kit requires that the cells under study are first lysed.

The appearance of the 3' OH ends can also be quantified as a measure of apoptosis in whole cells by an alternative method which does not require cell lysis. Such an alternative method for use in mixed-cell populations is called the TUNEL assay (terminal deoxynucleotidyl transferase-mediated dUTP nick end-labeling), also known as the bromodeoxyuridine terminal deoxynucleotidyl transferase assay. As an example of this kit,[39] L929 cells (3×10^6) were placed in a flask and PVA-coated SPION (200 mM/mL) added for 72 h. Identical cultures without added SPION were used as controls. Cells were then fixed with paraformaldehyde in phosphate-buffered saline (PBS), followed by ethanol fixation. The cells were then washed and reacted with the TdT enzyme (terminal deoxynucleotidyl transferase) and Br-dUTP (bromodeoxyuridinetriphosphate) in buffered solution at 37 °C for 60 min. Bromodeoxyuridine was covalently incorporated into the 3' DNA ends during this incubation. Cells were then thoroughly rinsed and incubated with a FITC (fluorescein isothiocyanate) labeled antibody directed to bromodeoxyuridine for 30 min. After washing away unbound antibody, immunostaining with the FITC labeled antibromodeoxyuridine antibody was indicative of the number of free 3' ends. The RNA of the cells was then digested and the total DNA stained by incubation with a solution containing RNase A plus propidium iodide. Staining of cells with propidium iodide allows normalizing FITC staining to the total amount of DNA in the cells. The stained cells were then analyzed by flow cytometry with an argon laser emitting at 488 nm. FITC fluorescence was observed at 520 nm and propidium iodide simultaneously at 623 nm. Flow cytometry result[39] indicates that the cells did not face apoptosis due to SPION exposure even at high molarities of 200 mM. The toxicity of SPION thorough flow cytometry method was reported by some groups.[54, 98, 124, 125]

5.2.4.2. Comet Assay

The single cell gel electrophoresis assay known as the Comet assay is a simple, rapid, and sensitive technique for analyzing and quantifying DNA damage in individual mammalian cells. Initially, there was much discussion about whether the assay could really distinguish apoptosis from necrosis, a topic which has been resolved in recent years with an improved protocol. The Comet assay can be performed by coating a glass slide with agarose and adding the test cells in a separate agarose layer on top.[126] Using this approach, successful magnetic hyperthermia treatment was confirmed after the direct injection of magnetic nanoparticles into an MX-1 tumor and application of an alternating magnetic field (frequency 400 kHz and amplitude 6.5 kA/m).[127] Tumor cells treated with magnetic nanoparticles showed prominent DNA damage visible as distinct comet trails. Oxidative DNA lesions in cultured A549 cells after exposure to 40 μg/ml and 80 μg/ml iron oxide nanoparticles for 4 h, showed statistically significant ($p<0.05$) increased levels compared to those of the control in the highest dose.[108]

Omidkhoda et al.[128] studied the adverse effects of SPION on the Mesenchymal stem cells (MSC) using the comet assay. MSC were grown *in vitro* and labeled with various doses of SPION for various time intervals. Results showed that SPION at various doses (50–250 μg/ml) in conjunction with protamine sulphate, used as a cationic transfection agent; and treatment times of 24–72 h did not statistically affect the frequency of apoptosis in labeled MSC ($p>0.05$).

5.3. IN VIVO TOXICITY

Magnetic nanoparticles can be screened in animals for preliminary *in vivo* toxicity, before treating patients or performing a clinical trial. Promising formulations should be submitted to clinical trial for full toxicological evaluation. SPION has undergone toxicity tests and have received FDA approval for use as MRI contrast agents. Other magnetic micro- and nanospheres were approved for clinical trials of cancer agents, but are not yet fully approved for patient use. Since enhancing MRI contrast is recognized as a one of the most promising applications of SPION, the early toxicity *in vivo* evaluation (i.e. animal tests) were examined with dextran-coated magnetic nanoparticles. The particles were entirely taken up by the reticuloendothelial system (RES), and gathered in spleen and liver. Neither apparent acute and subacute, nor mutagenic toxicity were observed for particle concentration of up to 250 mg of iron/kilogram of rat or 3 mmol Fe/kilogram of dog or rat.[86, 129-131] Considering suitable biocompatibility properties, the SPIONs are recognized as suitable agents for spleen and liver cancer imaging. No evidence of hepatic mitochondrial or microsomal lipid peroxidation or organelle dysfunction was found. It should be noted that the modified, cellular enzyme-linked immunosorbent assay (ELISA), named cellular magnetic-linked immunosorbent assay (C-MALISA), has been developed as an application of MRI for *in vitro* clinical diagnosis.[132] Overall, three contrast agents targeted to integrins were synthesized by grafting to SPION: (a) the CS1 (connecting segment-1) fragment of fibronectin, (b) the peptide GRGD, and (c) a nonpeptidic RGD mimetic. Following cell fixation on ELISA plates, incubation of Jurkat cells and rat mononuclear cells stimulated to activate their integrins with the contrast agents, rinsing, and digestion, the samples were analyzed by MRI. The apparent dissociation constants of the three contrast agents were estimated on the basis of the MRI measurement.[133] Full descriptions on the various applications of SPION on MRI were described elsewhere.[133]

Chertok et al. [134] explored the possibility of utilizing SPION as a drug delivery vehicle for minimally invasive, MRI monitored magnetic targeting of brain tumors. The *in vivo* effect of magnetic targeting on the extent and selectivity of nanoparticle accumulation in tumors of rats, harboring orthotopic 9L-gliosarcomas, was quantified with MRI. Animals were intravenously injected with nanoparticles (12 mg Fe/kg) under a magnetic field density of 0 T (control) or 0.4 T (experimental) applied for 30 min. Following their results, accumulation of iron oxide nanoparticles in gliosarcomas can be significantly enhanced by magnetic targeting and successfully quantified by MR imaging with no toxicity observation. *Yu et al.* [135] reported that the excellent passive tumor targeting efficiency of thermally cross-linked (TCL)-SPION allowed detection of tumors by MR imaging and at the same time delivery of sufficient amounts of anticancer drugs that in turn were released from the nanoparticles to exhibit anticancer activity. Consequently, doxorubicin@TCL-SPION showed exceptional antitumor effects without any systemic toxicity. SPION injection could diminish the accumulation in the liver and lungs, which are well known filters for nanoparticles.[24, 136, 137] These particles seemed to be moderately degraded, started to show up in red blood cells within 6 weeks, and then slowly cleared from the liver within about 3 months, which is confirmed by radiotracer studies with ^{59}Fe.[86, 130, 138] After animal studies, Feridex (Endorem^{TM}) was probed in humans and recognized as safe and efficacious.[139] The most frequent side effect was focused on back pain which was detected in nine patients (4%) and necessitated interruption of the infusion of ferumoxides in five of the patients. Although lumbar pain has been associated with administration of a variety of colloids and emulsions, the physiologic causes are unknown, since no significant changes in chemistry values, vital signs, and electrocardiographic findings were found. Limitations may also arise in extrapolating from animal models to humans. There are many physiological parameters to consider, ranging from variations in weight, blood volume, cardiac output, and circulation time to tumor volume/location/blood flow, complicating the extrapolation of data obtained in animal models.[140, 141] Low toxicity of SPION coated with dextran was observed by morphological studies. Minor variations in histology of both spleen and liver were observed at higher concentrations (i.e. 200× higher doses than that of used for MR imaging).[142] Ultra small magnetic nanoparticles have been shown to be effective, during MRI, in identification of lymph node metastases, thoracic and abdominal vasculature and of coronary arteries in rats and pigs, respectively.[143-145] Nanoparticles showed a good acceptance and safety profile with an LD_{50}> 17.9 mmol Fe/kg. Toxicity was observed only at very high exposure levels due to the massive iron overload after repeated injections. The contrast agent was not mutagenic but teratogenic in rats and rabbits. Regarding to its single-dose diagnostic agent for human MR imaging of lymph nodes, these contrast agents would be compatible. Another clinical application of ultra small magnetic nanoparticles is their usage in improving the delineation of brain tumor boundaries and quantify tumor volumes.[146, 147] Histology studies showed no pathological brain cell or myelin changes, making these particles (ferumoxtran-10) candidates for brain imaging contrast agents.[148] Various SPIONs as MRI contrast agents have been also used for magnetic cell labeling *ex-vivo*. The magnetically labeled cells were re-injected into patients for cellular imaging of cell therapy following transplantation, transfusion, or gene therapy.[103, 149] The non-toxicity of the approach was determined beforehand by comprehensive *in vitro* measurements such as the iron uptake of the cells, MTT and annexin V assays.[30] Magnetic cellular labeling with the magnetic nanoparticle complex had no short or long-term toxic effects on tumor or stem cells. For stem cell therapy, the extracted stem cells must be purified from tumor cells before reinjection. A very efficient method to do this is magnetic separation. The non-toxicity of the magnetic particles used for this separation approach was evaluated by at least one group.[150] They compared the well established Dynabeads to 16 different small magnetic nanoparticles. In their test, porous silica beads of 250 nm size fared best in terms of yield, purity, and viability of isolated epithelial tumor cells. *Lübbe et al.* [151] observed no obvious toxicity due to the intravenous injection of magnetic nanoparticles (diameter of 100 nm) into mice. The critical matter which would be important in toxicity response is the concentration of ferrofluid; therefore, due to their acute iron overload[152], high amounts of SPION showed toxic effect[153], however the lower amount of the same material was fully biocompatible.[154]

De Vries et al. [149] showed that *in vivo* magnetic resonance tracking of magnetically labeled cells is feasible in human for detecting very low numbers of dendritic cells in conjunction with detailed anatomical information. Autologous dendritic cells were labeled with a clinical superparamagnetic iron oxide formulation or In-oxine[111] and were co-injected intranodally in melanoma patients under ultrasound guidance. In contrast to scintigraphic imaging, magnetic resonance imaging (MRI) allowed assessment of the accuracy of dendritic cell delivery and of inter- and intra-nodal cell migration patterns. Following their results, MRI cell tracking using iron oxide appears clinically safe and well suited to monitor cellular therapy in humans.

Farrell et al.[68] demonstrated for the first time the effects of SPION labeling on chondrocyte behavior, illustrated its potential for *in vivo* tracking of implanted chondrocytes. The SPION labeling did not significantly affect any of parameters relative to unlabelled controls (i.e. cell proliferation). They also demonstrated SPION retention within the cells for the full duration of the experiment. Comprehensive studies for the application of SPION on MRI, drug delivery and cancer were presented as reviews elsewhere.[155-157] Finally, the last form of toxicity evaluation which is becoming increasingly popular is computer simulations processing.[158] Although this method is not routinely integrated in toxicology assessments, it is becoming a useful technique of looking at the toxicity of drugs even before their synthesis during drug discovery. *Dames et al.*[159] showed theoretically by computer-aided simulation, and for the first time experimentally in mice, that targeted aerosol delivery to the lung can be achieved with aerosol droplets comprising Superparamagnetic iron oxide nanoparticles in combination with a target-directed magnetic gradient field. They suggested that nanomagnetosols may be useful for treating localized lung disease, by targeting foci of bacterial infection or tumor nodules.

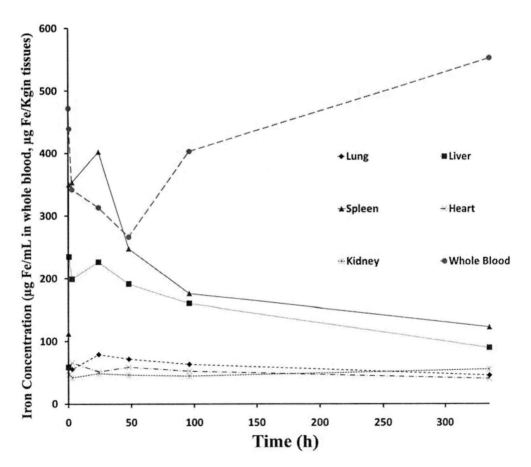

Figure 5.4. Iron concentration in different tissues at different time points after intravenous injection of SPION (data given from the table in Ref [163]).

5.4. PHARMACOKINETICS AND BIODISTRIBUTION

In order to commercialize SPION for human *in vivo* applications, the toxicity of particles should be explored by comprehensive *in vitro* cytotoxicity assays. After apoptosis, assays will give us some indication of toxicity, the toxic effects could be determined (which cell function is damaged). If the above tests show only minor or no effects in the concentrations to be used, then the SPION could be forwarded for some *in vivo* (animal) tests in a relevant (similar to human application) model. If that looks good, then they could be introduced to the FDA for approval for first human clinical trials. Two kind of SPIONs, consisting Ferumoxides (i.e. Endorem® in Europe, Feridex® in the USA and Japan, Advanced Magnetics, USA) coated with dextran, and Ferucarbutran (i.e. Resovist® in Europe and Japan, Schering, Germany) coated with carboxy dextran, have already been approved for clinical usage. These particles approved for application in medical imaging as a contrast agent.

In order to determine the biodistribution, clearance and biocompatibility of SPION (oleic acid-Pluronic-coated) and to ensure their safe clinical use, Jain *et al.*[160] injected the nanoparticles (at dose of 10 mg Fe/kg) to rats through the tail vein. They studied the changes in serum and tissue iron levels over 3 weeks after intravenous administration of nanoparticles to rats. Serum alanine aminotransferase (ALT), aspartate aminotransferase (AST), alkaline phosphatase (AKP) levels and total iron-binding capacity (TIBC) were also measured with time to assess the effect of SPION on liver function. Selected tissues were also analyzed for oxidative stress and studied histologically to determine biocompatibility of SPION. Results showed that serum iron levels gradually increased for up to 1 week but levels slowly declined thereafter. Biodistribution of iron in various body tissues changed with time but greater fraction of the injected iron localized in the liver and spleen than in the brain, heart, kidney, and lung. Magnetization measurements of the liver and spleen samples showed a steady decrease over 3 weeks; the authors claimed that increasing magnetization amounts is related to particle degradation. Serum showed a transient increase in ALT, AST, AKP levels, and TIBC over a period of 6-24 h following SPION injection. The increase in oxidative stress was tissue dependent, reaching a peak at about 3 days and then slowly declining thereafter. Histological analyses of liver, spleen, and kidney samples collected at 1 and 7 days showed no apparent abnormal changes. Similar investigation reported by *Ma et al.* [161-163]. Alginate coated nanoparticles with concentration of 109.5 µmol Fe/Kg have been injected to rats and their both biodistribution and pharmacokinetics were probed. Figure 5.4 shows the average iron content in the different organs.

REFERENCES

[1] Berkovski B, Bashtovoy V. *Magnetic fluids and applications, Handbook.* Begell House, Inc., New York. NY. 1996.
[2] Wang YX, Hussain SM, Krestin GP. *Eur. Radiol.* 2001;11: 2319-2331.
[3] DeRosa L, Montuoro A, Pandolfi A, Lanti T, Pescador L, Morara R, DeLaurenzi A. *Haematologica* 1991;76: 37-40.
[4] Häfeli US, W.; Teller, J.; Zborowski, M. *Plenum:* New York. 1997.
[5] Mornet S, Vasseur S, Grasset F, Duguet E. Magnetic nanoparticle design for medical diagnosis and therapy. *Journal of Materials Chemistry* 2004;14: 2161-2175.

[6] Park W, Owens JM. Future Directions in the Treatment of Oral Cancer. *Otolaryngologic Clinics of North America* 2006;39: 381-396.

[7] Yang J, Gunn J, Dave SR, Zhang M, Wang YA, Gao X. Ultrasensitive detection and molecular imaging with magnetic nanoparticles. *Analyst* 2008;133: 154-160.

[8] Corr SA, Rakovich YP, Gun'Ko YK. Multifunctional magnetic-fluorescent nanocomposites for biomedical applications. *Nanoscale Research Letters* 2008;3: 87-104.

[9] Zaitsev VS, Filimonov DS, Presnyakov IA, Gambino RJ, Chu B. Physical and chemical properties of magnetite and magnetite-polymer nanoparticles and their colloidal dispersions. *Journal of Colloid and Interface Science* 1999;212: 49-57.

[10] Mahmoudi M, Simchi A, Imani M, Stroeve P, Sohrabi A. *Templated growth of superparamagnetic iron oxide nanoparticles by temperature programming in the presence of poly(vinyl alcohol).* Thin Solid Films 2010;in press.

[11] Mahmoudi M, Milani AS, Stroeve P. Synthesis and Surface Architecture of Superparamagnetic Iron Oxide Nanoparticles for Application in Drug Delivery and Their Biological Response: *A Review. International Journal of Biomedical Nanoscience and Nanotechnology 2010*;in press.

[12] Mahmoudi M, Simchi A, Imani M, Milani AS, Stroeve P. Optimal Design and Characterization of Superparamagnetic Iron Oxide Nanoparticles Coated with Polyvinyl Alcohol for Targeted Delivery and Imaging. *Journal of Physical Chemistry B* 2008;112: 14470-14481.

[13] Bulte JWM, Kraitchman DL. Monitoring cell therapy using iron oxide MR contrast agents. *Current Pharmaceutical Biotechnology* 2004;5: 567-584.

[14] Bulte JWM, Kraitchman DL. Iron oxide MR contrast agents for molecular and cellular imaging. *NMR in Biomedicine* 2004;17: 484-499.

[15] LaConte L, Nitin N, Bao G. Magnetic nanoparticle probes. *Materials Today* 2005;8: 32-38.

[16] LaConte LEW, Nitin N, Zurkiya O, Caruntu D, O'Connor CJ, Hu X, Bao G. Coating thickness of magnetic iron oxide nanoparticles affects R 2 relaxivity. *Journal of Magnetic Resonance* Imaging 2007;26: 1634-1641.

[17] LaConte LEW, Nitin N, Zurkiya O, Hu X, Bao G. Engineering magnetic nanoparticle-based MRI contrast agents for molecular imaging. In: *Proceedings of the 2005 Summer Bioengineering Conference;* 2005. p. 651-652.

[18] Corot C, Robert P, Ideجبe JM, Port M. Recent advances in iron oxide nanocrystal technology for medical imaging. *Advanced Drug Delivery Reviews* 2006;58: 1471-1504.

[19] Thorek DLJ, Chen AK, Czupryna J, Tsourkas A. Superparamagnetic iron oxide nanoparticle probes for molecular imaging. *Annals of Biomedical Engineering* 2006;34: 23-38.

[20] Gupta AK, Naregalkar RR, Vaidya VD, Gupta M. *Recent advances on surface engineering of magnetic iron oxide nanoparticles and their biomedical applications. Nanomedicine* 2007;2: 23-39.

[21] Häfeli UO, Aue J, Damani J. *The biocompatibility and toxicity of magnetic particles.* In *Magnetic cell separation* (Zborowski, M. and Chalmers, J. J., eds.), pp. 163-223, Elsevier, Amsterdam 2007.

[22] Lewinski N, Colvin V, Drezek R. Cytotoxicity of nanopartides. *Small* 2008;4: 26-49.

[23] Mahmoudi M, Hofmann H, A. F. ??? 2010.
[24] Veranth JM, Kaser EG, Veranth MM, Koch M, Yost GS. Cytokine responses of human lung cells (BEAS-2B) treated with micron-sized and nanoparticles of metal oxides compared to soil dusts. *Particle and Fibre Toxicology* 2007;4.
[25] Ciapetti G, Granchi D, Arciola CR, Cenni E, Savarino L, Stea S, Montanaro L, Pizzoferrato A. In *vitro testing of cytotoxicity of materials. Biomaterials and Bioengineering Handbook* Marcel Dekker Inc., NewYork, NY 2000: 179-98.
[26] Pieters R, Huismans DR, Leyva A, Veerman AJP. Comparison of the rapid automated MTT-assay with a dye exclusion assay for chemosensitivity testing in childhood leukaemia. *British Journal of Cancer* 1989;59: 217-220.
[27] Devineni D, Klein-Szanto A, Gallo JM. Tissue distribution of methotrexate following administration as a solution and as a magnetic microsphere conjugate in rats bearing brain tumors. *Journal of Neuro-Oncology* 1995;24: 143-152.
[28] Gupta AK, Berry C, Gupta M, Curtis A. Receptor-mediated targeting of magnetic nanoparticles using insulin as a surface ligand to prevent endocytosis. *IEEE Transactions on Nanobioscience* 2003;2: 255-261.
[29] Gupta AK, Wells S. Surface-Modified Superparamagnetic Nanoparticles for Drug Delivery: Preparation, Characterization, and Cytotoxicity Studies. *IEEE Transactions on Nanobioscience* 2004;3: 66-73.
[30] Arbab AS, Bashaw LA, Miller BR, Jordan EK, Lewis BK, Kalish H, Frank JA. Characterization of biophysical and metabolic properties of cells labeled with superparamagnetic iron oxide nanoparticles and transfection agent for cellular MR imaging. *Radiology* 2003;229: 838-846.
[31] Gupta AK, Curtis ASG. Lactoferrin and ceruloplasmin derivatized superparamagnetic iron oxide nanoparticles for targeting cell surface receptors. *Biomaterials* 2004;25: 3029-3040.
[32] Fischer D, Li Y, Ahlemeyer B, Krieglstein J, Kissel T. In vitro cytotoxicity testing of polycations: *Influence of polymer structure on cell viability and hemolysis. Biomaterials* 2003;24: 1121-1131.
[33] Petri-Fink A, Chastellain M, Juillerat-Jeanneret L, Ferrari A, Hofmann H. Development of functionalized superparamagnetic iron oxide nanoparticles for interaction with human cancer cells. *Biomaterials* 2005;26: 2685-2694.
[34] Cengelli F, Maysinger D, Tschudi-Monnet F, Montet X, Corot C, Petri-Fink A, Hofmann H, Juillerat-Jeanneret L. Interaction of functionalized superparamagnetic iron oxide nanoparticles with brain structures. *Journal of Pharmacology and Experimental Therapeutics* 2006;318: 108-116.
[35] Mahmoudi M, Simchi A, Imani M, Hafeli UO. Superparamagnetic iron oxide nanoparticles with rigid cross-linked polyethylene glycol fumarate coating for application in imaging and drug delivery. *Journal of Physical Chemistry C* 2009;113: 8124-8131.
[36] Hussain SM, Hess KL, Gearhart JM, Geiss KT, Schlager JJ. In vitro toxicity of nanoparticles in BRL 3A rat liver cells. *Toxicology in Vitro* 2005;19: 975-983.
[37] Cheng FY, Su CH, Yang YS, Yeh CS, Tsai CY, Wu CL, Wu MT, Shieh DB. Characterization of aqueous dispersions of Fe3O4 nanoparticles and their biomedical applications. *Biomaterials* 2005;26: 729-738.

[38] Hafeli UO, Pauer GJ. In vitro and in vivo toxicity of magnetic microspheres. *Journal of Magnetism and Magnetic Materials* 1999;194: 76-82.

[39] Mahmoudi M, Shokrgozar MA, Simchi A, Imani M, Milani AS, Stroeve P, Vali H, Hafeli UO, Bonakdar S. Multiphysics Flow Modeling and in Vitro Toxicity of Iron Oxide Nanoparticles Coated with Poly(vinyl alcohol). *Journal of Physical Chemistry C* 2009;113: 2322-2331.

[40] Mahmoudi M, Simchi A, Imani M, Milani AS, Stroeve P. An in vitro study of bare and poly(ethylene glycol)-co-fumarate-coated superparamagnetic iron oxide nanoparticles: a new toxicity identification procedure. *Nanotechnology* 2009;20.

[41] Brunner TJ, Wick P, Manser P, Spohn P, Grass RN, Limbach LK, Bruinink A, Stark WJ. In vitro cytotoxicity of oxide nanoparticles: Comparison to asbestos, silica, and the effect of particle solubility. *Environmental Science and Technology* 2006;40: 4374-4381.

[42] Lee H, Lee E, Kim DK, Jang NK, Jeong YY, Jon S. Antibiofouling polymer-coated superparamagnetic iron oxide nanoparticles as potential magnetic resonance contrast agents for in vivo cancer imaging. *Journal of the American Chemical Society* 2006;128: 7383-7389.

[43] Raty JK, Liimatainen T, Wirth T, Airenne KJ, Ihalainen TO, Huhtala T, Hamerlynck E, Vihinen-Ranta M, Närvänen A, Ylä-Herttuala S, Hakumäki JM. Magnetic resonance imaging of viral particle biodistribution in vivo. *Gene Therapy* 2006;13: 1440-1446.

[44] Kim JS, Yoon TJ, Yu KN, Mi SN, Woo M, Kim BG, Lee KH, Sohn BH, Park SB, Lee JK, Cho MH. Cellular uptake of magnetic nanoparticle is mediated through energy-dependent endocytosis in A549 cells. *Journal of Veterinary Science* 2006;7: 321-326.

[45] Muller K, Skepper JN, Posfai M, Trivedi R, Howarth S, Corot C, Lancelot E, Thompson PW, Brown AP, Gillard JH. Effect of ultrasmall superparamagnetic iron oxide nanoparticles (Ferumoxtran-10) on human monocyte-macrophages in vitro. *Biomaterials* 2007;28: 1629-1642.

[46] Mykhaylyk O, Antequera YS, Vlaskou D, Plank C. Generation of magnetic nonviral gene transfer agents and magnetofection in vitro. *Nature protocols* 2007;2: 2391-2411.

[47] Lee KJ, An JH, Shin JS, Kim DH, Kim C, Ozaki H, Koh JG. Protective effect of maghemite nanoparticles on ultraviolet-induced photo-damage in human skin fibroblasts. *Nanotechnology* 2007;18.

[48] Martin AL, Bernas LM, Rutt BK, Foster PJ, Gillies ER. Enhanced cell uptake of superparamagnetic iron oxide nanoparticles functionalized with dendritic guanidines. *Bioconjugate Chemistry 2008*;19: 2375-2384.

[49] Lv G, He F, Wang X, Gao F, Zhang G, Wang T, Jiang H, Wu C, Guo D, Li X, Chen B, Gu Z. Novel nanocomposite of nano Fe3O4 and polylactide nanofibers for application in drug uptake and induction of cell death of leukemia cancer cells. *Langmuir* 2008;24: 2151-2156.

[50] Huang G, Diakur J, Xu Z, Wiebe LI. Asialoglycoprotein receptor-targeted superparamagnetic iron oxide nanoparticles. *International Journal of Pharmaceutics* 2008;360: 197-203.

[51] Chen BA, Dai YY, Wang XM, Zhang RY, Xu WL, Shen HL, Gao F, Sun Q, Deng XJ, Ding JH, Gao C, Sun YY, Cheng J, Wang J, Zhao G, Chen NN. Synergistic effect of the combination of nanoparticulate Fe3O4 and Au with daunomycin on K562/A02 cells. *International Journal of Nanomedicine* 2008;3: 343-350.

[52] Wang AZ, Bagalkot V, Vasilliou CC, Gu F, Alexis F, Zhang L, Shaikh M, Yuet K, Cima MJ, Langer R, Kantoff PW, Bander NH, Jon S, Farokhzad OC. Superparamagnetic iron oxide nanoparticle-aptamer bioconjugates for combined prostate cancer imaging and therapy. *ChemMedChem* 2008;3: 1311-1315.

[53] Horie M, Nishio K, Fujita K, Kato H, Nakamura A, Kinugasa S, Endoh S, Miyauchi A, Yamamoto K, Murayama H, Niki E, Iwahashi H, Yoshida Y, Nakanishi J. Ultrafine NiO particles induce cytotoxicity in vitro by cellular uptake and subsequent Ni(II) release. *Chemical Research in Toxicology* 2009;22: 1415-1426.

[54] Villanueva A, Cĺete M, Roca AG, Calero M, Veintemillas-Verdaguer S, Serna CJ, Del Puerto Morales M, Miranda R. The influence of surface functionalization on the enhanced internalization of magnetic nanoparticles in cancer cells. *Nanotechnology* 2009;20.

[55] Ge Y, Zhang Y, He S, Nie F, Teng G, Gu N. Fluorescence modified chitosan-coated magnetic nanoparticles for high-efficient cellular imaging. *Nanoscale Research Letters* 2009;4: 287-295.

[56] Mahmoudi M, Simchi A, Imani M, Shokrgozar MA, Milani AS, Hafeli UO, Stroeve P. A new approach for the in vitro identification of the cytotoxicity of superparamagnetic iron oxide nanoparticles. *Colloids and Surfaces B:* Biointerfaces, in press, doi:10.1016/j.colsurfb.2009.08.044 2009.

[57] Mahmoudi M, Simchi A, Imani M, Shokrgozar MA, Milani AS, Hafeli U, Stroeve P. A new approach for the in vitro identification of the cytotoxicity of superparamagnetic iron oxide nanoparticles. *Colloids and Surfaces B: Biointerfaces* 2010;75: 300-309.

[58] Worle-Knirsch JM, Pulskamp K, Krug HF. Oops They Did It Again! Carbon Nanotubes Hoax Scientists in Viability Assays. *Nanoletters* 2006;6: 1261-1268.

[59] Huang JH, Parab HJ, Liu RS, Lai TC, Hsiao M, Chen CH, Sheu HS, Chen JM, Tsai DP, Hwu YK. Investigation of the growth mechanism of iron oxide nanoparticles via a seed-mediated method and its cytotoxicity studies. *Journal of Physical Chemistry C* 2008;112: 15684-15690.

[60] Delcroix GJR, Jacquart M, Lemaire L, Sindji L, Franconi F, Le Jeune JJ, Montero-Menei CN. Mesenchymal and neural stem cells labeled with HEDP-coated SPIO nanoparticles: In vitro characterization and migration potential in rat brain. *Brain Research* 2009;1255: 18-31.

[61] Pawelczyk E, Arbab AS, Chaudhry A, Balakumaran A, Robey PG, Frank JA. In vitro model of bromodeoxyuridine or iron oxide nanoparticle uptake by activated macrophages from labeled stem cells: Implications for cellular therapy. *Stem Cells* 2008;26: 1366-1375.

[62] Walczak P, Ruiz-Cabello J, Kedziorek DA, Gilad AA, Lin S, Barnett B, Qin L, Levitsky H, Bulte JWM. Magnetoelectroporation: improved labeling of neural stem cells and leukocytes for cellular magnetic resonance imaging using a single FDA-approved agent. *Nanomedicine: Nanotechnology, Biology, and Medicine* 2006;2: 89-94.

[63] Au C, Mutkus L, Dobson A, Riffle J, Lalli J, Aschner M. Effects of nanoparticles on the adhesion and cell viability on astrocytes. *Biological Trace Element Research* 2007;120: 248-256.

[64] Wan S, Huang J, Guo M, Zhang H, Cao Y, Yan H, Liu K. Biocompatible superparamagnetic iron oxide nanoparticle dispersions stabilized with poly(ethylene

glycol)-oligo(aspartic acid) hybrids. *Journal of Biomedical Materials Research - Part A* 2007;80: 946-954.

[65] Lee CM, Jeong HJ, Kim EM, Cheong SJ, Park EH, Kim DW, Lim ST, Sohn MH. Synthesis and characterization of iron oxide nanoparticles decorated with carboxymethyl curdlan. *Macromolecular Research* 2009;17: 133-136.

[66] Lee CM, Jeong HJ, Kim SL, Kim EM, Kim DW, Lim ST, Jang KY, Jeong YY, Nah JW, Sohn MH. SPION-loaded chitosan-linoleic acid nanoparticles to target hepatocytes. *International Journal of Pharmaceutics* 2009;371: 163-169.

[67] Shi X, Thomas TP, Myc LA, Kotlyar A, Baker Jr JR. Synthesis, characterization, and intracellular uptake of carboxyl-terminated poly(amidoamine) dendrimer-stabilized iron oxide nanoparticles. *Physical Chemistry Chemical Physics* 2007;9: 5712-5720.

[68] Farrell E, Wielopolski P, Pavljasevic P, Kops N, Weinans H, Bernsen MR, van Osch GJVM. Cell labelling with superparamagnetic iron oxide has no effect on chondrocyte behaviour. *Osteoarthritis and Cartilage* 2009;17: 958-964.

[69] Arbab AS, Yocum GT, Wilson LB, Parwana A, Jordan EK, Kalish H, Frank JA. Comparison of transfection agents in forming complexes with ferumoxides, cell labeling efficiency, and cellular viability. *Molecular Imaging* 2004;3: 24-32.

[70] Polyak B, Fishbein I, Chorny M, Alferiev I, Williams D, Yellen B, Friedman G, Levy RJ. High field gradient targeting of magnetic nanoparticle-loaded endothelial cells to the surfaces of steel stents. *Proceedings of the National Academy of Sciences of the United States of America* 2008;105: 698-703.

[71] Chorny M, Polyak B, Alferiev IS, Walsh K, Friedman G, Levy RJ. Magnetically driven plasmid DNA delivery with biodegradable polymeric nanoparticles. *FASEB Journal* 2007;21: 2510-2519.

[72] Samanta B, Yan H, Fischer NO, Shi J, Jerry DJ, Rotello VM. Protein-passivated Fe3O4 nanoparticles: Low toxicity and rapid heating for thermal therapy. *Journal of Materials Chemistry* 2008;18: 1204-1208.

[73] Cree IA, Andreotti PE. Measurement of cytotoxicity by ATP-based luminescence assay in primary cell cultures and cell lines. *Toxicology in Vitro 1997*;11: 553-556.

[74] Weyermann J, Lochmann D, Zimmer A. A practical note on the use of cytotoxicity assays. *International Journal of Pharmaceutics* 2005;288: 369-376.

[75] Eirheim HU, Bundgaard C, Nielsen HM. Evaluation of different toxicity assays applied to proliferating cells and to stratified epithelium in relation to permeability enhancement with glycocholate. *Toxicology in Vitro* 2004;18: 649-657.

[76] Cheng CM, Chu PY, Chuang KH, Roffler SR, Kao CH, Tseng WL, Shiea J, Chang WD, Su YC, Chen BM, Wang YM, Cheng TL. Hapten-derivatized nanoparticle targeting and imaging of gene expression by multimodality imaging systems. *Cancer Gene Therapy* 2009;16: 83-90.

[77] Simko M, Hartwig C, Lantow M, Lupke M, Mattsson MO, Rahman Q, Rollwitz J. Hsp70 expression and free radical release after exposure to non-thermal radio-frequency electromagnetic fields and ultrafine particles in human Mono Mac 6 cells. *Toxicology Letters* 2006;161: 73-82.

[78] Krotz F, de Wit C, Sohn HY, Zahler S, Gloe T, Pohl U, Plank C. Magnetofection - A highly efficient tool for antisense oligonucleotide delivery in vitro and in vivo. *Molecular Therapy* 2003;7: 700-710.

[79] Rook GAW, Steele J, Umar S, Dockrell HM. A simple method for the solubilisation of reduced NBT, and its use as a colorimetric assay for activation of human macrophages by ³ζ-interferon. *Journal of Immunological Methods* 1985;82: 161-167.

[80] Muller K, Skepper JN, Tang TY, Graves MJ, Patterson AJ, Corot C, Lancelot E, Thompson PW, Brown AP, Gillard JH. Atorvastatin and uptake of ultrasmall superparamagnetic iron oxide nanoparticles (Ferumoxtran-10) in human monocyte-macrophages: Implications for magnetic resonance imaging. *Biomaterials* 2008;29: 2656-2662.

[81] Babic M, Horok D, Jendelova ‹P, Glogarov ‹ⁱK, Herynek V, Trchova M, Likavoanov ‹ⁱ K, Lesny P, Pollert E, Hajek M, Sykova E. Poly(N,N-dimethylacrylamide)-coated maghemite nanoparticles for stem cell labeling. *Bioconjugate Chemistry* 2009;20: 283-294.

[82] Chen G, Chen W, Wu Z, Yuan R, Li H, Gao J, Shuai X. MRI-visible polymeric vector bearing CD3 single chain antibody for gene delivery to T cells for immunosuppression. *Biomaterials* 2009;30: 1962-1970.

[83] Terrovitis J, Stuber M, Youssef A, Preece S, Leppo M, Kizana E, Schaζˆr M, Gerstenblith G, Weiss RG, Marbaجﻪn E, Abraham MR. Magnetic resonance imaging overestimates ferumoxide-labeled stem cell survival after transplantation in the heart. *Circulation* 2008;117: 1555-1562.

[84] Fotakis G, Timbrell JA. In vitro cytotoxicity assays: Comparison of LDH, neutral red, MTT and protein assay in hepatoma cell lines following exposure to cadmium chloride. *Toxicology Letters* 2006;160: 171-177.

[85] Olbrich C, Bakowsky U, Lehr CM, Muζˆller RH, Kneuer C. Cationic solid-lipid nanoparticles can efficiently bind and transfect plasmid DNA. *Journal of Controlled Release* 2001;77: 345-355.

[86] Weissleder R, Stark DD, Engelstad BL, Bacon BR, Compton CC, White DL, Jacobs P, Lewis J. Superparamagnetic iron oxide: Pharmacokinetics and toxicity. *American Journal of Roentgenology* 1989;152: 167-173.

[87] Jang M, Ghio AJ, Cao G. Exposure of BEAS-2B cells to secondary organic aerosol coated on magnetic nanoparticles. *Chemical Research in Toxicology* 2006;19: 1044-1050.

[88] Boutry S, Brunin S, Mahieu I, Laurent S, Vander Elst L, Muller RN. Magnetic labeling of non-phagocytic adherent cells with iron oxide nanoparticles: A comprehensive study. *Contrast Media and Molecular Imaging* 2008;3: 223-232.

[89] Apopa PL, Qian Y, Shao R, Guo NL, Schwegler-Berry D, Pacurari M, Porter D, Shi X, Vallyathan V, Castranova V, Flynn DC. Iron oxide nanoparticles induce human microvascular endothelial cell permeability through reactive oxygen species production and microtubule remodeling. *Particle and Fibre Toxicology* 2009;6.

[90] Pettibone JM, Adamcakova-Dodd A, Thorne PS, O'Shaughnessy PT, Weydert JA, Grassian VH. Inflammatory response of mice following inhalation exposure to iron and copper nanoparticles. *Nanotoxicology* 2008;2: 189-204.

[91] Lehmann J, Natarajan A, DeNardo GL, Ivkov R, Foreman AR, Catapano C, Mirick G, Quang T, Gruettner C, DeNardo SJ. Short communication: Nanoparticle thermotherapy and external beam radiation therapy for human prostate cancer cells. *Cancer Biotherapy and Radiopharmaceuticals* 2008;23: 265-271.

[92] Wang TW, Wu HC, Wang WR, Lin FH, Lou PJ, Shieh MJ, Young TH. The development of magnetic degradable DP-Bioglass for hyperthermia cancer therapy. *Journal of Biomedical Materials Research - Part A* 2007;83: 828-837.

[93] Terrovitis JV, Bulte JWM, Sarvananthan S, Crowe LA, Sarathchandra P, Batten P, Sachlos E, Chester AH, Czernuszka JT, Firmin DN, Taylor PM, Yacoub MH. Magnetic resonance imaging of ferumoxide-labeled mesenchymal stem cells seeded on collagen scaffolds - Relevance to tissue engineering. *Tissue Engineering* 2006;12: 2765-2775.

[94] Yang A, Cardona DL, Barile FA. In vitro cytotoxicity testing with fluorescence-based assays in cultured human lung and dermal cells. *Cell Biology and Toxicology* 2002;18: 97-108.

[95] De Clerck LS, Bridts CH, Mertens AM, Moens MM, Stevens WJ. Use of fluorescent dyes in the determination of adherence of human leucocytes to endothelial cells and the effect of fluorochromes on cellular function. *Journal of Immunological Methods* 1994;172: 115-124.

[96] Papadopoulos NG, Dedoussis GVZ, Spanakos G, Gritzapis AD, Baxevanis CN, Papamichail M. An improved fluorescence assay for the determination of lymphocyte-mediated cytotoxicity using flow cytometry. *Journal of Immunological Methods* 1994;177: 101-111.

[97] Xiu Ming W, Terasaki PI, Rankin Jr GW, Chia D, Hui Ping Z, Hardy S. A new microcellular cytotoxicity test based on calcein AM release. *Human Immunology* 1993;37: 264-270.

[98] Haugland RP, MacCoubrey IC, Moore PL. Dual-fluorescence cell viability assay using ethidium homodimer and calcein AM. In: *United States patent and trademark office granted patent;* 1994.

[99] Berkova Z, Kriz J, Girman P, Zacharovova K, Koblas T, Dovolilova E, Saudek F. Vitality of pancreatic islets labeled for magnetic resonance imaging with iron particles. *Transplantation Proceedings* 2005;37: 3496-3498.

[100] Hasnat MA, Uddin MM, Samed AJF, Alam SS, Hossain S. Adsorption and photocatalytic decolorization of a synthetic dye erythrosine on anatase TiO_2 and ZnO surfaces. *Journal of Hazardous Materials* 2007;147: 471-477.

[101] Toso C, Vallee JP, Morel P, Ris F, Demuylder-Mischler S, Lepetit-Coiffe M, Marangon N, Saudek F, James Shapiro AM, Bosco D, Berney T. Clinical magnetic resonance imaging of pancreatic islet grafts after iron nanoparticle labeling. *American Journal of Transplantation* 2008;8: 701-706.

[102] Wang SH, Shi X, Van Antwerp M, Cao Z, Swanson SD, Bi X, Baker Jr JR. Dendrimer-functionalized iron oxide nanoparticles for specific targeting and imaging of cancer cells. *Advanced Functional Materials* 2007;17: 3043-3050.

[103] Seo WS, Lee JH, Sun X, Suzuki Y, Mann D, Liu Z, Terashima M, Yang PC, McConnell MV, Nishimura DG, Dai H. FeCo/graphitic-shell nanocrystals as advanced magnetic-resonance-imaging and near-infrared agents. *Nature Materials* 2006;5: 971-976.

[104] Song EQ, Wang GP, Xie HY, Zhang ZL, Hu J, Peng J, Wu DC, Shi YB, Pang DW. Visual recognition and efficient isolation of apoptotic cells with fluorescent-magnetic-biotargeting multifunctional nanospheres. *Clinical Chemistry* 2007;53: 2177-2185.

[105] Dunning MD, Lakatos A, Loizou L, Kettunen M, Ffrench-Constant C, Brindle KM, Franklin RJM. Superparamagnetic iron oxide-labeled schwann cells and olfactory ensheathing cells can be traced in vivo by magnetic resonance imaging and retain functional properties after transplantation into the CNS. *Journal of Neuroscience* 2004;24: 9799-9810.

[106] Evgenov NV, Medarova Z, Pratt J, Pantazopoulos P, Leyting S, Bonner-Weir S, Moore A. In vivo imaging of immune rejection in transplanted pancreatic islets. *Diabetes* 2006;55: 2419-2428.

[107] Wu YJ, Muldoon LL, Varallyay C, Markwardt S, Jones RE, Neuwelt EA. In vivo leukocyte labeling with intravenous ferumoxides/protamine sulfate complex and in vitro characterization for cellular magnetic resonance imaging. *American Journal of Physiology - Cell Physiology* 2007;293: C1698-C1708.

[108] Reichardt W, Dürr C, von Elverfeldt D, Jüttner E, Gerlach UV, Yamada M, Smith B, Negrin RS, Zeiser R. Impact of mammalian target of rapamycin inhibition on lymphoid homing and tolerogenic function of nanoparticle-labeled dendritic cells following allogeneic hematopoietic cell transplantation. *Journal of immunology* (Baltimore, Md.: 1950) 2008;181: 4770-4779.

[109] Zhang Z, Van Den Bos EJ, Wielopolski PA, De Jong-Popijus M, Bernsen MR, Duncker DJ, Krestin GP. In vitro imaging of single living human umbilical vein endothelial cells with a clinical 3.0-T MRI scanner. *Magnetic Resonance Materials in Physics, Biology and Medicine* 2005;18: 175-185.

[110] Zurkiya O, Chan AWS, Hu X. MagA is sufficient for producing magnetic nanoparticles in mammalian cells, making it an MRI reporter. *Magnetic Resonance in Medicine* 2008;59: 1225-1231.

[111] Berry CC, Wells S, Charles S, Curtis ASG. Dextran and albumin derivatised iron oxide nanoparticles: Influence on fibroblasts in vitro. Biomaterials 2003;24: 4551-4557.

[112] Lappalainen K, Jaaskelainen I, Syrjanen K, Urtti A, Syrjanen S. Comparison of cell proliferation and toxicity assays using two cationic liposomes. *Pharmaceutical Research* 1994;11: 1127-1131.

[113] Karlsson HL, Cronholm P, Gustafsson J, Möller L. Copper oxide nanoparticles are highly toxic: A comparison between metal oxide nanoparticles and carbon nanotubes. *Chemical Research in Toxicology* 2008;21: 1726-1732.

[114] Pittet MJ, Swirski FK, Reynolds F, Josephson L, Weissleder R. Labeling of immune cells for in vivo imaging using magnetofluorescent nanoparticles. *Nature protocols* 2006;1: 73-79.

[115] Gojova A, Guo B, Kota RS, Rutledge JC, Kennedy IM, Barakat AI. Induction of inflammation in vascular endothelial cells by metal oxide nanoparticles: Effect of particle composition. *Environmental Health Perspectives* 2007;115: 403-409.

[116] Matuszewski L, Persigehl T, Wall A, Sehwindt W, Tombach B, Fobker M, Poremba C, Ebert W, Heindel W, Bremer C. Cell tagging with clinically approved iron oxides: Feasibility and effect of lipofection, particle size, and surface coating on labeling efficiency. *Radiology* 2005;235: 155-161.

[117] Mahmoudi M, Simchi A, Vali H, Imani M, Shokrgozar MA, Azadmanesh K, Azari F. Cytotoxicity and cell cycle effects of bare and polyvinyl alcohol coated iron oxide nanoparticles in mouse fibroblasts. *Advanced Biomaterials,* in press 2009.

[118] Mahmoudi M, Simchi A, Imani M. Cytotoxicity of uncoated and polyvinyl alcohol coated superparamagnetic iron oxide nanoparticles. *Journal of Physical Chemistry C* 2009;113: 9573-9580.

[119] Verdijk P, Scheenen TWJ, Lesterhuis WJ, Gambarota G, Veltien AA, Walczak P, Scharenborg NM, Bulte JWM, Punt CJA, Heerschap A, Figdor CG, De Vries IJM. Sensitivity of magnetic resonance imaging of dendritic cells for in vivo tracking of cellular cancer vaccines. *International Journal of Cancer* 2007;120: 978-984.

[120] Sundstrom JB, Mao H, Santoianni R, Villinger F, Little DM, Huynh TT, Mayne AE, Hao E, Ansari AA. Magnetic Resonance Imaging of Activated Proliferating Rhesus Macaque T Cells Labeled with Superparamagnetic Monocrystalline Iron Oxide Nanoparticles. *Journal of Acquired Immune Deficiency Syndromes* 2004;35: 9-21.

[121] Kamei K, Mukai Y, Kojima H, Yoshikawa T, Yoshikawa M, Kiyohara G, Yamamoto TA, Yoshioka Y, Okada N, Seino S, Nakagawa S. Direct cell entry of gold/iron-oxide magnetic nanoparticles in adenovirus mediated gene delivery. *Biomaterials* 2009;30: 1809-1814.

[122] Reinhold WC, Kouros-Mehr H, Kohn KW, Maunakea AK, Lababidi S, Roschke A, Stover K, Alexander J, Pantazis P, Miller L, Liu E, Kirsch IR, Urasaki Y, Pommier Y, Weinstein JN. Apoptotic susceptibility of cancer cells selected for camptothecin resistance: Gene expression profiling, functional analysis, and molecular interaction mapping. *Cancer Research* 2003;63: 1000-1011.

[123] Lohakan M, Junchaichanakun P, Boonsang S, Pintavirooj C. A computational model of magnetic drug targeting in blood vessel using finite element method. In: *ICIEA 2007: 2007 Second IEEE Conference on Industrial Electronics and Applications;* 2007. p. 231-234.

[124] Berry CC, Wells S, Charles S, Aitchison G, Curtis ASG. Cell response to dextran-derivatised iron oxide nanoparticles post internalisation. *Biomaterials 2004*;25: 5405-5413.

[125] Zhao M, Beauregard DA, Loizou L, Davletov B, Brindle KM. Non-invasive detection of apoptosis using magnetic resonance imaging and a targeted contrast agent. *Nature Medicine 2001;*7: 1241-1244.

[126] Singh NP, McCoy MT, Tice RR, Schneider EL. A simple technique for quantitation of low levels of DNA damage in individual cells. *Experimental Cell Research 1988;*175: 184-191.

[127] Hilger I, Rapp A, Greulich KO, Kaiser WA. Assessment of DNA damage in target tumor cells after thermoablation in mice. *Radiology 2005*;237: 500-506.

[128] Omidkhoda A, Mozdarani H, Movasaghpoor A, Fatholah AAP. Study of apoptosis in labeled mesenchymal stem cells with superparamagnetic iron oxide using neutral comet assay. *Toxicology in Vitro 2007*;21: 1191-1196.

[129] Bacon BR, Stark DD, Park CH, Saini S, Groman EV, Hahn PF, Compton CC, Ferrucci Jr JT. Ferrite particles: A new magnetic resonance imaging contrast agent. Lack of acute or chronic hepatotoxicity after intravenous administration. *Journal of Laboratory and Clinical Medicine* 1987;110: 164-171.

[130] Fahlvik AK, Holtz E, Schroder U, Klaveness J. Magnetic starch microspheres, biodistribution and biotransformation. A new organ-specific contrast agent for magnetic resonance imaging. *Investigative Radiology 1990;*25: 793-797.

[131] Kawamura Y, Endo K, Watanabe Y, Saga T, Nakai T, Hikita H, Kagawa K, Konishi J. Use of magnetic particles as a contrast agent for MR imaging of the liver. *Radiology* 1990;174: 357-360.

[132] Burtea C, Laurent S, Roch A, Vander Elst L, Muller RN. C-MALISA (cellular magnetic-linked immunosorbent assay), a new application of cellular ELISA for MRI. *Journal of Inorganic Biochemistry* 2005;99: 1135-1144.

[133] Laurent S, Forge D, Port M, Roch A, Robic C, Vander Elst L, Muller RN. Magnetic iron oxide nanoparticles: Synthesis, stabilization, vectorization, physicochemical characterizations and biological applications. *Chemical Reviews* 2008;108: 2064-2110.

[134] Chertok B, Moffat BA, David AE, Yu F, Bergemann C, Ross BD, Yang VC. Iron oxide nanoparticles as a drug delivery vehicle for MRI monitored magnetic targeting of brain tumors. *Biomaterials* 2008;29: 487-496.

[135] Yu MK, Jeong YY, Park J, Park S, Kim JW, Min JJ, Kim K, Jon S. Drug-loaded superparamagnetic iron oxide nanoparticles for combined cancer imaging and therapy in vivo. *Angewandte Chemie - International Edition* 2008;47: 5362-5365.

[136] Lee H, Mi KY, Park S, Moon S, Jung JM, Yong YJ, Kang HW, Jon S. Thermally cross-linked superparamagnetic iron oxide nanoparticles: Synthesis and application as a dual imaging probe for cancer in vivo. *Journal of the American Chemical Society* 2007;129: 12739-12745.

[137] Jalilian AR, Panahifar A, Mahmoudi M, Akhlaghi M, Simchi A. Preparation and biological evaluation of [67Ga]-labeled- superparamagnetic nanoparticles in normal rats. *Radiochimica Acta* 2009;97: 51-56.

[138] Lawaczeck R, Bauer H, Frenzel T, Hasegawa M, Ito Y, Kito K, Miwa N, Tsutsui H, Vogler H, Weinmann HJ. Magnetic iron oxide particles coated with carboxydextran for parenteral administration and liver contrasting: *Pre-clinical profile of SH U555A. Acta Radiologica* 1997;38: 584-597.

[139] Ros PR, Freeny PC, Harms SE, Seltzer SE, Davis PL, Chan TW, Stillman AE, Muroff LR, Runge VM, Nissenbaum MA, Jacobs PM. Hepatic MR imaging with ferumoxides: A multicenter clinical trial of the safety and efficacy in the detection of focal hepatic lesions. *Radiology* 1995;196: 481-488.

[140] Lubbe AS, Alexiou C, Bergemann C. Clinical applications of magnetic drug targeting. *Journal of Surgical Research* 2001;95: 200-206.

[141] Lubbe AS, Bergemann C, Brock J, McClure DG. Physiological aspects in magnetic drug-targeting. *Journal of Magnetism and Magnetic Materials* 1999;194: 149-155.

[142] Okon EE, Pouliquen D, Pereverzev AE, Kudryavtsev BN, Jallet P. To the problem of toxicity of magnetite-dextran particles: The morphologic study. *Tsitologiya* 2000;42: 365-366.

[143] Wagner S, Schnorr J, Pilgrimm H, Hamm B, Taupitz M. Monomer-coated very small superparamagnetic iron oxide particles as contrast medium for magnetic resonance imaging: Preclinical in vivo characterization. *Investigative Radiology* 2002;37: 167-177.

[144] Harisinghani MG, Barentsz J, Hahn PF, Deserno WM, Tabatabaei S, Van de Kaa CH, De la Rosette J, Weissleder R. Noninvasive detection of clinically occult lymph-node metastases in prostate cancer. *New England Journal of Medicine* 2003;348: 2491-2499.

[145] Bourrinet P, Bengele HH, Bonnemain B, Dencausse A, Idee JM, Jacobs PM, Lewis JM. Preclinical safety and pharmacokinetic profile of ferumoxtran-10, an ultrasmall

superparamagnetic iron oxide magnetic resonance contrast agent. *Investigative Radiology* 2006;41: 313-324.

[146] Neuwelt EA, Várallyay P, Bagó AG, Muldoon LL, Nesbit G, Nixon R. Imaging of iron oxide nanoparticles by MR and light microscopy in patients with malignant brain tumours. *Neuropathology and Applied Neurobiology* 2004;30: 456-471.

[147] Enochs WS, Harsh G, Hochberg F, Weissleder R. Improved delineation of human brain tumors on MR images using a long- circulating, superparamagnetic iron oxide agent. *Journal of Magnetic Resonance Imaging* 1999;9: 228-232.

[148] Muldoon LL, Sandor M, Pinkston KE, Neuwelt EA. Imaging, distribution, and toxicity of superparamagnetic iron oxide magnetic resonance nanoparticles in the rat brain and intracerebral tumor. *Neurosurgery*. 2005;57: 785-796; discussion 785-796.

[149] De Vries IJM, Lesterhuis WJ, Barentsz JO, Verdijk P, Van Krieken JH, Boerman OC, Oyen WJG, Bonenkamp JJ, Boezeman JB, Adema GJ, Bulte JWM, Scheenen TWJ, Punt CJA, Heerschap A, Figdor CG. Magnetic resonance tracking of dendritic cells in melanoma patients for monitoring of cellular therapy. *Nature Biotechnology* 2005;23: 1407-1413.

[150] Sieben S, Bergemann C, Lübbe A, Brockmann B, Rescheleit D. Comparison of different particles and methods for magnetic isolation of circulating tumor cells. *Journal of Magnetism and Magnetic* Materials 2001;225: 175-179.

[151] Lubbe AS, Bergemann C, Huhnt W, Fricke T, Riess H, Brock JW, Huhn D. Preclinical experiences with magnetic drug targeting: Tolerance and efficacy. *Cancer Research* 1996;56: 4694-4701.

[152] Paruta S, Horl WH. Iron and infection. *Kidney International* 1999;55: 125-130.

[153] Van Hecke P, Marchal G, Decrop E, Baert AL. Experimental study of the pharmacokinetics and dose response of ferrite particles used as a contrast agent in MRI of the normal liver of the rabbit. *Investigative Radiology* 1989;24: 397-399.

[154] Lubbe AS, Bergemann C, Riess H, Schriever F, Reichardt P, Possinger K, Matthias M, Dörken B, Herrmann F, Gürtler R, Hohenberger P, Haas N, Sohr R, Sander B, Lemke AJ, Ohlendorf D, Huhnt W, Huhn D. Clinical experiences with magnetic drug targeting: A phase I study with 4'-epidoxorubicin in 14 patients with advanced solid tumors. *Cancer Research* 1996;56: 4686-4693.

[155] Sun C, Lee JSH, Zhang M. Magnetic nanoparticles in MR imaging and drug delivery. *Advanced Drug Delivery Reviews* 2008;60: 1252-1265.

[156] Rogers WJ, Meyer CH, Kramer CM. Technology Insight: In vivo cell tracking by use of MRI. *Nature Clinical Practice Cardiovascular Medicine* 2006;3: 554-562.

[157] Ferrari M. Cancer nanotechnology: Opportunities and challenges. *Nature Reviews Cancer* 2005;5: 161-171.

[158] Yu H, Adedoyin A. ADME-Tox in drug discovery: Integration of experimental and computational technologies. *Drug Discovery Today* 2003;8: 852-861.

[159] Dames P, Gleich B, Flemmer A, Hajek K, Seidl N, Wiekhorst F, Eberbeck D, Bittmann I, Bergemann C, Weyh T, Trahms L, Rosenecker J, Rudolph C. Targeted delivery of magnetic aerosol droplets to the lung. *Nature Nanotechnology* 2007;2: 495-499.

[160] Jain TK, Reddy MK, Morales MA, Leslie-Pelecky DL, Labhasetwar V. Biodistribution, clearance, and biocompatibility of iron oxide magnetic nanoparticles in rats. *Molecular Pharmaceutics* 2008;5: 316-327.

[161] Ma HL, Qi XR, Ding WX, Maitani Y, Nagai T. Magnetic targeting after femoral artery administration and biocompatibility assessment of superparamagnetic iron oxide nanoparticles. *Journal of Biomedical Materials Research - Part* A 2008;84: 598-606.

[162] Ma Hl, Qi Xr, Maitani Y, Nagai T. Preparation and characterization of superparamagnetic iron oxide nanoparticles stabilized by alginate. *International Journal of Pharmaceutics* 2007;333: 177-186.

[163] Ma HL, Xu YF, Qi XR, Maitani Y, Nagai T. Superparamagnetic iron oxide nanoparticles stabilized by alginate: Pharmacokinetics, tissue distribution, and applications in detecting liver cancers. *International Journal of Pharmaceutics* 2008;354: 217-226.

Chapter 6

APPLICATION OF SPIONs[*]

6.1. SPIONs AS MAGNETIC RESONANCE IMAGING CONTRAST AGENT

Introduction

MR contrast agents exhibit properties of being paramagnetic or superparamagnetic. These agents induce the change in the nuclear magnetic resonance relaxation (NMR) times of the water protons in solution or tissue, known as T1, T2 and T2*. The spin-lattice or longitudinal relaxation time or T1 represents the exponential recovery of the proton spin to align with the external magnetic field. The spin-spin or transverse relaxation time or T2 is the exponential loss of coherence among the spins oriented at an angle to the static magnetic field due to interactions of the spins. The T2* (T2 star) is the loss of phase coherence of the spins in the external magnetic field and is the combination of magnetic field inhomogeneities and T2.

Paramagnetic Agents

Paramagnetism refers to the ability of a metal such as manganese, gadolinium or iron to interact with water protons through dipole-dipole interaction with direct inner sphere effects, resulting in a shortening of NMR relaxation times, and is usually associated with enhancement (increase in signal intensity) on T1 weighted images. Gadolinium chelates (i.e., GdDTPA, GdDOTA or GdDO3A) and manganese chloride are paramagnetic contrast agents used in experimental and clinical studies. These agents tend to shorten T1 relaxation time greater than the T2 and T2* of tissues. Fe^{3+} molecules also behave like other paramagnetic contrast agent. However, investigators have used newer ultra small monocrystalline ironoxide nanoparticles (USPIO) as an alternative to the traditional, gadolinium based, T1 contrast agents in different disease models and in human [1]. Monocrystalline ironoxides

---

[*] Branislava Janic, PhD, from Cellular and Molecular Imaging Laboratory, Department of Radiology, Henry Ford Hospital, 1 Ford Place 2F, Box- 82, Detroit, MI 48202 is contributed in this chapter.

nanoparticles with smaller size (<30 nm) behave like paramagnetic agents (positive contrast) at lower concentration on T1-weighted sequences because of less susceptible effect.

Superparamagnetic Agents

Superparamagnetism is primarily associated with iron oxide crystals (magnetite, Fe_3O_4, maghemite, γFe_2O_3, or other ferrites) and occurs when crystals containing regions of unpaired spins are large enough to be considered a single-magnetic domain with a net magnetic dipole that is larger than the sum of the field generated by the unpaired electrons in the crystal. In the presence of an external magnetic field, the magnetic domains reorient and align resulting in a large magnetic moment due to the magnetic ordering exceeding the one that would occur from paramagnetic ions such as manganese, iron and gadolinium in aqueous solution. This induced magnetic moment causes a distortion of the local magnetic fields when the SPIO nanoparticles are placed in a MR scanner and imaged, resulting in a loss of signal because of magnetic susceptibility and causing a blooming artifact on the T2* weighted image [2,3,4,5]. The SPIO nanoparticles do not retain residual magnetization once the external field is removed. SPIO nanoparticles shorten T1, T2 and T2* relaxation time properties of water or tissue when present in high enough concentrations [6,7]. In general, (U)SPIO nanoparticles will alter the T2/T2* of the surrounding tissue compared to the T1 relaxation times in part due to field gradients surrounding the nanoparticles, resulting in a rapid dephasing of the protons in the environment. (U)SPIO nanoparticles effect on MRI signal intensities depend on various factors including particle size, hydrodynamic radius, concentration of particles within the voxel, image acquisition parameters, and whether the MR contrast agent is in solution or compartmentalized within a cell [8,9,10]. Long echo time T2 weighted spin echo pulse sequences or T2* weighted gradient echo MR pulse sequences are usually used to detect the presence of (U)SPIO nanoparticles within tissues and these agents usually appear as hypointensities with or without associated susceptibility artifacts on the images.

Various methods are used to prepare SPIO nanoparticles, resulting in nanoparticles with a wide range of physiochemical differences including core size (e.g., ultrasmall (U)SPIO), shape, mono or oligocrystalline composition, and outer coating that may alter the ability to use these agents to label cells. The basic chemistry behind the formation of superparamagnetic iron oxide nanoparticles is a mixture of ferrous and ferric iron salts at alkaline pH with a coating (dextran or other types of coatings) that is actively stirred or sonicated resulting in magnetite containing various ratios of Fe_3O_4 to Fe_2O_3 in the crystals [7,11]. The typical synthesis of SPIO nanoparticles is represented by the chemical formula:

$$Fe^{2+} + Fe^{3+} + OH^- + [Coating] \rightarrow Magnetite + coat$$

Superparamagnetism is a through space or outer sphere dephasing effect on local water protons due to alterations in local magnetic field gradients (magnetic susceptibility effect). SPIO nanoparticles shorten the T1, T2 and T2* relaxation times of water or tissue when present at high concentrations. Various superparamagnetic iron oxides (SPIO) based contrast agents have been approved for clinical use. Based on the size of core, hydrodynamic diameter, and surface coating, SPIO based contrast agents can be used either as T2 or T1 agents. SPIO with larger size and higher R2 relaxivities are used as T2 contrast agents (Table 6.1). Generally, SPIO shortens T2 or T2* and causes darkening of enhanced areas.

The size of the (U)SPIO nanoparticles depends on the surface coating used and the surface coating will determine if the particle is monocrystalline (ferumoxtran) or consists of multiplecrystalline or oligocrystalline such as ferumoxides [7]. Surface coatings on (U)SPIO nanoparticles may be variously sized and surface charged molecules, including dextran and modified cross-linked dextran, dendrimers, starches, citrate, or viral particles [2,5,7,12,13,14,15,16,17, 18,19,20,21,22,23,24,25,26,27,28,29]. The coating is usually added during formation of the Fe_3O_4 to Fe_2O_3 crystals and this allows SPIO nanoparticles to exist in a colloidal suspension in aqueous solutions. For several clinically approved SPIO nanoparticles (e.g., ferumoxides, ferumoxtran and ferucarbotran), the coating is dextran that is attached through electrostatic interaction to the iron core by hydrogen bonds between some of the dextran hydroxyl groups and the surface oxide groups of the iron core [7]. The unattached dextran tails covers the rest of the iron crystals and contributes to most of the hydrodynamic diameter of the (U)SPIO nanoparticles [7]. For SPIO nanoparticles (e.g., ferumoxides or ferucarbotran) the dextran coating links multiple iron oxide crystals together and they have a hydrodynamic diameter of about between 60 and 200 nm [7].

Table 6.1 shows the current ironoxide nanoparticle based contrast agents used in clinics and are under development for human use.

Current Clinical Use of (U)SPIONs as Contrast Agents

Oral Contrast Agents

Oral SPIO contrast agents are larger than the injectable agents. Larger SPIO used as oral contrast agents for bowel delineation may precipitate on storage. Currently two agents are approved for clinical use (Ferumoxil (Gastromark/Lumirem) and Ferristene (OMP). Oral agents are coated with non-biodegradable and insoluble materials, such as siloxane (ferumoxil) or polystyrene (OMP), and suspended in food additives to prevent absorption and aggregation when ingested [11,30]. Animal studies showed less that 1% absorption of iron ingested and exclusive excretion through feces with 48 hours [30]. These agents homogeneously distribute throughout the intestine to decrease the signal intensity on T2 or T2* images. The agents are usually administered with large volume of water 30-60 minutes before MRI study. Results from clinical studies have been published [30,31,32,33] and there are reported increased confidence levels in identifying lesions in the abdomen, compared to that of precontrast MR images [34,35,36]. Investigators also showed improved diagnosis of carcinoma of pancreatic head when dynamic contrast enhanced MRI was used following administration of oral magnetic particles [37]. The investigators concluded that two effects, which facilitated the better diagnosis on MRI were: (1) the delineation of the pancreas from the duodenum by the negative contrast medium, and (2) the enhancement pattern of pancreatic tumors by gadolinium-enhanced dynamic MRI compared to normal tissue within the early enhancement after contrast injection. Oral SPIOs also improved the diagnosis of lymphoma and enlarged lymphnodes [36]. However, due to large particle size and susceptible effects, investigators have also recommended the use of small TE (echo time) with faster image sequences [38]. However, these agents are still not popular contrast agents within the MRI communities.

Table 6.1. Superparamagnetic Ironoxides Nanoparticles (SPIONs) that are clinically used

Generic name	Commercial name	Hydrodynamic diameter of the particles	Core Crystal size (TEM)	R1 relaxivity values (mM.s)$^{-1}$	R2 relaxivity values (mM.s)$^{-1}$	Contrast effect
Ferumoxsil (SPIO)	Gastromark/ Lumirem	300-nm	8.4± 2.5nm	3.2±.09	72.0± 12-0	T2*, Predominantly negative enhancement
Ferristene	Oral Magnetic Particles/ Aboscan	3.5 μm	50nm			T2*, Predominantly negative enhancement
Ferumoxides (SPIO)	Feridex I.V./Endorem	80-150nm	4.8± 1.9nm	23.7±1.2	107.0± 11.0	T2*, Predominantly negative enhancement
Ferucarbotran (SPIO)	Resovist/ Cliavist	62nm	4.2nm	25.4	151.0	T2/T1, Predominantly negative enhancement
PEG-feron (Feruglose)	Clariscan	20nm	5-7nm	20	35	T1/T2, predominantly negative enhancement
Ferumoxtran-10 (USPIO)	Combidex/ Sinerem	20-40nm	4.9± 1.5	22.7±0.2	53.1± 3.3	T1/T2, predominantly negative enhancement
Ferucarbotran (USPIO)	Supravist	20-25nm	3-5 nm	10.7-13.2	38-44	T1, Positive enhancement
Ferumoxytol (USPIO)	Faraheme	30±0.2nm	6.4±0.4	15-38	83-89	T1, positive enhancement

Injectable Contrast Agents

Ferumoxides (Feridex I.V./Endorem)

Ferumoxides (Feridex, Berlex, NJ) is a dextran coated larger SPIO (80-150nm) that has been approved by food and drug administration (FDA) in United States and by it's counter parts in Europe and Japan to be used as a MRI liver contrast agent (negative contrast). Once administered intravenously, SPIOs are taken up by reticuloendothelial system (RES) in liver and spleen. Due to its larger size and short blood half-life (~8 minutes) ferumoxides particles do not localize in the RES lymphnodes. However, after local interstitial injection, they can be localized into RES of lymphnode. SPIO particles taken up by the Kupffer cells in liver make the normal liver parenchyma dark on T2-weighted (T2WI) or T* weighted (T2*WI) images. Liver lesions such as hepatocellular carcinoma (HCC), metastasis, and cysts are devoid of Kuppffer cells and can easily be detected by T2WI and T2*WI on MRI [39,40,41,42]. By manipulating parameters of T2WI sequence different cystic lesions can be differentiated on MRI by administering SPIO [39] (Figure 6.1). Other lesion, such as focal nodular hyperplasia (FNH), can easily be differentiated from other solid hepatic lesions using ferumoxides injection [43,44,45]. However, investigators do not use SPIO routinely for diagnosing FNH.

On dynamic enhance MRI or CT, FNH show typical features, such as central scar, and they are diagnostic. Because of possible dose dependant acute hypotensive reactions that are probably due to microvascular embolization and lumbar pain, ferumoxides is not recommended for rapid bolus injection [46]. Intravenous administration is done as a slow drip over 30 minutes. MR images are obtained between 0.5 to 6 hours. The recommended dose is 15 µmol Fe/kg in 100 ml of 5% glucose with a biphasic infusion regimen (2ml/min for 10 minutes and 4ml/min over next 20 minutes) [11]. There were significant changes in the following parameters: protein level, serum iron, transferrin and ferritin levels, and transferrin saturation coefficient. Ferumoxides are contraindicated in patients with known allergy or hypersensitivity to dextran or to any of the other components, and should be used with caution in patients with hemosiderosis and hemochromatosis. One of the advantages of SPIO based contrast agent is that the agents are biodegradable and undergo disintegration in the endosomal-lysosomal pathways to take part in normal cellular iron metabolisms [47]. Animal studies during early stage of preclinical trails, demonstrated biphasic blood clearance of intravenously administered radioactive ferumoxides; the fast component cleared with a half-life of 10 minutes and the slower component cleared with a half-life of 92 minutes [48] an the half-lives were not dependent on injected doses. The relative uptake in the liver, spleen, and kidneys was 57%, 2.9%, and 2.0% of the ID, respectively. Animal studies also showed the incorporation of radioactive iron in the red blood cells (RBC), indicating incorporation of administered iron into normal metabolic pathways [49]. Post ferumoxides MR images are helpful in the diagnosis and detection of hepatic lesions mostly on T2 based sequences [50,51,52,53,54,55] and the diagnosis are as good as double-phase contrast enhanced spiral computed tomography (CT) or CT during arterial portography (CTAP) [56,57,58]. Unfortunately, the company has discontinued the production of ferumoxide.

Ferucarbotran/SHU555A (Resovist/Cliavist)

Ferucarbotran/SHU555A (Resovist/Cliavist) is another SPIO with little smaller size (62nm), which is coated with carboxydextran rather than dextran and has been approved for clinical use in Europe, Australia and Japan. Unlike feridex, resovist does not show side effects after rapid intravenous injection [59,60,61,62,63,64]. Due to smaller size and stronger T1 relaxation properties, resovist can be used as dynamic contrast enhanced agent similar to that of gadolinium based contrast agents [11,65,66,67,68,69]. During its retention phase, which is around 30 minutes following IV administration, liver images can be obtained with T2W sequences and different lesions can be detected. For standard T2WI of liver 8µmol Fe/Kg dose is recommended. Strong T1 enhancement can be achieved with a dose of 10µmol Fe/Kg. Bolus injection can be given within 3 sec of diluted resovist in 10 ml of saline [11,65,70]. Similar to feridex enhanced images, most of the liver lesions will show high signal intensity or increased lesion-to-liver contrast-to-noise ratio (CNR) due to absence of RES [70]. Recent human studies from different parts of the world have proved the effectiveness and superiority of resovist in detecting different lesions in the liver [71,72,73,74,75,76,77,78]. Recent reports also indicated that a slow infusion with T1-weighted 3D GRE as well as with fat sat T1WI image sequences are better in detecting hypervascular hepatic lesion on dynamic enhanced MRI when using resovist [79-80]. Especially for smaller hepatocellular carcinoma (HCC) (< or = 1 cm), resovist enhanced MRI showed higher sensitivity in detecting HCC than triple phase multi-detector CT [81].

Ferumoxtran-10 (USPIO) (Combidex/Senerem)

This is ultrasmall SPIO with a hemodynamic diameter of 20-40nm with complete dextran coating. Because of it's almost neutral surface charge and smaller size, combidex has much longer half-life. The blood pool plasma relaxation half-life is said to be almost 24 hours [11,82]. Due to longer half-life and strong T1 relaxation properties, combidex can be used as blood pool agent during early phase of intravenous administration. In the later phases, especially after 24 hours of administration, combidex is used to evaluate RES system in the body. It is most effective in evaluating lymph node metastasis [83,84,85,86], disruption of blood brain barrier [1,87,88,89,90], tumor vascular densities [91,92] and monocyte-macrophage accumulation in atherosclerotic plaques and in organ rejection [93,94,95,96,97,98,99]. A dose of 13.8 to 29.2 µmol Fe/Kg is used for MR angiography, however, larger dose 29.2 to 44.7 µmol Fe/Kg is recommended for lymphography [11]. Combidex is not recommended for bolus injection and should be injected over 30 minutes. It is to note that Combidex is not yet approved by FDA for clinical use. The most commonly reported treatment-related adverse events are back pain, pruritus, headache, and urticaria [100]. In a multicenter clinical trial, MRI scanning following administration of combidex detected pathological mediastinal lymphnodes in primary lung cancer patients with a sensitivity of 92% to 100% and specificity of 37.5% to 80% [101,102]. Combidex enhanced MRI also showed similar sensitivity and specificity in determining mediastinal staging in patients with non-small cell lung carcinoma compared to PET [103]. Post Combidex MRI for evaluating pelvic metastases from gynecological and prostate cancers is shown to be effective and better than MDCT or unenhanced MRI as both of these techniques rely on the size of lymph node [104], and current clinical trial results indicate that post Combidex MR lymphangiography could be sufficient to determine the characteristics of lymph nodes and its pathologies [105,106,107]. However, a certain level of experience may be required when interpreting Combidex enhanced images alone. A very recent meta analysis also showed the diagnostic performance of USPIO enhanced MRI in detecting LN metastases in different body regions, however, the authors recommended that the use of USPIO-enhanced MRI in clinical practice still needs to be investigated [108]. Figure 6.2 shows the enhancement of implanted glioma in rat model 24 hours after intravenous administration of Combidex.

PEG-feron (Feruglose/Clariscan)

This USPIO has an ironoxide core of 5-7 nm with a hemodynamic diameter of 20 nm. The nanoparticles are coated with carbohydrate-polyethelene glycol and have been developed for MR angiography and perfusion studies. Following bolus administration, vascular half-life of Clariscan is shown to be dose dependent and is usually 3-4 hours. [11]. Recently, the use of clariscan in clinical trail has been discontinued. Clariscan has been used as blood pool agent in 3D cine MR cardiac imaging and MR angiography [109,110,111,112,113]. The agent has been shown to be safe in human with a bolus injection up to 172 µmol Fe/kg. The optimal dose for MR angiography ranges from 50-100 µmol Fe/Kg [11].

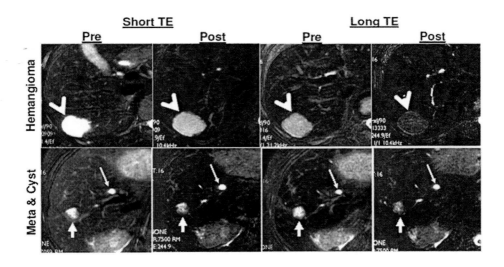

Figure 6.1. Advantage of using ferumoxides enhanced MRI with different effective echo times to differentiate different cystic lesions. T2-weighted images (T2WI) before and after administration of ferumoxides with fast spin echo short (80-90 ms) and long (180-250 ms) TE acquired by a 1.5 tesla MRI systems. Markedly low signal intensity on post ferumoxides images are observed with hemangioma (upper panel, arrowheads) with longer TE. Similar changes are not observed in metastasis (lower panel, short arrows) and in cyst (lower panel, long arrows).

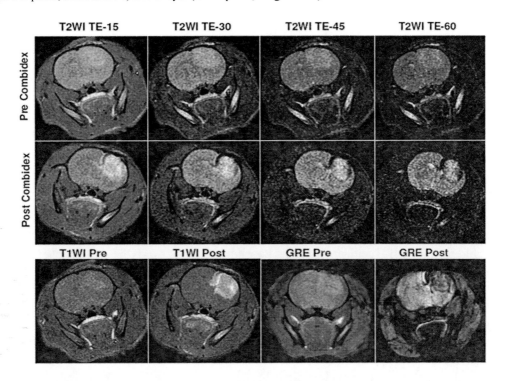

Figure 6.2. Signal intensity changes on T2WI, TWI and GRE images in implanted glioma (U251) in rat 24 hours following administration of Combidex. Note the effect of TE on the signal intensity changes on T2WI (TE = 15 vs TE = 60). At lower TE, T2WI show almost similar high signal intensity to that of T1WI. GRE image mostly show susceptible effect.

Ferucarbotran (USPIO)/SHU555C (Supravist)

This is another ultrasmall SPIO, which is under clinical trail as positive enhancer blood pool agent that can be administered as an IV bolus with a dose up to 80μmol Fe/Kg. This agent is smaller than combidex and resovist (~ 20 nm hydrodynamic diameter) and coated with carboxydextran. This agent is suitable for first pass and steady state MR angiography [9,114,115,116]. Recent studies showed dose dependent increase in the signal intensity in myocardium during first pass and equilibrium phases [116,117]. The agent showed wide range of safety margin during IV administration. Supravist is also tried in multiple sclerosis (MS) patients and investigators have concluded that USPIO enhanced MRI may give more insight about the inflammation in MS lesions [118,119]

Ferumoxytol (USPIO) (Feraheme)

Ferumoxytol is an FDA approved iron oxides nanoparticles (USPIO) to be used in iron deficiency anemia in adult chronic kidney diseases [120,121,122]. Ferumoxytol is approved as an intravenously administered agent with two 510 mg IV injections given 3-8 days apart. Initial clinical trials showed 0.2% of serious hypersensitivity reactions and 1.9% of patients exhibited hypotension following IV ferumoxytol [123,124,125]. Investigators are considering using ferumoxytol as alternative to Gadolinium based contrast enhancement in patients with chronic kidney diseases, to avoid nephrogenic systemic fibrosis (NSF) [1,126]. In recent published studies, both gadolinium based contrast agent and feromoxytol showed similar patterns of enhancement in brain tumors [1], but ferumoxytol showed better enhancement and lesion detection in case of acute disseminated encephalomyelitis (ADEM). First-pass and equilibrium phase MR angiography can be obtained after injecting ferumoxytol [127,128]. Ferumoxytol has a long blood half-life (10-14 hours) [129]. It would be interesting to see the real potential of the agents in future.

6.2. SPIONs FOR MAGNETIC LABELING OF CELLS

Introduction

Current and future diagnostic and therapeutic approaches for various diseases rely on the ability to *in vivo* visualize and track endogenous or transplanted cellular elements. Over the past few decades, superparamagnetic iron oxide nanoparticles (SPIONs) have proven to be a valuable tool for MRI cellular imaging. Labeling cells with SPIONs enables detection of single or clusters of labeled cells within target tissues following either direct implantation or intravenous injection [130,131]. For cellular imaging of transplanted cells SPIONs labeling is performed *in vitro*, by incubating labeling agent with cells for a given time period. This is usually achieved by adding labeling reagent to the serum free culture media containing the cells of interest, and allowing for labeling to take place for a given time period. In addition, several mechanical approaches such as gene gun or electroporation, have been used to effectively introduce nanoparticles into cells. The gene gun fires nanoparticles directly into cells in culture, driving the particles through the cell membrane or directly into the nucleus. Unfortunately, long-term effects on cellular functional properties are not known [132]. Moreover, this technique has limitations regarding the efficiency, potential tissue damages induced by nanoparticles impact and small area of coverage [133]. Since less traumatic methods to label cells with SPIONs are available, it is unlikely that gene gun approach will be used in the future.

To visualize cells or monitor the behavior of certain cell population within the host/patient, labeling SPIONs agent is administered systemically and the process of cell labeling is performed *in vivo*. Although most of cell labeling is performed in cell culture, cells such as peripheral macrophages have been often labeled *in vivo* using iron oxide particles. Wu et al. took the advantage of macrophages phagocytosis properties and applied intravenous administration of MPIOs in animal model of transplant organ rejection [134]. After 24h, *in vivo* MRI revealed dark areas due to macrophage accumulation at the sites of the host immune response to the transplant. Neural progenitor cells have also been used as a target for *in vivo* labeling. Shapiro et al. used this strategy to label neural stem cells in the subventricular zone in rat model by injecting MPIOs directly into the lateral cerebral ventricles. Five weeks after injection MRI analysis showed hypointense areas along the rostral migratory pathway, through which neuronal precursors migrate from subventricular zone of the brain to the main olfactory bulb, where they differentiate into interneurons. At the same time, immunohistochemistry of the same brain structures confirmed the presence of nanoparticles in neuronal stem cells, migrating neuroblasts and mature neurons [135].

SPIONs and its Modification for Labeling

Efficient cell labeling with SPIONs depends on nanoparticle's physicochemical properties including core size, shape, mono, or oligocrystalline composition, outer coating and surface charge, and on cellular properties such as nuclear/cytoplasm ratio, cell surface charge and the mode of nanoparticle transport across the cell membrane. [136,137,138,139]. In addition, successful cell labeling also depends on total iron content in cell culture media and labeling incubation time. Cellular mechanisms for internalization of SPIONs may include active process of phagocytosis or various endocytic pathways such as pinocytosis, caveolin or clathrin -dependent or -independent endocytosis, and the one involved in the particular transmembrane transport will depend on the cell type [140,141,142]. It is known that in phagocytic cells larger [143] and negatively charged [19] particles are more readily internalized. On the other hand, pinocytosis is more important for internalization of small particles such as USPIO. In addition, cells such as stem and progenitor cells that do not exhibit prominent phagocytic activity utilize endocytic pathways for SPIONs incorporation, and magnetic labeling of these cells has often resulted in low efficiency. Since these cells are extremely important and widely used in cell tracking studies, to increase their labeling efficiency researchers have continuously been improving currently available labeling procedures

With the development of cellular MRI, various dextran-coated forms of SPIO nanoparticles that were previously approved as clinical MR contrast agents, have been tried as cell labeling agents as well. For example, nanonparticles such as ferumoxides, ferumoxtran and ferucarbotran were used in clinical trials as blood pool agents or contrast agents for lymphangiography [53,144,145,146]. However, achieving the sufficient intracellular concentrations of these agents for cellular probe to be effectively detected by MRI, proved to be challenging. To improve labeling efficiency, different strategies have been used, with varying degree of success. Straightforward approach to increase the concentration of SPIONs in cell culture media during labeling procedure did not usually result in the increase in intracellular SPIONs in labeled cells [147,148]. However, modifying SPIONs agent itself to better accommodate and/or exploit cellular transmembrane transport and endocytosis pathways has proven more successful, with various different strategies employed. To reduce the repelling forces between negatively charged cell membrane and nanoparticle, cationic-coated SPIONs were designed that, through electrostatic interactions, attach to the cell membrane, and are more readily incorporated into the cells. Cations such as carboxypropyl trimethyl ammonium and citrate cation- modifications of USPIO exhibited advantage over the neutral carboxydextran-coated USPIO preparation in magnetic labeling of macrophage cell line [19].

Dendrimers are another group of molecules also used as SPIO modifying agents. Unlike classical polymers, dendrimers have a high degree of molecular uniformity, narrow molecular weight distribution, specific size and shape characteristics, and a highly- functionalized terminal surface. Previous applications of dendrimers in cell transfection methodologies demonstrated that branched, synthetic polymers nonspecifically bound to cell membrane and stimulated endocytosis. Numerous studies were performed to characterize dendrimers as MRI contrast agent modifiers [149,150,151,152]. One of the agents successfully used to generate cellular probes is magnetodendrimer (MD-100) that was synthesized by applying new class of macromolecules, polyamidoamines (PAMAM), and used as a coating for iron oxide nanoparticles. Bulte et al. showed that at the intracellular concentration of 9 and 14 pg iron/cell, labeled cells exhibited an *ex vivo* nuclear magnetic resonance (NMR) relaxation rate (1/T2) as high as 24-39 s^{-1}/mM iron. Labeled cells were also unaffected with regard to viability, proliferation and differentiation capacity. In addition, magnetically labeled oligodendroglial progenitors were detected, *in vivo,* six weeks after transplantation indicating the feasibility of the use of this agent for long term cellular tracking [5].

Modifying dextran coat of SPIO nanoparticles is yet another approach used to enhance cellular incorporation of SPIONs. This approach involves cross-linking the dextran strands (CLIO) and covalent attachment of biomolecules such as peptides and antibodies. This method utilizes specific signals through receptors and/or signal molecules to induce endocytosis. To facilitate incorporation into cell, human immunodeficiency virus transactivation transcription protein (HIV-Tat) that contains a membrane translocating signal was coupled to iron oxide nanoparticles [153]. Many groups used CLIO-HIV-Tat labeled cells as *in vivo* cellular probes and it was shown that homing of labeled lymphocytes to liver and spleen could be visualized by MRI [154]. *In vivo* MRI also successfully tracked CLIO-HIV-Tat labeled T cells in adoptive transfer experiments in autoimmune diabetes and melanoma mouse models [155,156,157]. Although the use of biomolecules as modifying agents carries the risk of affecting biological cellular functions, there were no reports indicating such an effect in the case of CLIO-HIV-Tat SPIONs.

To improve labeling efficiency in cells, monoclonal antibodies can be covalently attached to dextran moiety to facilitate receptor-mediated endocytosis. One of the examples is mouse anti-Tfr monoclonal antibody OX-26 that was conjugated to dextran coated monocrystalline iron oxide and used for labeling of oligodendrocytes and neuronal progenitors [158,159]. These studies showed that when transplanted, magnetically labeled progenitors preserved their migratory and myelinating capacity *in vivo*. Cells were easily detected by MR imaging and this allowed for three-dimensional mapping of the achieved extent of myelination.

Ahrens et al. used similar strategy, where they induced endocytosis of the labeling complex by targeting cell surface accessory molecule expressed by dentritic cells (DCs) with an anti-CD-11c monoclonal antibody conjugated to SPIO particles. This approach resulted in highly efficient particle uptake by the DCs, with no adverse effects on immunological function. In addition, labeled cells could be visualized by *in vivo* MRI for several days [160].

Viruses and viral proteins have also been explored as a tool for modifying SPIO nanoparticles. The hemagglutinating virus of Japan envelope (HVJ-E) has no charge and utilizes membrane fusion activity to deliver internalized materials. Therefore, encapsulation of SPION nanoparticles by HVJ-E was used by several groups to label microglial and neuronal progenitor cells in culture [14,161,162]. The HVJ-E-SPION particles proved to efficiently label cells; however, the lack of commercially available HVJ-E modified nanoparticles limits the use of this agent.

One of the newly re-discovered nanoparticles, micron sized iron oxide particles (MPIOs), have been extensively used for cellular MRI in various experimental models. One of the main advantages of using MPIOs is that they can be commercially obtained in a form that allows further chemical modifications to suit the particular research needs. The size of MPIOs is between 0.3 to .5 microns and they contain more than 60% magnetite in polymer coating that can include a fluorescent marker as well, enabling dual, MRI and fluorescent detection of labeled cells. MPIOs are available with their surface containing terminal amine or carboxyl groups that would allow further attachment of peptides, ligands and antibodies specific to the cellular target. It has been shown that MPIO did not alter viability, proliferation and differentiation capacity of labeled hematopoietic and mesenchymal stem cells [163].

Transfection Agents Mediated Modification of SPIONs for Cell Labeling

The use of transfection agents is yet another way to improve labeling efficiency of cells. Various commercially available transfection agents have been used to magnetically label cells and the common mechanism for most of them involve incorporation of SPIONs into liposomes that can potentially fuse with and cross the cell membrane, and intracellularly release the incorporated nanoparticle. The main drawback of using most of the commercially available transfection agents is dose dependent cytotoxicity. However, recent studies utilized commercially available, FDA approved Protamine Sulfate, as a transfection agent in combination with also FDA approved Ferumoxide [148,164,165]. Mixture of these two FDA approved agents generated complex that is very efficient in labeling stem cells [137,166]. Ferumoxides are dextran-coated colloidal iron oxide nanoparticles that magnetically saturates at low fields and have an extremely high NMR T2 relaxivity. Changes in R2 (R2=1/T2) are linear with respect to the iron concentration. Protamine sulfate is an FDA-approved drug containing >60 percent arginine and is used for treatment of heparin anticoagulation therapy overdose. Intracellular incorporation of ferumoxides-protamine sulfate (FePro) complex occurs *via* macropinocytosis and can be imaged at clinically relevant MRI fields using standard imaging techniques. Very recent study optimized the conditions for achieving the efficient FePro labeling in different cell types [165], and showed that in doses sufficient for efficient labeling, FePro does not affect physiology of various cell types [164,167]. The overview of SPIO labeling strategies used in different cell types is outlined in Table 6.2. (Modified with permission from Dr. Kiessling) [168].

Table 6.2. Iron oxide particle type, size, coating, and the cells used for MR cell-labeling

Particles	Size	Coating	Cell type	References
MPIO	0.9-5.8 μm	Divinyl benzene/styrene	Hepatocytes, embryonic fibroblasts, MSC	[139]
Feridex (SPIO)	80-150nm	Dextran	Monocyte, THP-1 cell line, Cord blood derived AC133+ progenitor cells	[164,165,169]
Ferumoxides and MION-46L	▫150nm, 8-20 nm	Dextran	Lymphocytes, MSC, CG-4 cells, cervix cancer cells	[147]
SHU 555a SHU 555c	45–65nm, <40nm	Dextran	Progenitor cells, fibroblasts, hepatocarcinoma cells	[143]
Magneticpoly-saccharides	▫50nm	Dextran	Hematopoietic, progenitor cells, dendritic cells	[160,170]
CLIO	▫45nm	Dextran	Lymphocytes, CD34+ progenitor cells	[153,171]
Anionic maghemite nanoparticles	▫35nm	DMSA	Macrophages, HeLa	[29,172]
Ferumoxtran	20-40nm	Dextran	Embryonic stem cells, muscle stem cells, MSC	[173,174,175]
AquaMag-100BMS180549	▫35nm	Dextran	T-cells	[2]
MION-46L	8–20 nm	Dextran	Oligodendrocytes, neural precursor cells, T-cells	[158,159,176]
MION	8–20 nm	Dextran	C6 tumor cell	[177]
MION	12–14 nm	PEG phospholipid	Fibroblasts, macrophages	[178,179]
VSOP-C125	▫8nm	Citrate	Macrophages	[19]
MD-100	7–8 nm	Dendrimers	Stem cells, neural progenitor cells, olfactory ensheathing glia, muscle stem cells	[5,180,181]

Optimization of Cell Labeling with Simple Incubation

Both modified and unmodified SPIONs have been used to label cells by simple incubation method. To efficiently label cells, modified or non-modified SPIONs are added to cell cultures at varying concentrations and incubated for few hours to few days. However, due to toxic effect of iron oxides and transfection agents, long term incubation is not usually recommended. Investigators are continually trying to optimize the labeling of different types of cells based on incubation time and concentration of SPIONs. Our group has developed a technique to make ferumoxides (FDA approved agent) transfection agents complexes to facilitate cellular uptake by endocytosis [136,137,147,166]. Very recently, instead of commonly used cellular transfection agents, such as lipofectamine, we introduced the use of protamine sulfate (FDA approved agent) to generate ferumoxides-protamine sulfate (FePro) complexes for efficient labeling of different mammalian cells, including stem cells and T-lymphocytes [137,182]. These labeled cells have been used in different animal models and tracked by both, high strength and clinical strength MRI systems [131,183,184]. One of the advantages of labeling cells using FDA approved agents is the possibility of clinical trial without facing major toxicity issues related to contrast and transfection agents. We have recently reported modification of our method to efficiently label cells using ferumoxides and protamine sulfate within 4 hours [165]. The following is the step-by-step methods to label suspension as well as adherent cells using our modified method:

For Suspension Cells (such as Hematopoietic stem cells, T-cells, etc)
Materials Needed
1. RPMI-1640 medium with L-glutamine, MEM non-essential amino acid and sodium pyruvate
2. Ferumoxides (Feridex, 11.2 mg/Fe per ml, Berlex Laboratory, NJ)
3. Protamine sulfate (American Pharmaceuticals partners, IL)
4. 15 or 50 ml centrifuge tubes
5. 6-well plates (each well has around 10 cm^2 growth surface)

Procedure:
1. Count the cells, put into 15 ml or 50 ml tubes (based on cell numbers)
2. Centrifuge and decant the supernatant
3. Add serum free RPMI-1640 to the cell suspension, mix and centrifuge it again (*this is important to get rid of all serum*).
4. Decant the supernatant and add serum free media again to *make 4×10^6 cells per ml of the serum free media*
3. Add *100 µg of Feridex (9 µl from the bottle)* for *every ml of cell suspension* and mix well by gentle pipetting
4. Then *add 3 µg* of freshly prepared (in distilled water) protamine sulfate for *every ml of cell suspension* and mix well by gentle pipetting.
5. Transfer the cells to 6-well plate at a concentration of 1×10^6 per 1 cm^2 area. For example add 2.5 ml cell suspension (10×10^6 cells) to a single well in a 6-well plate. (*The number of cells per cm^2 is very important*)
6. Incubate at 37°C for 15 minutes
7. Then add equal volume of complete media (containing serum) to the cell suspension. Complete media should contain serum and other growth factors for respective cells.
8. Incubate further for 4 hours.
9. Collect and wash the cells.
10. Determine the labeling efficiency and functional status of labeled cells.

Note: If higher amount of iron per cell is required, initial concentration of ferumoxides per ml of cell suspension may be increased to 200μg/ml (18μl of feridex), however, the ratio of protamine sulfate must be kept same. Therefore, for 200μg/ml of ferumoxides, 6μg of protamine sulfate should be added.

For Adherent Cells (such as Mesenchymal Stem Cells, Neural stem cells, etc)

Materials Needed
1. RPMI-1640 medium with L-glutamine, MEM non-essential amino acid and sodium pyruvate
2. Ferumoxides (Feridex, 11.2 mg/Fe per ml, Berlex Laboratory, NJ)
3. Protamine sulfate (American Pharmaceuticals partners, IL)
4. 15 or 50 ml centrifuge tubes
5. Cell culture flasks or plates

Procedure:
1. *Adherent Cell Labeling (MSC) in T75 culture flask*
   a) Culture cells to 80-90% confluence
   b) Remove old culture media completely and add 10 ml serum-free RPMI-1640 and wash the cells
   c) Remove the washing serum free RPMI 1640 and add 5 ml of serum free media
   d) Add *500 μg (45μl of feridex from the bottle) of Feridex (100μg/ml)* to the cells (in the flask containing 5 ml of serum free media) and mix thoroughly.
   e) Add *15 μg* of freshly prepared *Protamine sulfate (3μg/ml)* to the cells (in the flask containing 5 ml media plus Feridex) and mix thoroughly.
   f) Incubate at *37°C for 15 minutes.*
   g) Then add equal volume (5 ml) of complete media (containing serum) to the cell. *Complete media should contain serum and other growth factors for respective cells.*
   h) Incubate further for *4 hours.*
   i) Collect and wash the cells.
   j) Determine the labeling efficiency and functional status of labeled cells.

Note: The number of cells in T75 flask may not be 20 million; therefore, compared to suspension cells, the added amount of iron per cell is higher, which will in turn facilitate uptake of more iron to the adherent cells.

Figure 6.3 shows different types of labeled cells using ferumoxides-protamine sulfate and new 4 hour labeling method.

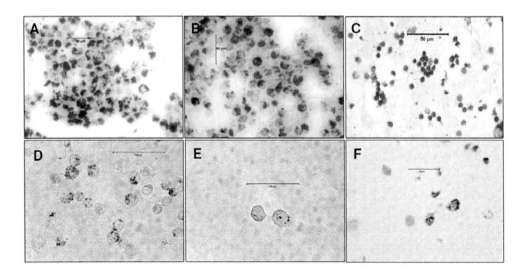

Figure 6.3. Representative images of different types of magnetically labeled cells. (A) U251 human glioma cells; Prussian blue staining, bar represents 50μm. (B) Umbilical cord blood derived immature dendritic cells; Prussian blue staining, bar represents 50μm. (C) Cytotoxic T-cells; DAB enhanced Prussian blue staining, bar represents 50μm. (D) Neural stem cells (C17.2); Prussian blue staining, bar represents 100μm. (E) Umbilical cord derived mesenchymal stem cells; Prussian blue staining, bar represents 100μm. (F) Hematopoietic stem cells (CD34+/AC133+); Prussian blue staining, bar represents 20μm.

Assay to Determine the Toxicity of Labeling

Whatever the approach is when labeling cells the most important parameters that have to be taken into account are the effects of nanoparticles on cell viability and functional status. For successful translation of any of the labeling technique from bench to bedside, it is essential to determine short- and long-term cytotoxic effect of the agent in use. Ions contained in nanoparticles and/or transfection agents can be toxic to cells. Ferrous ions from intracellularly incorporated SPIONs can initiate a Fenton reaction through Haber-Weiss chemistry and generate free radicals that in turn can induce oxidative cell injury and DNA damage [185,186,187].

The simplest, commonly used method to assess cell viability is Trypan blue exclusion assay that is usually employed at the beginning of labeling procedure to determine the percentage of viable cells by bright field microscopy. This assay is based on the premise that viable cells preserve cell membrane integrity and therefore will exclude the dye from entering the cell. The same principle is applied when using Propidium iodide (PI), but this approach has the advantage because it can be coupled with fluorescence based detection method, such as flow cytometry that provide for accurate quantification. Although widely used, trypan blue and PI assays can detect only dead cells, and lack the ability to identify cells that are undergoing apoptosis. Changes in apoptosis rate can be a very good indicator of a possible, long-term adverse effect and therefore an important parameter to assess when SPIONs labeling cells. The commonly used sensitive apoptotic assay is flow cytometric method that employs labeling with Annexin V in combination with 7-AAD or PI. One of the earliest changes in cells undergoing apoptosis is translocation of the membrane phosphatidyl serine

(PS) from the inner to the outer leaflet of plasma membrane. Annexin V is a Ca2 + dependent phospholipid-binding protein that has a high affinity for PS, and binds to cells with exposed PS [188]. Numerous previous studies compared Annexin V to other methods and confirmed that, Annexin V is an extremely sensitive probe for detecting cells that are in the early stages of apoptosis [189,190,191]. Annexin V is often used in conjunction with 7-AAD, an intercalative nucleic acid dye whose level of DNA incorporation depends on the extent of the loss of plasma membrane integrity and thus identifies dead cells. Investigators have found that in labeling methods described earlier, cell viability or rate of apoptosis of labeled cells does not differ significantly from unlabeled control cells. Short- and long-term effects of SPIONs on proliferation capacity of labeled cells can be assessed by tritiated-thymidine uptake, 5-bromo-2-deoxyuridine (BrdU) incorporation, or MTT 3-[4,5-dimethylthiazol-2-yl]-2,5-diphenyl tetrazolium bromide assay (MTT assay) (Roche Molecular Biochemicals, Indianapolis, IN). Proliferation assays are usually preformed as a pulse chase SPIO labeling experiment with determinations made at different time points [136]. Reactive oxygen species (ROS) can be detected by using CM-H2DCFDA probe (Molecular Probes Inc., Eugene, OR). In reaction with ROS, non-fluorescent CM-H2DCFDA form fluorescent esters that can be detected by flow cytometry, fluorescent or confocal microscopy.

Various other aspects of functional integrity in labeled cells can be analyzed. Cellular functional assay should be performed after labeling and compared with that of control, non-labeled cells. The specific approach is usually determined by the type of cells used. For example, in the case of T-cells, it is extremely important that SPIONs labeleing of cells does not activate or change one population of T-cells (such as T helper cells) to another (such as cytotoxic T-cells). When labeling immune cells, analysis of phenotypic cell surface marker expression (i.e., CD markers such as CD4, CD8, CD11a, CD19, CD25) and cytokine/chemokine production is warranted before and after labeling. For stem cells, primary and lineage markers should be assessed before and after labeling. Determining differentiation capacity in labeled cells is especially important for stem cells. For example, labeled mesenchymal stem cells (MSC) are analyzed for their ability to differentiate to adipogenic, chondrogenic, and osteogenic lineages while labeled neuronal stem cells (NSC) for their capacity to differentiate to neuronal and/or glial cell types. For hematopoietic stem cells (HSC), differentiation to colony-forming units (CFUs) should be determined for labeled and unlabeled cells. Labeled endothelial progenitor cells (EPC) can be assessed for their capacity to give rise to colony-forming units, endothelial cells, and cord-like structures in matrigel systems. It is also possible to determine differentiation capacity of magnetically labeled cells by tracking the *in vivo* migration and tissue/organ incorporation of labeled stem cells [192]. Various groups analyzed effects of SPIONs on different cell types and to date, little or no changes in cells viability, apoptotic rate, ROS formation, proliferation capacity, metabolic activity, functional and differentiation capacity were reported for cells labeled with clinically approved SPIONs. For example, Pawelczyk et al. analyzed the effects of Fepro on MSCs and macrophages. They demonstrated a transient decrease in transferrin receptor gene and protein expression, while ferritin gene and protein levels either remained stable or increased in response to the iron load [193]. These results indicate that Fe-Pro labeling of cells elicited appropriate and expected physiological changes of iron metabolism or storage. In another study, Janic et al examined the effect of SPIONs labeling on THP-1 macrophage like cell line where they demonstrated that intracellular incorporation of FePro complexes did not alter overall immunological properties of THP-1 cells [164]. Previous studies indicated cytotoxic capacity of TAT protein [194,195] that raised the concerns if TAT was to be used as a SPIONs modifying agent. However, recent publication indicated no short-term loss of cell viability or accumulation of CLIO-Tat in the nucleus [20]. With designing new labeling agents or new modifications to the existing labeling agents, new studies are needed to thoroughly assess their effects on all the aspects of cellular physiology.

6.3. SPIONS AND MAGNETICALLY LABELED CELLS FOR MULTIMODAL REGIMENS

SPIONs for Multimodal Imaging

SPIONs was originally developed as MRI contrast agents, however, due to complexity of investigations and separation of exogenous from endogenous iron (due to bleeding during tumor implantation or bleed into tumor from necrosis) it becomes essential to use another complimentary imaging or detection modality to confirm the accumulation of administered SPIONs. The complimentary imaging modality could be nuclear medicine scanning or detection, optical imaging, fluorescent imaging, etc. Most effective dual modality SPIONs would be to use radioactive iron to make the SPIONs. In fact, investigators have used radioactive iron (^{59}Fe) based SPIONs to determine the pharmacokinetic of AMI25 (ferumoxides) and other SPIONs [48,49,196,197]. However, this strategy is not used in humans. Another way of making SPIONs as dual contrast agents would be to modify the surface moiety of the nanoparticles and attach either fluorescent or other optical imaging probes or dye. Josephson et al. had developed a near-infrared fluorescent (NIFR) SPIONs [198]. The probes are prepared by conjugation of arginyl peptides to cross-linked iron oxide amine (amino-CLIO), either by a disulfide linkage or a thioether linker, followed by the attachment of the indocyanine dye Cy5.5. The NIRF of disulfide-linked conjugate was activated by DTT, while the NIRF of thioether-linked conjugate was activated by trypsin. The investigators had injected the probes subcutaneously and followed the migration to the adjacent lymphnodes. Axillary and brachial lymph nodes were darkened on MR images and easily delineated by NIRF imaging. Investigators also proposed to use similar NIRF SPIONs as a preoperative magnetic resonance imaging contrast agent and intraoperative optical probe [199] as well as probes for sentential lymphnodes detection [200]. To determine specific events in the tissue or area of interest SPIONs can be tagged with specific ligands and fluorescent dye. Effectiveness of annexin V and fluorescent probe Cy5.5 tagged small SPIONs (CLIO) was determined to detect the sites of apoptosis in animal model of myocardial infarction and compared with FITC labeled annexin V [201].

By using the method of encapsulation and tagging, Xie et al. have reported triple functional SPIONs (HAS-IONPs-^{64}Cu-DOTA-Cy5.5) [202]. To assess the biophysical characteristics of this novel nanosystem, the HSA coated iron oxide nanoparticles (IONPs) (HSA-IONPs) were further labeled with (64)Cu-DOTA and Cy5.5, and tested in a subcutaneous U87MG xenograft mouse model. *In vivo* positron emission tomography (PET)/near-infrared fluorescence (NIRF)/magnetic resonance imaging (MRI) tri-modality imaging, and *ex vivo* analyses and histological examinations were conducted to investigate the *in vivo* behavior of the nanoparticles. The nanoparticles showed a prolonged circulation half-life and increased accumulation in lesions with high extravasation rate, and low uptake of the particles by macrophages at the tumor area. Flexman et al. have used hemagglutinating virus of Japan envelopes (HVJ-Es) to encapsulate IONPs and a positron emitter (^{18}F-floride) and the nanoparticles were injected in rats to determine the biodistribution with or without placing a permanent magnet on the head. The investigators have observed significant differences in distribution of nanoparticles by PET scanning in two groups of rats (with or without magnetic targeting).

Multimodal imaging can also be used to determine the effectiveness of targeted delivery of SPIONs. Similar to the use of SPIONs as drug delivery vehicles (see next section) Meng et al. [203] have developed SPIONs tagged with FITC and chlorotoxin and used to target glioma cells *in vitro*. Both MRI and optical images showed effective uptake of SPIONs to the cells. SPIONs carrying gene or DNA are becoming current trends to make effective SPIONs based delivery system for the treatment of different diseases. Similar strategy can be utilized to deliver reporter gene to the sites of SPIONs accumulation, which can be determined by MRI (for SPIONs) and other complementary imaging modalities, such as optical or nuclear medicine imaging.

Cells for Multimodal Imaging

Multimodal imaging engages multiple contrast agents that can be detected by different imaging modalities. Usual multimodal approach would involve simultaneous use of two different contrast agents that can be detected by different imaging modalities, which are complementary to each other and each one provides an advantage that can compensate for drawbacks and limitations of the other technique. Optical fluorescent and nuclear medicine imagining are the techniques most commonly combined with MRI, when performing *in vivo* multimodal imaging. Optical imaging that is used in conjunction with fluorescent agents provides high sensitivity, but has the limitations with regard to limited tissue penetration due to the endogenous absorbers (hemoglobin, water and lipids) and scattering [201,204,205,206]. However, when combined with MRI, improved depth penetration, high spatial resolution and morphological information is provided by MRI [207]. Two principle methods are used to fluorescently label cells for *in vivo* tracking: 1) labeling with exogenous dyes and 2) genetic manipulation that modifies cells to trans-genetically express heterologous fluorescent proteins or enzymes that would activate a fluorescent probes. The most prominent class of exogenous fluorescent probes are lipophilic, near infra red cyanine dyes that when added to cell suspension bind to cell membrane phospholipid bi-layer [208]. Currently, numerous red fluorochromes are commercially available with the continuous addition of newly designed, improved reagents. The best known and commonly used are DiI, DiO, DiD and CM-DiI that have been proven effective for *in vitro* cell labeling and *in vivo* cell tracking [209]. These dyes demonstrated low cytotoxicity, high labeling efficiency and relatively high resistance to intercellular transfer [210]. However, the disadvantage is that the dye gets diluted if cells are proliferating and after several cell divisions signal may not be detectable anymore [168]. On the other hand, transgenic expression of fluorescent protein such as green fluorescent protein (GFP), is an excellent choice for labeling proliferating or trans-differentiating cells [211,212,213]. Nuclear medicine modalities like SPECT and PET are limited in their spatial resolution and tissue contrast ratio and therefore can be matched with MRI that can provide more detailed anatomical information. For *in vivo* cell tracking, double labeling of cells that can give efficient image when combining nuclear medicine with MRI, usually include combination of radiotracer and SPIONs labeling or SPIONs labeling of genetically modified cells that express reporter gene whose product interacts with radio probe in order to generate PET or SPECT signal. As radiotracers, ^{111}In agents are used to label cells in combination with SPIONs labeling. However when combining nuclear medicine and MRI modalities, reporter gene approaches have many advantages over direct and indirect cell labeling methods. Stable transfection of cells ensure for long term expression of the reporter gene that does not dilute out in proliferating cells. Furthermore, over time accumulated divided cells can generate increased signal that can be detected with repeated imaging. In addition, the signals detected prove the *in vivo* presence of viable cells. One of the most widely used reporter genes for PET imaging is wild-type herpes simplex virus type 1 thymidine kinase (HSV1-tk) and it's HSV1-sr39tk mutant. This enzyme efficiently phosphorylates purine and pyrimidine analogs and has been very successfully used with radio labeled reporter probes such as ^{124}I-2'-fluoro-2'-deoxy-1-β-D-β-arabinofuranosyl-5-iodouracil (FIAU), ^{18}F-2'-fluoro-2'-deoxy-1-β-D-β-arabinofuranosyl-5-ethyluracil (FEAU) and ^{18}F-9-(4- ^{18}F-fluoro-3-hydroxymethyl-butyl) guanine (FHBG) [214,215,216,217]. Reporter gene commonly used with SPECT imaging is sodium iodide symporter (NIS) that has been widely used in imaging applications in conjunction with ^{99m}Tc-pertechnetate or ^{124}I and in anticancer therapy with ^{131}I and ^{188}Re. By active transport via transgenictially encoded and expressed NIS channel, cells can take up radioactive probe and subsequently be monitored by gamma camera or SPECT scanners. Availability of human reporter gene has been a major advantage in exploiting NIS gene as an imaging and therapeutic tool. Various studies utilized viral vectors to stably express NIS in cells [218,219,220,221,222]. However, when tracking the NIS expressing cells in whole body imaging applications, due to the presence of endogenous NIS regions like thyroid, stomach and radioactivity in bladder, may result in background signals. One of the advantages of using multimodal imaging in cell tracking would be to determine the status of administered cells. For example, cells labeled with SPIONs alone can be tracked by MRI, however this will not provide information on the functional status of the administered cells. If a bimodal contrast agent was used to transfect genes into the cells, the expression of the gene product can be detected by another complementary imaging modalities such as optical imager or nuclear medicine techniques indicative of the functional status of the cells. One of the examples of successful application of bimodal imaging in cell tracking is recent study where endothelial progenitor cells were used as cellular probes [223]. Magnetically labeled transgenic endothelial progenitor cells (EPC) were used to determine the migration, incorporation and expression of gene product in a mouse model of breast cancer. EPCs were genetically modified by using human sodium iodide symporter (hNIS) reporter gene that enabled to assess of the functional status by SPECT, while labeling with FDA approved agents, ferumoxides and protamine sulfate enabled tracing by MRI. Figure 6.4 shows the migration and accumulation of intravenously administered magnetically labeled, transgenic (carrying hNIS gtene) EPC to the subcutaneously implanted breast cancer in mouse model. The cell migration was detected both by MRI and SPECT systems.

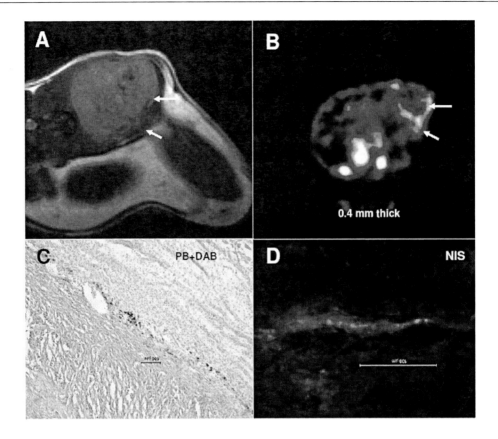

Figure 6.4. Accumulation of magnetically labeled, transgenic AC133+ progenitor cells around the implanted tumor. MRI shows low signal intensity areas at the margin of the tumor (A), which are at the corresponding sites of iron positive cells detected by Prussian blue staining (C). Trans-axial sections of SPECT study (B) indicate the accumulated transgenic AC133+ cells detected by T-99m. The SPECT study also proves the migration and homing of AC133+ cells at the margin of the tumors (seen on MRI). Immunohistochemistry shows the accumulation of hNIS positive cells at the corresponding sites, as detected by labeled secondary antibodies (D).

6.4. SPIONs as Drug Delivery Vehicles

Introduction

Nanoparticle mediated drug delivery is a new exciting approaches to combat malignant disease as well as to restore ischemic or infracted tissues. One of the advantages using iron based nanoparticle as drug delivery system is its property of para- or super-magnetism that can be scanned by non-invasive magnetic resonance imaging (MRI). The following criteria should be applied for any nanoparticles considered as drug delivery systems: 1) Hydrodynamic diameter should be small enough to extravasate through leaky vessels, however, size should not be as such to pass through normal endothelial lining, 2) surface charge should be optimized (near neutral) to make the particle not available for non-specific binding to plasma protein or any other tissues, 3) blood half-life should be long enough to recirculate through the organ of interest and accumulate at specific sites, 4) coating of the SPIONs should be functionalized to make the tagging easy with any deliverable drugs or peptide, 5) overall tagging chemistry should be easy, 6) the SPIONs should be biodegradable, and 7) payload should be released to the specific sites only. To release drugs to the site of interest, an enzymatic cleavable linkage can be inserted between the SPIONs surface and the drugs or payload. Legumain is one of the enzymes that can be utilized to cleave specific peptide sequences and legumain is highly expressed in different malignant cells, but very negligible expressed in normal cells [224,225].

Active Targeting Using Ligand Substrate Mechanisms, Antigen-Antibody or Receptors

In this strategy, iron containing nanoparticles are developed with the surface coating modified or functionalized to be attached to: a) antibody, to target specific antigen or receptors [226,227,228], b) different peptides, to target specific sites on cell membrane or block specific receptors [229], c) chemotherapeutic agents, to deliver it to the cells [230], d) folic acid [231,232], or e) hormones [233]. All these modification is for the facilitated uptake to the target tissues or cells. Dedicated iron oxide nanoparticle based chemistry laboratories with a dedicated chemist are essential to develop functionalized SPIO to target specific tumors and deliver drugs.

Antibody mediated targeting of tumors is not new and addition of SPIO enable investigators to determine the exact site of migration and accumulation of nanoparticles tagged with antibodies. One disadvantage of antibody mediated targeting of cancer is non-specific binding of antibody and formation of host antibody against administered antibody. Nuclear medicine based immunoscintigraphy was applied in early 90's but non-specific uptake to other non-target tissues and formation of HAMA (human anti-mouse antibody) discouraged the techniques [234,235,236,237]. However, recent development of antibody technology (making of monobody, smaller fraction of antibody) revived the investigations [238,239,240]. Antibody mediated targeting and drug delivery depends on the expression of specific antigen on the surface of the cells. Malignant tumors usually express higher amount of transferrin receptor (CD71) or epidermal growth factor receptor (EGFR). Successful antibody mediated SPIONs based drug delivery systems have been developed and used in various experiments [226,227,241,242,243,244].

Targeting cell membrane receptors using specific ligands is an attractive way of SPIONs based drug delivery. It is important to understand the basic mechanisms of ligand receptor interaction that can be used to deliver the therapeutic agents into the cells. Folate receptor has been targeted by oncologist to deliver chemotherapeutic agents to prevent DNA synthesis. Folic acid is an essential nutrient for the development of DNA and survival of malignant tumors [245,246,247,248]. Most of the malignant tumors, especially hematopoietic, express folic acid receptor many fold higher than that of normal cells, therefore, injecting SPIONs tagged with folate are thought to target malignant cells [232,247,249,250]. Investigators have made different SPIONs based delivery system that targets folic acid receptors [231,248,249]. Liong et al. has developed a SPIONs based delivery system that targets folic acid receptors and deliver paclitaxel and camptothecin in pancreatic cancer cells [230]. SPIONs was delivered into the cells through folic acid receptor and once inside SPIONs released the chemotherapeutic agents and initiated cells toxicity. Similar strategy has also been adopted by other investigators to target hormone receptors [251,252]. Peptides that target cell receptors have also been developed to targeted delivery of drugs using SPIONs [229,253]. Cyclic RDG peptide is a ligand for αvβ3 integrin. αvβ3 integrin is expressed in endothelial cells lining the neovasculatures of malignant tumors and cyclic RDG is important for angiogenesis. Analog of RDG peptide, Cilengitide, is being used in clinical trials to control angiogenesis in tumors, especially glioma. Cilengitide actively binds to and inhibit the activities of αvβ3 and αvβ5, thereby inhibit endothelial cell-cell interaction and angiogenesis [254,255,256].

Enhanced Permeability Retention (EPR) Effect

Tumor growth depends on angiogenesis[257,258,259]. Vascular endothelial growth factor (VEGF) is produced and released by tumor cells undergoing hypoxia and it causes active angiogenesis. These newly formed blood vessels are usually abnormal in morphology and architecture. The lining of the vessels is incomplete and enables macromolecules to pass through and accumulate into tumors. EPR is the property by which administered macromolecules such as smaller SPIONs will accumulate in the tumor tissues [260,261,262,263]. To be able to accumulate, the size of the SPIO must be optimal and the blood half-life long enough. Currently available SPIONs such as Combidex, Supravist and Femoxytol accumulate in tumors by EPR effect due to VEGF induced increased permeability of tumor vasculatures (*figure 6.2 shows an example of epr effects*). The EPR property of tumors has been utilized to deliver SPIONs loaded with different anticancer agents [264,265]. Delivery of SPIONs with payloads to the tumors using EPR effect is a passive mechanism and SPIONs with long blood half-life SPIONs will accumulate within interstitial spaces of the tumors tissues, however, to penetrate tumor cell membrane to release the payloads into cytosol or enodsome-lysome, either membrane penetrable peptide or facilitated phagocytic mechanisms should be utilized.

Convection-Enhanced Delivery

Convection is the process of movement of molecules in fluids and the process does not happen in solids. Convection-enhanced drug delivery (CED) was first proposed and introduced by the investigators at the National Institutes of Health (NIH) in USA [266]. The process utilizes continuous injection under positive pressure to disperse the therapeutic agents in fluids. The CED method was developed for the treatment of brain tumors due to limited transfer of drugs through blood brain barrier (BBB). Saito et al. has utilized CED method to determine the distribution of liposome as drug delivery vehicles in rat brain tumor and monitor the migration and distribution by *in vivo* MRI [267]. For *in vivo* imaging, gadolinium agent Gadodiamide was entrapped into liposomes and tagged with a fluorescent dye. The distribution was confirmed by histology and fluorescent microscopy. The authors also used the same liposomal composition to deliver a chemotherapeutic agent, Doxil, to brain tumor. Similar approach has been utilized by Perlstein et al. [268] to deliver monodispersed maghemite nanopartciles (MNPs) into normal rat brain and showed that CED can be used to deliver SPIONs and attached payloads to malignant tumors or to the site of interests.

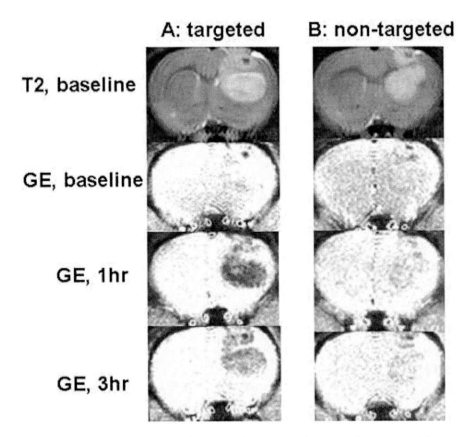

Figure 6.5. Representative subset of kinetic series of MRI scans demonstrating nanoparticle accumulation in 9L gliosarcoma (A) with and (B) without magnetic targeting. The spin echo T2-weighted baseline images illustrate the tumor location clearly observable as a hyper-intense lesion. GE baseline images were acquired before the nanoparticle injection, while 1-hr and 3-hr images were acquired 1 and 3 hours after nanoparticle administration, respectively. (Reprinted from Publication Biomaterials. 2008; 29(4): 487–496, by Beata Chertok, Bradford A. Moffat, Allan E. David, et al. "Iron Oxide Nanoparticles as a Drug Delivery Vehicle for MRI Monitored Magnetic Targeting of Brain Tumors", Copyright (2010), with permission from Elsevier).

Magnet Activated Delivery

Magnet activated separation of SPIONs (SPIONs attached with cells, proteins, etc) is well known method in cell and molecular biology laboratories. In general, biological samples tagged with SPIONs are separated from rest of the samples using strong magnet. These methods are mostly used in *in vitro* situations. Our group first published the results on enhanced delivery and retention of SPIONs labeled cells in rat liver after intravenous administration of SPIO labeled mesenchymal stem cells (MSC) [269]. In the experiments a group of rats wore jackets with a magnet placed over the liver during and 29 days after the intravenous administration of SPIO labeled MSC. The number of accumulated MSC, iron content (Fe μg/gm liver tissues) and MRI signal intensity changes were compared with another group of rats that received SPIONs labeled MSC with no magnet placed over liver. Similar magnet activated enhanced delivery and retention of SPIO labeled administered cells to the site of interest have been reported by other groups [270,271,272,273,274]. A proof of principle study on magnet activated delivery of SPIONs, loaded with drugs, to brain tumor in rats has been reported by investigators at University of Michigan, USA [275,276]. The investigators have used a magnetic field density of 0 Tesla (for control group) and 0.4 Tesla (for experimental group) for 30 minutes during and following IV administration of SPIONs (12 mg Fe/Kg). MR images were acquired before and after the administration of SPIONs. Image analysis revealed 5-fold increased in the accumulation of SPIONs in the tumors that were exposed to magnetic field during injection (Figure 6.5). The investigators are currently utilizing the method to deliver different payloads to the tumors. Other investigators also used magnetic targeting for site-specific drug delivery [277,278].

Drug Delivery by Direct Injection

Conventional intra-articular injection of corticosteroid for the treatment of arthritis suffers from limitations such as crystal formation and rapid clearance of drug from the joints. Investigators have used SPIONs based microparticles containing dexamethons or functionalized PVA-vinyl coated SPIONs and injected in to athritic joints [279,280]. The use of SPIONs was to extend the retention of injected particles by an external magnetic field [281]. Butoescu et al. used an antigen-induced mouse arthritis model and SPIONs embedded microparticles containing dexamethason [282,283]. The agents were injected directly into the joint and a magnetic field was applied to retain the agents for longer time. The results showed extended retention of SPIONs based microparticles in the joint with remission of joint inflammation.

6.5. SPIONS FOR HYPERTHERMIA TREATMENT

Hyperthermia (also called thermal therapy or thermotherapy) is a type of cancer treatment in which body tissue or tumor is exposed to high temperatures or tumor temperature is raised to about 42.5°C (108°F) for about 45 to 60 minutes. Research has shown that high temperatures can damage and kill cancer cells with minimal injury to normal tissues [284,285]. Hyperthermia can kill cancer cells and damage proteins and structures within cells [286], and may decrease the size of the tumors. Heat improves the blood circulation to the tumors and makes them susceptible to radiation therapy. In recent years investigators have gained interest in SPIONs for its properties (strong magnetic moments) to induce or generate temperatures when exposed to an alternating magnetic field [287,288,289]. The induction of heat following exposure to magnetic field is called selective ferromagnetic induced hyperthermia or magnetic fluid hyperthermia (MFH) [290,291,292]. Investigators have used dextran coated SPIO [293,294], lanthanum magnetite particles with inserted silver ions to control Curie temperature [288,295,296] in the hyperthermia range of interest, aminosilan shell magnetite [297], magnetite cationic liposomes [298,299], and thermosensitive magnetoliposomes [300], thermosensitive magnetoliposomes [300] and other formulations [301] for SPIONs based hyperthermia treatment of tumors. Based on the *in vitro* preliminary studies Hofmann-Amtenbrink *et al.* made embolization formulations of magnetic nanoparticles embedded in silica and administered in preformed subcutaneous colon carcinoma in mouse model. The tumor was then exposed to alternating magnetic field strength between 6 to 12 MT with a frequency of 141 kHz and showed the temperature increase up to 44.5°C with the tumor. Two days after exposure, animals were euthanized and histology analysis of tissue sections revealed heat induced necrosis in the tumors [281]. Maelnikov et al. [295] showed the effectiveness of a new class of nanoparticles based on silver-doped manganites La1–xAgxMnO3þd. Silver-doped perovskite manganites particles demonstrated the effect of adjustable Curie temperature for cellular hyperthermia. The magnetic relaxation properties of the particles are comparable with that of SPIO, therefore they were able to monitor the particle movement and retention by MRI. Thus, the new material combines the MRI contrast enhancement capability with targeted hyperthermia treatment. Rise of temperatures during exposure to alternating magnetic fields and the effects on tumor cell killing are also depend on the size and coating of SPIONs. Jordan et al. [297] have used two different SPIONs, carboxydextrancoated DDM128 P6 (Schering AG, Berlin, Germany) with an average core diameter of 3 nm and aminosilanecoated MFL AS (MagForce Nanotechnologies AG, Berlin, Germany) with an average core diameter of 15 nm, for alternating magnetic field induced hyperthermia treatment in rat glioma. The authors concluded that due to surface coating and size of the SPIONs, the distribution of the particles and the rise of temperatures in the tumors were different. The dextran coated particles spread into surrounding tissues soon after intratumor injection and there was only 2°C rise in core temperature, whereas aminosilanecoated coated magnetic particles remained within the tumors with a substantial temperature rise that was above the temperature needed to kill tumor cells. The overall survival of animals was 4.5 fold higher in group that received aminosilane-coated nanopartcles compared to control group or group treated with dextran coated SPIONs. Again, it is important to achieve high concentration of SPIONs within the site of interest for optimal rise in temperature following alternating magnetic fields. In this respect, targeted delivery by modifying the surface coating of SPIONs is essential, as discussed in the section of drug delivery. Targeted drug delivery (cytotoxic) followed by hyperthermia induced killing of tumor cells could be the future prospect for SPIONs technology. The agents will act as both therapeutic and imaging probes.

6.6. SPIONs FOR ENHANCED TRANSFECTION OF DNA (MAGNETOFECTION)

Magnetofection is a technique that utilizes self-assembling complexes of enhancers like cationic lipids with plasmid DNA or small interfering RNA (siRNA) that are associated with magnetic nanoparticles. These complexes are then concentrated at the surface of cultured cells by applying a permanent inhomogeneous magnetic field [302,303,304,305,306,307]. This technique delivers nanoparticles directly to the cytoplasm and it results in a considerable increase in transfection efficiency compared to transfections done with nonmagnetic gene vectors. For a while, it presented itself as a promising new approach for introducing foreign oligonuclotides to the cell. Although it did not compensate for all the drawbacks of any given standard gene transfer method *in vitro*, its major potential is considered to be the extraordinarily rapid and efficient transfection at low vector doses and the possibility of remotely controlled vector targeting *in vivo* [304]. This technique proved useful if used in adherent cells where it delivered genes rapidly with very low amount of vectors used. Krotz et al demonstrated that magnetofection potentiated gene delivery in cultured endothelial cell [308]. Similar reports came from the studies on mouse embryonic stem cells [309], neurons [310] and respiratory epithelial cells [305]. Although, recent report indicated efficient use of this method for viral gene transfer in human leukemia K562 suspension cell line [302], additional studies are needed to determine magnetofection applicability in labeling suspension cells or cell with small cytoplasm to nuclear ratio (such as T-cells and hematopoietic stem cells). One of the major deterrents for the use of magnetofection is direct delivery to the cell cytoplasm and possible cytotoxicity following the release of iron into the cytoplasm or nucleus. Moreover, detailed work on magnetofection toxicity and nuclear uptake has not been reported yet.

6.7. SPIONs for Cell/Protein Separation

Beside their application as MRI contrast agents, SPIONs are used in cell and protein separation techniques that utilize SPIONs in the form of magnetic beads. At the present time, majority of laboratories involved with cell isolation and protein purification projects use commercially available magnetic beads that, with their speed, ease of use and affordability, became a popular choice for cell/protein separation. Magnetic beads products can be purchased from a number of different companies that make their products available in various shapes and forms and under the distinct trademark names. Therefore, the fine details on the exact chemical composition of these products are usually proprietary information. Nevertheless, the basic operating concepts are common for majority of products. Superparamagnetic properties of the beads come from the Iron (III) oxide (Fe_2O_3) particles deposited in the core that is coated with polystyrene shell. This shell provide a surface containing hydrophobic groups that facilitate physical absorption of molecules, such as antibodies, and hydroxyl groups that can covalently bind molecules such as streptavidin, lectins, and peptides. With this design, the beads act as magnets when exposed to a magnetic field, without residual magnetization when magnetic field is removed. The size of superparamagnetic beads is 20-100 nm and they are therefore called Micro Beads. While, larger size beads (0.5-5µm) can also be used, their application is associated with non-specific cell selection and adverse effects on cell viability. Initially, small nanoparticles were not utilized for cell separation protocols due to their small magnetic moment that, in context of conventional magnetic fields, resulted in long separation times [311,312]. However, novel studies of the time introduced new designs using small nanoparticles for cell separation that ultimately decreased the need for large magnetic spheres. Molday and MacKenzie described novel application of immunospecific iron dextran microbeads with high gradient magnetic cell separation technique [313,314,315]. This protocol was further modified by Miltenyi *et al.*, who added fluorescent labeling to the process and laid out the platform for combining magnetic based cell separation and fluorescent sorting [316]. Since first introduced for cell separation purposes, several types of magnetic microbeads have been available, but most widely used was combination of iron oxide and polysaccharide, and in early 1980s' for companies such as Miltenyi Biotec it became the focal point of research.

General concept for cell separation is simple and starts with incubating the mono-disperse suspension of cells with magnetic beads designed to target specific cell population, usually with specific monoclonal antibody covalently attached to the bead surface. Next, cells are washed, to remove the excess reagent, and passed through the column that contains ferromagnetic matrix and is exposed to a strong magnetic field. Cells labeled with microbeads are retained, while non-labeled pass through the column and can be collected. Labeled cells are released after removing the column from the magnet. Thus, with this approach both positive and negative cell selection can be performed, with relatively high purity. Since micro beads are very small and are made of biodegradable elements, they do not have to be removed from cells that can immediately be used for further experiments. As mentioned earlier, a variety of ready to use and custom-made products are available through different companies. However, the most widely used are MACS® Magnetic Cell Separation and similar products from M microbeads from Miltenyi Biotec, DYNAL® magnetic beads, and immunomagnetic based products from STEMCELL Technologies®,

In addition to cell separation, magnetic beads have also become widely used for protein purification. While the basics of magnetisms applied are the same as for cell separation technique, modifications and molecular tags are designed to suit protein affinity purification. Functional magnetic beads with different surface groups such as amine, carboxy, aldehyde, epoxy, hydrazide etc. that can be easily used to couple proteins and antibodies are commercially available from most of the manufacturers that produce cell separation systems. In addition to magnetic beads, complete magnetic purification systems can also be commercially obtained and the most commonly used are specifically designed for fast purifications of recombinantly expressed fusion proteins such as His- and GST-tagged proteins, IgG antibodies and removal of *E.coli* protein-reactivated antibodies. However, new technologies are introduced constantly and new highly sensitive methods and complex arrays are continually being added to the range of magnetic protein chromatography protocols.

6.8. SPIONs as Catalyst

In addition to the described applications, iron oxide nanoparticles (IONPs) could also function as artificial peroxidase [317]. IONPs are more stable and posses an unchanged catalytic activity over a wide range of temperature and pH [317]. In an investigation Yu et al. [318] have reported the impact of coating on the peroxidase activity of IONPs. Nanoparticles with six different coating structures, unmodified IONPs, citrate and glycine-modified, polylysine, poly-ethyleneimine, carboxymethyl dextran, and heparin-coated IONPs were synthesized and characterized by FTIR, TGA, TEM, size, zeta potential, and SQUID; and evaluated for peroxidase activity. The peroxidase-like activity of iron oxide nanoparticles was found dependent on the surface attributes of nanoparticles. Anionic nanoparticles had a high affinity for 3,3',5,5'-tetramethylbenzidine (TMB) substrate and exhibited a high catalytic activity.

It is already known that catalytic properties of bulky or nanoscale ironoxides differ significantly. For instance, relatively large iron oxide nanoparticles (~13—15 nm) obtained at a high (10—15%) iron content on silica gel are reducible with hydrogen on heating to metallic iron. At the same time, fine iron oxide nano clusters (<5 nm in size), which are formed on the support at a low (down to 5%) iron content, cannot virtually be reduced under similar conditions. This is caused by a stronger interaction between nanoclusters of a smaller size and the support surface. In addition to the particle size, the structure of the support, its surface, and porosity exert a substantial effect on the formation of catalytically active iron [319,320].

Gold-iron nanoparticles were also synthesize by Horvath *et al.* and showed the catalytic activity by CO oxidation [321]. Similar findings on CO oxidation were also reported by Herzing, *et al.* where they used active gold nanocluster on iron oxide [322]. Different composition of iron based catalysts are in use in the environmental sciences and in industries to clean up hormones, arsenic and other contaminants [323].

6.9. SPIONS FOR TISSUE REPAIR

SPIONs have no specific role in tissue repair. However, modification of SPIONs to carry different payloads may help combat different diseases. Moreover, by using alternating magnetic field, accumulated SPIONs can generate heat to kill tumor cells and the technique can be used to combat malignant tumors. In addition, SPIONs labeled stem cells can be delivered to the site of interest by external magnet to enhance the effect of stem cell therapy.

REFERENCES

[1] Neuwelt EA, Hamilton BE, Varallyay CG, Rooney WR, Edelman RD, et al. (2009) Ultrasmall superparamagnetic iron oxides (USPIOs): a future alternative magnetic resonance (MR) contrast agent for patients at risk for nephrogenic systemic fibrosis (NSF)? *Kidney Int* 75: 465-474.

[2] Yeh TC, Zhang W, Ildstad ST, Ho C (1993) Intracellular labeling of T-cells with superparamagnetic contrast agents. *Magn Reson Med* 30: 617-625.

[3] Bulte JW, De Jonge MW, Kamman RL, Go KG, Zuiderveen F, et al. (1992) Dextran-magnetite particles: contrast-enhanced MRI of blood-brain barrier disruption in a rat model. *Magnetic Resonance in Medicine* 23: 215-223.

[4] Bulte JW, Brooks RA, Moskowitz BM, Bryant LH, Jr., Frank JA (1999) Relaxometry and magnetometry of the MR contrast agent MION-46L. *Magn Reson Med* 42: 379-384.

[5] Bulte JW, Douglas T, Witwer B, Zhang SC, Strable E, et al. (2001) Magnetodendrimers allow endosomal magnetic labeling and in vivo tracking of stem cells. *Nature Biotechnology* 19: 1141-1147.

[6] Josephson L, Lewis J, Jacobs P, Hahn PF, Stark DD (1988) The effects of iron oxides on proton relaxivity. *Magn Reson Imaging* 6: 647-653.

[7] Jung CW (1995) Surface properties of superparamagnetic iron oxide MR contrast agents: ferumoxides, ferumoxtran, ferumoxsil. *Magn Reson Imaging* 13: 675-691.

[8] Bulte JW, Arbab AS, Douglas T, Frank JA (2004) Preparation of magnetically labeled cells for cell tracking by magnetic resonance imaging. *Methods Enzymol* 386: 275-299.

[9] Taupitz M, Schmitz S, Hamm B (2003) [Superparamagnetic iron oxide particles: current state and future development]. *Rofo* 175: 752-765.
[10] Modo M, Hoehn M, Bulte JW (2005) Cellular MR imaging. *Mol Imaging* 4: 143-164.
[11] Wang YX, Hussain SM, Krestin GP (2001) Superparamagnetic iron oxide contrast agents: physicochemical characteristics and applications in MR imaging. *Eur Radiol* 11: 2319-2331.
[12] Ittrich H, Lange C, Dahnke H, Zander AR, Adam G, et al. (2005) [Labeling of mesenchymal stem cells with different superparamagnetic particles of iron oxide and detectability with MRI at 3T]. *Rofo* 177: 1151-1163.
[13] Mikhaylova M, Kim do, K, Bobrysheva, N, Osmolowsky, M, Semenov, V, Tsakalakos, T, Muhammed, M. (2004) Superparamagnetism of magnetite nanoparticles: dependence on surface modification. *Langmuir* 20: 2472-2477.
[14] Toyoda K, Tooyama I, Kato M, Sato H, Morikawa S, et al. (2004) Effective magnetic labeling of transplanted cells with HVJ-E for magnetic resonance imaging. *Neuroreport* 15: 589-593.
[15] Hogemann D, Josephson L, Weissleder R, Basilion JP (2000) Improvement of MRI probes to allow efficient detection of gene expression. *Bioconjug Chem* 11: 941-946.
[16] Hawrylak N, Ghosh P, Broadus J, Schlueter C, Greenough WT, et al. (1993) Nuclear magnetic resonance (NMR) imaging of iron oxide-labeled neural transplants. *Exp Neurol* 121: 181-192.
[17] Yeh TC, Zhang W, Ildstad ST, Ho C (1995) In vivo dynamic MRI tracking of rat T-cells labeled with superparamagnetic iron-oxide particles. *Magn Reson Med* 33: 200-208.
[18] Shen TT, Bogdanov A, Jr., Bogdanova A, Poss K, Brady TJ, et al. (1996) Magnetically labeled secretin retains receptor affinity to pancreas acinar cells. *Bioconjug Chem* 7: 311-316.
[19] Fleige G, Seeberger F, Laux D, Kresse M, Taupitz M, et al. (2002) In vitro characterization of two different ultrasmall iron oxide particles for magnetic resonance cell tracking. *Invest Radiol* 37: 482-488.
[20] Kaufman CL, Williams M, Ryle LM, Smith TL, Tanner M, et al. (2003) Superparamagnetic iron oxide particles transactivator protein-fluorescein isothiocyanate particle labeling for in vivo magnetic resonance imaging detection of cell migration: uptake and durability. *Transplantation* 76: 1043-1046.
[21] Koch AM, Reynolds F, Kircher MF, Merkle HP, Weissleder R, et al. (2003) Uptake and metabolism of a dual fluorochrome Tat-nanoparticle in HeLa cells. *Bioconjug Chem* 14: 1115-1121.
[22] Ho C, Hitchens TK (2004) A non-invasive approach to detecting organ rejection by MRI: monitoring the accumulation of immune cells at the transplanted organ. *Curr Pharm Biotechnol* 5: 551-566.
[23] Song H, Choi, JS, Huh, YM, Kim, S, Jun, YW, Suh, JS, Cheon, J. (2005) Surface modulation of magnetic nanocrystals in the development of highly efficient magnetic resonance probes for intracellular labeling. *J Am Chem Soc* 127: 9992-9993.
[24] Schulze E, Ferrucci, JT Jr, Poss, K, Lapointe, L, Bogdanova, A, Weissleder, R. (1995) Cellular uptake and trafficking of a prototypical magnetic iron oxide label in vitro. *Invest Radiol* 30: 604-610.

[25] Bulte JW, Douglas T, Witwer B, Zhang SC, Lewis BK, et al. (2002) Monitoring stem cell therapy in vivo using magnetodendrimers as a new class of cellular MR contrast agents. *Academic Radiology 9 Suppl* 2: S332-335.

[26] Smirnov P, Gazeau F, Lewin M, Bacri JC, Siauve N, et al. (2004) In vivo cellular imaging of magnetically labeled hybridomas in the spleen with a 1.5-T clinical MRI system. *Magn Reson Med* 52: 73-79.

[27] Riviere C, Boudghene FP, Gazeau F, Roger J, Pons JN, et al. (2005) Iron oxide nanoparticle-labeled rat smooth muscle cells: cardiac MR imaging for cell graft monitoring and quantitation. *Radiology* 235: 959-967.

[28] Brillet PY, Gazeau F, Luciani A, Bessoud B, Cuenod CA, et al. (2005) Evaluation of tumoral enhancement by superparamagnetic iron oxide particles: comparative studies with ferumoxtran and anionic iron oxide nanoparticles. *Eur Radiol* 15: 1369-1377.

[29] Wilhelm C, Billotey C, Roger J, Pons JN, Bacri JC, et al. (2003) Intracellular uptake of anionic superparamagnetic nanoparticles as a function of their surface coating. *Biomaterials* 24: 1001-1011.

[30] Rinck PA, Smevik O, Nilsen G, Klepp O, Onsrud M, et al. (1991) Oral magnetic particles in MR imaging of the abdomen and pelvis. *Radiology* 178: 775-779.

[31] Oksendal AN, Jacobsen TF, Gundersen HG, Rinck PA, Rummeny E (1991) Superparamagnetic particles as an oral contrast agent in abdominal magnetic resonance imaging. *Invest Radiol* 26 Suppl 1: S67-70; discussion S71.

[32] Van Beers B, Grandin C, Jamart J, Demeure R, Jacobsen TF, et al. (1992) Magnetic resonance imaging of lower abdominal and pelvic lesions: assessment of oral magnetic particles as an intestinal contrast agent. *Eur J Radiol* 14: 252-257.

[33] Van Beers BE, Grandin C, De Greef D, Lundby B, Pringot J (1996) Ferristene as intestinal MR contrast agent. Distribution and safety of a fast ingestion procedure with oral metoclopramide. *Acta Radiol* 37: 676-679.

[34] MacVicar D, Jacobsen TF, Guy R, Husband JE (1993) Phase III trial of oral magnetic particles in MRI of abdomen and pelvis. *Clin Radiol* 47: 183-188.

[35] Oksendal AN, Bach-Gansmo T, Jacobsen TF, Eide H, Andrew E (1993) Oral magnetic particles. Results from clinical phase II trials in 216 patients. *Acta Radiol* 34: 187-193.

[36] Jacobsen TF, Laniado M, Van Beers BE, Dupas B, Boudghene FP, et al. (1996) Oral magnetic particles (ferristene) as a contrast medium in abdominal magnetic resonance imaging. *Acad Radiol* 3: 571-580.

[37] Tervahartiala P, Kivisaari L, Lamminen A, Maschek A, Wohling H, et al. (1997) Dynamic fast-gradient echo MR imaging of pancreatic tumours. *Eur J Radiol* 25: 74-80.

[38] Boudghene FP, Bach-Gansmo T, Grange JD, Lame S, Nantois C, et al. (1993) Contribution of oral magnetic particles in MR imaging of the abdomen with spin-echo and gradient-echo sequences. *J Magn Reson Imaging* 3: 107-112.

[39] Arbab AS, Ichikawa T, Sou H, Araki T, Nakajima H, et al. (2002) Ferumoxides-enhanced double-echo T2-weighted MR imaging in differentiating metastases from nonsolid benign lesions of the liver. *Radiology* 225: 151-158.

[40] Scott J, Ward J, Guthrie JA, Wilson D, Robinson PJ (2000) MRI of liver: a comparison of CNR enhancement using high dose and low dose ferumoxide infusion in patients with colorectal liver metastases. *Magn Reson Imaging* 18: 297-303.

[41] Tanimoto A (2001) [Liver-specific MR contrast agents: current status and prospects]. *Nippon Igaku Hoshasen Gakkai Zasshi - Nippon Acta Radiologica* 61: 525-533.

[42] Takahama K, Amano Y, Hayashi H, Ishihara M, Kumazaki T (2003) Detection and characterization of focal liver lesions using superparamagnetic iron oxide-enhanced magnetic resonance imaging: comparison between ferumoxides-enhanced T1-weighted imaging and delayed-phase gadolinium-enhanced T1-weighted imaging. *Abdom Imaging* 28: 525-530.

[43] Grazioli L, Morana G, Kirchin MA, Caccia P, Romanini L, et al. (2003) MRI of focal nodular hyperplasia (FNH) with gadobenate dimeglumine (Gd-BOPTA) and SPIO (ferumoxides): an intra-individual comparison. *J Magn Reson Imaging* 17: 593-602.

[44] Frohlich JM (2004) MRI of focal nodular hyperplasia (FNH) with gadobenate dimeglumine (Gd-BOPTA) and SPIO (ferumoxides): an intra-individual comparison.[comment]. *Journal of Magnetic Resonance Imaging* 19: 375-376; author reply 376.

[45] Paley MR, Mergo PJ, Torres GM, Ros PR (2000) Characterization of focal hepatic lesions with ferumoxides-enhanced T2-weighted MR imaging. *AJR Am J Roentgenol* 175: 159-163.

[46] Stark DD, Weissleder R, Elizondo G, Hahn PF, Saini S, et al. (1988) Superparamagnetic iron oxide: clinical application as a contrast agent for MR imaging of the liver. *Radiology* 168: 297-301.

[47] Weissleder R, Stark DD, Engelstad BL, Bacon BR, Compton CC, et al. (1989) Superparamagnetic iron oxide: pharmacokinetics and toxicity. *AJR Am J Roentgenol* 152: 167-173.

[48] Majumdar S, Zoghbi SS, Gore JC (1990) Pharmacokinetics of superparamagnetic iron-oxide MR contrast agents in the rat. *Invest Radiol* 25: 771-777.

[49] Pouliquen D, Le Jeune JJ, Perdrisot R, Ermias A, Jallet P (1991) Iron oxide nanoparticles for use as an MRI contrast agent: pharmacokinetics and metabolism. *Magn Reson Imaging* 9: 275-283.

[50] Abe Y, Yamashita Y, Namimoto T, Tang Y, Takahashi M (2000) The value of fast and ultrafast T2-weighted MR imaging sequences in hepatic enhancement with ferumoxides: comparison with conventional spin-echo sequence. *Radiat Med* 18: 97-105.

[51] Reimer P, Jahnke N, Fiebich M, Schima W, Deckers F, et al. (2000) Hepatic lesion detection and characterization: value of nonenhanced MR imaging, superparamagnetic iron oxide-enhanced MR imaging, and spiral CT-ROC analysis. *Radiology* 217: 152-158.

[52] Bluemke DA, Paulson EK, Choti MA, DeSena S, Clavien PA (2000) Detection of hepatic lesions in candidates for surgery: comparison of ferumoxides-enhanced MR imaging and dual-phase helical CT. *AJR Am J Roentgenol* 175: 1653-1658.

[53] Bluemke D, Weber, TM, Rubin, D, de Lange, EE, Semelka, R, Redvanly, RD, Chezmar, J, Outwater, E, Carlos, R, Saini, S, Holland, GA, Mammone, JF, Brown, JJ, Milestone, B, Javitt, MC, Jacobs, P. (2003) Hepatic MR imaging with ferumoxides: multicenter study of safety and effectiveness of direct injection protocol. *Radiology* 228: 457-464.

[54] Kanematsu M, Itoh K, Matsuo M, Maetani Y, Ametani F, et al. (2001) Malignant hepatic tumor detection with ferumoxides-enhanced MR imaging with a 1.5-T system: comparison of four imaging pulse sequences. *J Magn Reson Imaging* 13: 249-257.

[55] Winter PM, Caruthers SD, Yu X, Song SK, Chen J, et al. (2003) Improved molecular imaging contrast agent for detection of human thrombus. *Magnetic Resonance in Medicine* 50: 411-416.

[56] Muller RD, Vogel K, Neumann K, Hirche H, Barkhausen J, et al. (1999) SPIO-MR imaging versus double-phase spiral CT in detecting malignant lesions of the liver. *Acta Radiol* 40: 628-635.

[57] Kondo H, Kanematsu M, Hoshi H, Murakami T, Kim T, et al. (2000) Preoperative detection of malignant hepatic tumors: comparison of combined methods of MR imaging with combined methods of CT. *AJR Am J Roentgenol* 174: 947-954.

[58] Choi D, Kim S, Lim J, Lee W, Jang H, et al. (2001) Preoperative detection of hepatocellular carcinoma: ferumoxides-enhanced mr imaging versus combined helical CT during arterial portography and CT hepatic arteriography. *AJR Am J Roentgenol* 176: 475-482.

[59] Lawaczeck R, Bauer H, Frenzel T, Hasegawa M, Ito Y, et al. (1997) Magnetic iron oxide particles coated with carboxydextran for parenteral administration and liver contrasting. Pre-clinical profile of SH U555A. *Acta Radiol* 38: 584-597.

[60] Kopp AF, Laniado M, Dammann F, Stern W, Gronewaller E, et al. (1997) MR imaging of the liver with Resovist: safety, efficacy, and pharmacodynamic properties. *Radiology* 204: 749-756.

[61] Hirohashi S, Ichikawa T, Tanimoto A, Isobe Y, Hachiya J, et al. (2003) [Dose investigation of superparamagnetic iron oxide (SPIO) SH U 555 A in liver MR imaging]. *Nippon Igaku Hoshasen Gakkai Zasshi* 63: 539-550.

[62] Kehagias DT, Gouliamos AD, Smyrniotis V, Vlahos LJ (2001) Diagnostic efficacy and safety of MRI of the liver with superparamagnetic iron oxide particles (SH U 555 A). *J Magn Reson Imaging* 14: 595-601.

[63] Reimer P, Schuierer G, Balzer T, Peters PE (1995) Application of a superparamagnetic iron oxide (Resovist) for MR imaging of human cerebral blood volume. *Magn Reson Med* 34: 694-697.

[64] Onishi H, Murakami T, Kim T, Hori M, Hirohashi S, et al. (2009) Safety of ferucarbotran in MR imaging of the liver: a pre- and postexamination questionnaire-based multicenter investigation. *J Magn Reson Imaging* 29: 106-111.

[65] Reimer P, Muller M, Marx C, Wiedermann D, Muller R, et al. (1998) T1 effects of a bolus-injectable superparamagnetic iron oxide, SH U 555 A: dependence on field strength and plasma concentration--preliminary clinical experience with dynamic T1-weighted MR imaging. *Radiology* 209: 831-836.

[66] Reimer P, Tombach B (1998) Hepatic MRI with SPIO: detection and characterization of focal liver lesions. *Eur Radiol* 8: 1198-1204.

[67] Ichikawa T, Arbab AS, Araki T, Touyama K, Haradome H, et al. (1999) Perfusion MR imaging with a superparamagnetic iron oxide using T2-weighted and susceptibility-sensitive echoplanar sequences: evaluation of tumor vascularity in hepatocellular carcinoma. *AJR Am J Roentgenol* 173: 207-213.

[68] Arbab AS, Ichikawa T, Araki T, Toyama K, Nambu A, et al. (2000) Detection of hepatocellular carcinoma and its metastases with various pulse sequences using superparamagnetic iron oxide (SHU-555-A). *Abdom Imaging* 25: 151-158.

[69] Mintorovitch J, Shamsi K (2000) Eovist Injection and Resovist Injection: two new liver-specific contrast agents for MRI. *Oncology (Williston Park)* 14: 37-40.

[70] Reimer P, Rummeny EJ, Daldrup HE, Balzer T, Tombach B, et al. (1995) Clinical results with Resovist: a phase 2 clinical trial. *Radiology* 195: 489-496.

[71] Santoro L, Grazioli L, Filippone A, Grassedonio E, Belli G, et al. (2009) Resovist enhanced MR imaging of the liver: does quantitative assessment help in focal lesion classification and characterization? *J Magn Reson Imaging* 30: 1012-1020.

[72] Saito K, Sugimoto K, Nishio R, Araki Y, Moriyasu F, et al. (2009) Perfusion study of liver lesions with superparamagnetic iron oxide: distinguishing hepatocellular carcinoma from focal nodular hyperplasia. *Clin Imaging* 33: 447-453.

[73] Heilmaier C, Lutz AM, Bolog N, Weishaupt D, Seifert B, et al. (2009) Focal liver lesions: detection and characterization at double-contrast liver MR Imaging with ferucarbotran and gadobutrol versus single-contrast liver MR imaging. *Radiology* 253: 724-733.

[74] Rief M, Wagner M, Franiel T, Bresan V, Taupitz M, et al. (2009) Detection of focal liver lesions in unenhanced and ferucarbotran-enhanced magnetic resonance imaging: a comparison of T2-weighted breath-hold and respiratory-triggered sequences. *Magn Reson Imaging* 27: 1223-1229.

[75] Motosugi U, Ichikawa T, Nakajima H, Sou H, Sano M, et al. (2009) Imaging of small hepatic metastases of colorectal carcinoma: how to use superparamagnetic iron oxide-enhanced magnetic resonance imaging in the multidetector-row computed tomography age? *J Comput Assist Tomogr* 33: 266-272.

[76] Kim YK, Kim CS, Han YM (2009) Detection of small hepatocellular carcinoma: comparison of conventional gadolinium-enhanced MRI with gadolinium-enhanced MRI after the administration of ferucarbotran. *Br J Radiol* 82: 468-484.

[77] Kim HS, Choi Y, Song IC, Moon WK (2009) Magnetic resonance imaging and biological properties of pancreatic islets labeled with iron oxide nanoparticles. *NMR Biomed* 22: 852-856.

[78] Kim YK, Han YM, Kim CS (2009) Comparison of diffuse hepatocellular carcinoma and intrahepatic cholangiocarcinoma using sequentially acquired gadolinium-enhanced and Resovist-enhanced MRI. *Eur J Radiol* 70: 94-100.

[79] Grazioli L, Bondioni MP, Romanini L, Frittoli B, Gambarini S, et al. (2009) Superparamagnetic iron oxide-enhanced liver MRI with SHU 555 A (RESOVIST): New protocol infusion to improve arterial phase evaluation--a prospective study. *J Magn Reson Imaging* 29: 607-616.

[80] Chou CT, Chen RC, Chen WT, Lii JM (2009) Detection of hepatocellular carcinoma by ferucarbotran-enhanced magnetic resonance imaging: the efficacy of accumulation phase fat-suppressed T1-weighted imaging. *Clin Radiol* 64: 22-29.

[81] Kim SJ, Kim SH, Lee J, Chang S, Kim YS, et al. (2008) Ferucarbotran-enhanced 3.0-T magnetic resonance imaging using parallel imaging technique compared with triple-phase multidetector row computed tomography for the preoperative detection of hepatocellular carcinoma. *J Comput Assist Tomogr* 32: 379-385.

[82] McLachlan SJ, Morris MR, Lucas MA, Fisco RA, Eakins MN, et al. (1994) Phase I clinical evaluation of a new iron oxide MR contrast agent. *J Magn Reson Imaging* 4: 301-307.

[83] Anzai Y, Blackwell KE, Hirschowitz SL, Rogers JW, Sato Y, et al. (1994) Initial clinical experience with dextran-coated superparamagnetic iron oxide for detection of lymph node metastases in patients with head and neck cancer. *Radiology* 192: 709-715.

[84] Harisinghani MG, Saini S, Hahn PF, Weissleder R, Mueller PR (1998) MR imaging of lymph nodes in patients with primary abdominal and pelvic malignancies using ultrasmall superparamagnetic iron oxide (Combidex). Academic Radiology 5 Suppl 1: S167-169.

[85] Narayanan P, Iyngkaran T, Sohaib SA, Reznek RH, Rockall AG (2009) Magnetic resonance lymphography: a novel technique for lymph node assessment in gynecologic malignancies. *Cancer Biomark* 5: 81-88.

[86] Narayanan P, Iyngkaran T, Sohaib SA, Reznek RH, Rockall AG (2009) Pearls and pitfalls of MR lymphography in gynecologic malignancy. Radiographics 29: 1057-1069; *discussion* 1069-1071.

[87] Neuwelt EA, Varallyay P, Bago AG, Muldoon LL, Nesbit G, et al. (2004) Imaging of iron oxide nanoparticles by MR and light microscopy in patients with malignant brain tumours. *Neuropathol Appl Neurobiol* 30: 456-471.

[88] Taschner CA, Wetzel SG, Tolnay M, Froehlich J, Merlo A, et al. (2005) Characteristics of ultrasmall superparamagnetic iron oxides in patients with brain tumors. *AJR Am J Roentgenol* 185: 1477-1486.

[89] Muldoon LL, Sandor M, Pinkston KE, Neuwelt EA (2005) Imaging, distribution, and toxicity of superparamagnetic iron oxide magnetic resonance nanoparticles in the rat brain and intracerebral tumor. *Neurosurgery* 57: 785-796; *discussion* 785-796.

[90] Manninger SP, Muldoon LL, Nesbit G, Murillo T, Jacobs PM, et al. (2005) An exploratory study of ferumoxtran-10 nanoparticles as a blood-brain barrier imaging agent targeting phagocytic cells in CNS inflammatory lesions. *AJNR Am J Neuroradiol* 26: 2290-2300.

[91] Christoforidis GA, Yang M, Kontzialis MS, Larson DG, Abduljalil A, et al. (2009) High resolution ultra high field magnetic resonance imaging of glioma microvascularity and hypoxia using ultra-small particles of iron oxide. *Invest Radiol* 44: 375-383.

[92] Hyodo F, Chandramouli GV, Matsumoto S, Matsumoto K, Mitchell JB, et al. (2009) Estimation of tumor microvessel density by MRI using a blood pool contrast agent. *Int J Oncol* 35: 797-804.

[93] Klug G, Bauer L, Bauer WR (2008) Patterns of USPIO deposition in murine atherosclerosis. *Arterioscler Thromb Vasc Biol* 28: E157; author reply E158-159.

[94] Klug K, Gert G, Thomas K, Christan Z, Marco P, et al. (2009) Murine atherosclerotic plaque imaging with the USPIO Ferumoxtran-10. *Front Biosci* 14: 2546-2552.

[95] Tang TY, Muller KH, Graves MJ, Li ZY, Walsh SR, et al. (2009) Iron oxide particles for atheroma imaging. *Arterioscler Thromb Vasc Biol* 29: 1001-1008.

[96] Truijers M, Futterer JJ, Takahashi S, Heesakkers RA, Blankensteijn JD, et al. (2009) In vivo imaging of the aneurysm wall with MRI and a macrophage-specific contrast agent. *AJR Am J Roentgenol* 193: W437-441.

[97] Turner GH, Olzinski AR, Bernard RE, Aravindhan K, Boyle RJ, et al. (2009) Assessment of macrophage infiltration in a murine model of abdominal aortic aneurysm. *J Magn Reson Imaging* 30: 455-460.

[98] Wu YL, Ye Q, Sato K, Foley LM, Hitchens TK, et al. (2009) Noninvasive evaluation of cardiac allograft rejection by cellular and functional cardiac magnetic resonance. *JACC Cardiovasc Imaging* 2: 731-741.

[99] Epstein FH (2009) Heterogeneity of acute heart transplant rejection can be visualized by cellular and functional cardiac magnetic resonance. *JACC Cardiovasc Imaging* 2: 742-743.

[100] Bernd H, De Kerviler E, Gaillard S, Bonnemain B (2009) Safety and tolerability of ultrasmall superparamagnetic iron oxide contrast agent: comprehensive analysis of a clinical development program. Invest Radiol 44: 336-342.

[101] Nguyen BC, Stanford W, Thompson BH, Rossi NP, Kernstine KH, et al. (1999) Multicenter clinical trial of ultrasmall superparamagnetic iron oxide in the evaluation of mediastinal lymph nodes in patients with primary lung carcinoma. *J Magn Reson Imaging* 10: 468-473.

[102] Pannu HK, Wang KP, Borman TL, Bluemke DA (2000) MR imaging of mediastinal lymph nodes: evaluation using a superparamagnetic contrast agent. *J Magn Reson Imaging* 12: 899-904.

[103] Kernstine KH, Stanford W, Mullan BF, Rossi NP, Thompson BH, et al. (1999) PET, CT, and MRI with Combidex for mediastinal staging in non-small cell lung carcinoma. *Ann Thorac Surg* 68: 1022-1028.

[104] Harisinghani MG, Saksena M, Ross RW, Tabatabaei S, Dahl D, et al. (2005) A pilot study of lymphotrophic nanoparticle-enhanced magnetic resonance imaging technique in early stage testicular cancer: a new method for noninvasive lymph node evaluation. *Urology* 66: 1066-1071.

[105] Heesakkers RA, Hovels AM, Jager GJ, van den Bosch HC, Witjes JA, et al. (2008) MRI with a lymph-node-specific contrast agent as an alternative to CT scan and lymph-node dissection in patients with prostate cancer: a prospective multicohort study. *Lancet Oncol* 9: 850-856.

[106] Barentsz JO, Futterer JJ, Takahashi S (2007) Use of ultrasmall superparamagnetic iron oxide in lymph node MR imaging in prostate cancer patients. *Eur J Radiol* 63: 369-372.

[107] Harisinghani MG, Saksena MA, Hahn PF, King B, Kim J, et al. (2006) Ferumoxtran-10-enhanced MR lymphangiography: does contrast-enhanced imaging alone suffice for accurate lymph node characterization? *AJR Am J Roentgenol* 186: 144-148.

[108] Wu L, Cao Y, Liao C, Huang J, Gao F (2010) Diagnostic performance of USPIO-enhanced MRI for lymph-node metastases in different body regions: *A meta-analysis.* Eur J Radiol.

[109] Rohl L, Ostergaard L, Simonsen CZ, Vestergaard-Poulsen P, Sorensen L, et al. (1999) NC100150-enhanced 3D-SPGR MR angiography of the common carotid artery in a pig vascular stenosis model. Quantification of stenosis and dose optimization. *Acta Radiol* 40: 282-290.

[110] Alley MT, Napel S, Amano Y, Paik DS, Shifrin RY, et al. (1999) Fast 3D cardiac cine MR imaging. *J Magn Reson Imaging* 9: 751-755.

[111] Kellar KE, Fujii DK, Gunther WH, Briley-Saebo K, Spiller M, et al. (1999) 'NC100150', a preparation of iron oxide nanoparticles ideal for positive-contrast MR angiography. *MAGMA* 8: 207-213.

[112] Ahlstrom KH, Johansson LO, Rodenburg JB, Ragnarsson AS, Akeson P, et al. (1999) Pulmonary MR angiography with ultrasmall superparamagnetic iron oxide particles as a blood pool agent and a navigator echo for respiratory gating: pilot study. *Radiology* 211: 865-869.

[113] Panting JR, Taylor AM, Gatehouse PD, Keegan J, Yang GZ, et al. (1999) First-pass myocardial perfusion imaging and equilibrium signal changes using the intravascular contrast agent NC100150 injection. *J Magn Reson Imaging* 10: 404-410.

[114] Amemiya S, Akahane M, Aoki S, Abe O, Kamada K, et al. (2009) Dynamic contrast-enhanced perfusion MR imaging with SPIO: a pilot study. *Invest Radiol* 44: 503-508.

[115] Bjerner T, Johansson L, Ericsson A, Wikstrom G, Hemmingsson A, et al. (2001) First-pass myocardial perfusion MR imaging with outer-volume suppression and the intravascular contrast agent NC100150 injection: preliminary results in eight patients. *Radiology* 221: 822-826.

[116] Reimer P, Bremer C, Allkemper T, Engelhardt M, Mahler M, et al. (2004) Myocardial perfusion and MR angiography of chest with SH U 555 C: results of placebo-controlled clinical phase i study. *Radiology* 231: 474-481.

[117] Tombach B, Reimer P, Bremer C, Allkemper T, Engelhardt M, et al. (2004) First-pass and equilibrium-MRA of the aortoiliac region with a superparamagnetic iron oxide blood pool MR contrast agent (SH U 555 C): results of a human pilot study. *NMR in Biomedicine* 17: 500-506.

[118] Vellinga MM, Vrenken H, Hulst HE, Polman CH, Uitdehaag BM, et al. (2009) Use of ultrasmall superparamagnetic particles of iron oxide (USPIO)-enhanced MRI to demonstrate diffuse inflammation in the normal-appearing white matter (NAWM) of multiple sclerosis (MS) patients: an exploratory study. *J Magn Reson Imaging* 29: 774-779.

[119] Vellinga MM, Oude Engberink RD, Seewann A, Pouwels PJ, Wattjes MP, et al. (2008) Pluriformity of inflammation in multiple sclerosis shown by ultra-small iron oxide particle enhancement. *Brain* 131: 800-807.

[120] Provenzano R, Schiller B, Rao M, Coyne D, Brenner L, et al. (2009) Ferumoxytol as an intravenous iron replacement therapy in hemodialysis patients. *Clin J Am Soc Nephrol* 4: 386-393.

[121] Rosner MH, Bolton WK (2009) Ferumoxytol for the treatment of anemia in chronic kidney disease. *Drugs Today (Barc)* 45: 779-786.

[122] Schwenk MH (2010) Ferumoxytol: a new intravenous iron preparation for the treatment of iron deficiency anemia in patients with chronic kidney disease. *Pharmacotherapy* 30: 70-79.

[123] Singh A, Patel T, Hertel J, Bernardo M, Kausz A, et al. (2008) Safety of ferumoxytol in patients with anemia and CKD. *Am J Kidney Dis* 52: 907-915.

[124] Spinowitz BS, Kausz AT, Baptista J, Noble SD, Sothinathan R, et al. (2008) Ferumoxytol for treating iron deficiency anemia in CKD. *J Am Soc Nephrol* 19: 1599-1605.

[125] Spinowitz BS, Schwenk MH, Jacobs PM, Bolton WK, Kaplan MR, et al. (2005) The safety and efficacy of ferumoxytol therapy in anemic chronic kidney disease patients. *Kidney Int* 68: 1801-1807.

[126] Varallyay CG, Muldoon LL, Gahramanov S, Wu YJ, Goodman JA, et al. (2009) Dynamic MRI using iron oxide nanoparticles to assess early vascular effects of antiangiogenic versus corticosteroid treatment in a glioma model. *J Cereb Blood Flow Metab* 29: 853-860.

[127] Ersoy H, Jacobs P, Kent CK, Prince MR (2004) Blood pool MR angiography of aortic stent-graft endoleak. *AJR Am J Roentgenol* 182: 1181-1186.

[128] Harisinghani M, Ross RW, Guimaraes AR, Weissleder R (2007) Utility of a new bolus-injectable nanoparticle for clinical cancer staging. *Neoplasia* 9: 1160-1165.

[129] Li W, Tutton S, Vu AT, Pierchala L, Li BS, et al. (2005) First-pass contrast-enhanced magnetic resonance angiography in humans using ferumoxytol, a novel ultrasmall superparamagnetic iron oxide (USPIO)-based blood pool agent. *J Magn Reson Imaging* 21: 46-52.

[130] Arbab AS, Janic B, Knight RA, Anderson SA, Pawelczyk E, et al. (2008) Detection of migration of locally implanted AC133+ stem cells by cellular magnetic resonance imaging with histological findings. *FASEB J* 22: 3234-3246.

[131] Arbab AS, Pandit SD, Anderson SA, Yocum GT, Bur M, et al. (2006) Magnetic resonance imaging and confocal microscopy studies of magnetically labeled endothelial progenitor cells trafficking to sites of tumor angiogenesis. *Stem Cells* 24: 671-678.

[132] Zhang RL, Zhang L, Zhang ZG, Morris D, Jiang Q, et al. (2003) Migration and differentiation of adult rat subventricular zone progenitor cells transplanted into the adult rat striatum. *Neuroscience* 116: 373-382.

[133] Davidson JM, Krieg T, Eming SA (2000) Particle-mediated gene therapy of wounds. *Wound Repair Regen* 8: 452-459.

[134] Wu YL, Ye Q, Foley LM, Hitchens TK, Sato K, et al. (2006) In situ labeling of immune cells with iron oxide particles: an approach to detect organ rejection by cellular MRI. *Proc Natl Acad Sci U S A* 103: 1852-1857.

[135] Shapiro EM, Gonzalez-Perez O, Manuel Garcia-Verdugo J, Alvarez-Buylla A, Koretsky AP (2006) Magnetic resonance imaging of the migration of neuronal precursors generated in the adult rodent brain. *Neuroimage* 32: 1150-1157.

[136] Arbab AS, Bashaw LA, Miller BR, Jordan EK, Lewis BK, et al. (2003) Characterization of biophysical and metabolic properties of cells labeled with superparamagnetic iron oxide nanoparticles and transfection agent for cellular MR imaging. *Radiology* 229: 838-846.

[137] Arbab AS, Yocum GT, Kalish H, Jordan EK, Anderson SA, et al. (2004) Efficient magnetic cell labeling with protamine sulfate complexed to ferumoxides for cellular MRI. *Blood* 104: 1217-1223.

[138] Arbab AS, Yocum GT, Wilson LB, Parwana A, Jordan EK, et al. (2004) Comparison of transfection agents in forming complexes with ferumoxides, cell labeling efficiency, and cellular viability. *Mol Imaging* 3: 24-32.

[139] Shapiro EM, Skrtic S, Koretsky AP (2005) Sizing it up: cellular MRI using micron-sized iron oxide particles. *Magn Reson Med* 53: 329-338.

[140] Conner SD, Schmid SL (2003) Regulated portals of entry into the cell. *Nature* 422: 37-44.
[141] Medina-Kauwe LK, Xie J, Hamm-Alvarez S (2005) Intracellular trafficking of nonviral vectors. *Gene Ther* 12: 1734-1751.
[142] Mukherjee S, Ghosh RN, Maxfield FR (1997) Endocytosis. *Physiol Rev* 77: 759-803.
[143] Sun R, Dittrich J, Le-Huu M, Mueller MM, Bedke J, et al. (2005) Physical and biological characterization of superparamagnetic iron oxide- and ultrasmall superparamagnetic iron oxide-labeled cells: a comparison. *Invest Radiol* 40: 504-513.
[144] Harisinghani M, Saini, S, Weissleder, R, Hahn, PF, Yantiss, RK, Tempany, C, Wood, BJ, Mueller, PR. (1999) MR lymphangiography using ultrasmall superparamagnetic iron oxide in patients with primary abdominal and pelvic malignancies: radiographic-pathologic correlation. *AJR Am J Roentgenol* 172: 1347-1351.
[145] Harisinghani M, Barentsz, J, Hahn, PF, Deserno, WM, Tabatabaei, S, van de Kaa, CH, de la Rosette, J, Weissleder, R. (2003) Noninvasive detection of clinically occult lymph-node metastases in prostate cancer. *N Engl J Med* 343: 2491-2499.
[146] Harisinghani MG, Saini S, Weissleder R, Halpern EF, Schima W, et al. (1997) Differentiation of liver hemangiomas from metastases and hepatocellular carcinoma at MR imaging enhanced with blood-pool contrast agent Code-7227. *Radiology* 202: 687-691.
[147] Frank JA, Miller BR, Arbab AS, Zywicke HA, Jordan EK, et al. (2003) Clinically applicable labeling of mammalian and stem cells by combining superparamagnetic iron oxides and transfection agents. *Radiology* 228: 480-487.
[148] Frank JA, Zywicke H, Jordan EK, Mitchell J, Lewis BK, et al. (2002) Magnetic intracellular labeling of mammalian cells by combining (FDA-approved) superparamagnetic iron oxide MR contrast agents and commonly used transfection agents. *Acad Radiol 9 Suppl* 2: S484-487.
[149] Sato N, Kobayashi H, Hiraga A, Saga T, Togashi K, et al. (2001) Pharmacokinetics and enhancement patterns of macromolecular MR contrast agents with various sizes of polyamidoamine dendrimer cores. *Magn Reson Med* 46: 1169-1173.
[150] Bryant LH, Jr., Brechbiel MW, Wu C, Bulte JW, Herynek V, et al. (1999) Synthesis and relaxometry of high-generation (G = 5, 7, 9, and 10) PAMAM dendrimer-DOTA-gadolinium chelates. *J Magn Reson Imaging* 9: 348-352.
[151] Kobayashi H, Kawamoto S, Jo SK, Bryant HL, Jr., Brechbiel MW, et al. (2003) Macromolecular MRI contrast agents with small dendrimers: pharmacokinetic differences between sizes and cores. *Bioconjug Chem* 14: 388-394.
[152] Yan GP, Hu B, Liu ML, Li LY (2005) Synthesis and evaluation of gadolinium complexes based on PAMAM as MRI contrast agents. *J Pharm Pharmacol* 57: 351-357.
[153] Josephson L, Tung CH, Moore A, Weissleder R (1999) High-efficiency intracellular magnetic labeling with novel superparamagnetic-Tat peptide conjugates. *Bioconjug Chem* 10: 186-191.
[154] Dodd CH, Hsu HC, Chu WJ, Yang P, Zhang HG, et al. (2001) Normal T-cell response and in vivo magnetic resonance imaging of T cells loaded with HIV transactivator-peptide-derived superparamagnetic nanoparticles. Journal *of Immunological Methods* 256: 89-105.

[155] Moore A, Grimm J, Han B, Santamaria P (2004) Tracking the recruitment of diabetogenic CD8+ T-cells to the pancreas in real time. *Diabetes* 53: 1459-1466.

[156] Moore A, Sun PZ, Cory D, Hogemann D, Weissleder R, et al. (2002) MRI of insulitis in autoimmune diabetes. *Magn Reson Med* 47: 751-758.

[157] Kircher MF, Allport JR, Graves EE, Love V, Josephson L, et al. (2003) In vivo high resolution three-dimensional imaging of antigen-specific cytotoxic T-lymphocyte trafficking to tumors. *Cancer Res* 63: 6838-6846.

[158] Bulte JW, Ben-Hur T, Miller BR, Mizrachi-Kol R, Einstein O, et al. (2003) MR microscopy of magnetically labeled neurospheres transplanted into the Lewis EAE rat brain. *Magn Reson Med* 50: 201-205.

[159] Bulte JW, Zhang S, van Gelderen P, Herynek V, Jordan EK, et al. (1999) Neurotransplantation of magnetically labeled oligodendrocyte progenitors: magnetic resonance tracking of cell migration and myelination. *Proceedings of the National Academy of Sciences of the United States of America* 96: 15256-15261.

[160] Ahrens ET, Feili-Hariri M, Xu H, Genove G, Morel PA (2003) Receptor-mediated endocytosis of iron-oxide particles provides efficient labeling of dendritic cells for in vivo MR imaging. *Magnetic Resonance in Medicine* 49: 1006-1013.

[161] Miyoshi S, Flexman, JA, Cross, DJ, Maravilla, KR, Kim, Y, Anzai, Y, Oshima, J, Minoshima, S. (2005) Transfection of Neuroprogenitor Cells with Iron Nanoparticles for Magnetic Resonance Imaging Tracking: Cell Viability, Differentiation, and Intracellular Localization. *Mol Imaging iol* 4: 1-10.

[162] Song Y, Morikawa S, Morita M, Inubushi T, Takada T, et al. (2006) Magnetic resonance imaging using hemagglutinating virus of Japan-envelope vector successfully detects localization of intra-cardially administered microglia in normal mouse brain. *Neurosci Lett* 395: 42-45.

[163] Hinds KA, Hill JM, Shapiro EM, Laukkanen MO, Silva AC, et al. (2003) Highly efficient endosomal labeling of progenitor and stem cells with large magnetic particles allows magnetic resonance imaging of single cells. *Blood* 102: 867-872.

[164] Janic B, Iskander AS, Rad AM, Soltanian-Zadeh H, Arbab AS (2008) Effects of ferumoxides-protamine sulfate labeling on immunomodulatory characteristics of macrophage-like THP-1 cells. *PLoS ONE* 3: e2499.

[165] Janic B, Rad AM, Jordan EK, Iskander AS, Ali MM, et al. (2009) Optimization and validation of FePro cell labeling method. *PLoS One* 4: e5873.

[166] Arbab AS, Bashaw LA, Miller BR, Jordan EK, Bulte JW, et al. (2003) Intracytoplasmic tagging of cells with ferumoxides and transfection agent for cellular magnetic resonance imaging after cell transplantation: methods and techniques. *Transplantation* 76: 1123-1130.

[167] Arbab AS, Yocum GT, Rad AM, Khakoo AY, Fellowes V, et al. (2005) Labeling of cells with ferumoxides-protamine sulfate complexes does not inhibit function or differentiation capacity of hematopoietic or mesenchymal stem cells. *NMR Biomed* 18: 553-559.

[168] Kiessling F (2008) Noninvasive cell tracking. *Handb Exp Pharmacol*: 305-321.

[169] Zelivyanskaya ML, Nelson JA, Poluektova L, Uberti M, Mellon M, et al. (2003) Tracking superparamagnetic iron oxide labeled monocytes in brain by high-field magnetic resonance imaging. *J Neurosci Res* 73: 284-295.

[170] Daldrup-Link HE, Rudelius M, Oostendorp RA, Settles M, Piontek G, et al. (2003) Targeting of hematopoietic progenitor cells with MR contrast agents. *Radiology* 228: 760-767.

[171] Lewin M, Carlesso N, Tung CH, Tang XW, Cory D, et al. (2000) Tat peptide-derivatized magnetic nanoparticles allow in vivo tracking and recovery of progenitor cells. *Nat Biotechnol* 18: 410-414.

[172] Billotey C, Wilhelm C, Devaud M, Bacri JC, Bittoun J, et al. (2003) Cell internalization of anionic maghemite nanoparticles: quantitative effect on magnetic resonance imaging. *Magn Reson Med* 49: 646-654.

[173] Hoehn M, Kustermann, E, Blunk, J, Wiedermann, D, Trapp, T, Wecker, S, Focking, M, Arnold, H, Hescheler, J, Fleischmann, BK, Schwindt, W, Buhrle, C. (2002) Monitoring of implanted stem cell migration in vivo: a highly resolved in vivo magnetic resonance imaging investigation of experimental stroke in rat. *Proc Natl Acad Sci U S A* 99: 16267-16272.

[174] Kraitchman DL, Heldman AW, Atalar E, Amado LC, Martin BJ, et al. (2003) In vivo magnetic resonance imaging of mesenchymal stem cells in myocardial infarction. *Circulation* 107: 2290-2293.

[175] Cahill KS, Germain S, Byrne BJ, Walter GA (2004) Non-invasive analysis of myoblast transplants in rodent cardiac muscle. *Int J Cardiovasc Imaging* 20: 593-598.

[176] Anderson SA, Shukaliak-Quandt J, Jordan EK, Arbab AS, Martin R, et al. (2004) Magnetic resonance imaging of labeled T-cells in a mouse model of multiple sclerosis. *Ann Neurol* 55: 654-659.

[177] Weissleder R, Cheng HC, Bogdanova A, Bogdanov A, Jr. (1997) Magnetically labeled cells can be detected by MR imaging. *J Magn Reson Imaging* 7: 258-263.

[178] Nitin N, LaConte LE, Zurkiya O, Hu X, Bao G (2004) Functionalization and peptide-based delivery of magnetic nanoparticles as an intracellular MRI contrast agent. *J Biol Inorg Chem* 9: 706-712.

[179] Moore A, Weissleder, R, Bogdanov, A, Jr. (1997) Uptake of dextran-coated monocrystalline iron oxides in tumor cells and macrophages. *J Magn Reson Imaging* 7: 1140-1145.

[180] Lee IH, Bulte JW, Schweinhardt P, Douglas T, Trifunovski A, et al. (2004) In vivo magnetic resonance tracking of olfactory ensheathing glia grafted into the rat spinal cord. *Exp Neurol* 187: 509-516.

[181] Walter GA, Cahill KS, Huard J, Feng H, Douglas T, et al. (2004) Noninvasive monitoring of stem cell transfer for muscle disorders. *Magn Reson Med* 51: 273-277.

[182] Pittet MJ, Swirski FK, Reynolds F, Josephson L, Weissleder R (2006) Labeling of immune cells for in vivo imaging using magnetofluorescent nanoparticles. *Nat Protoc* 1: 73-79.

[183] Arbab AS, Rad AM, Iskander AS, Jafari-Khouzani K, Brown SL, et al. (2007) Magnetically-labeled sensitized splenocytes to identify glioma by MRI: a preliminary study. *Magn Reson Med* 58: 519-526.

[184] Arbab AS, Frank JA (2008) Cellular MRI and its role in stem cell therapy. *Regen Med* 3: 199-215.

[185] Gutteridge JM, Halliwell B (1982) The role of the superoxide and hydroxyl radicals in the degradation of DNA and deoxyribose induced by a copper-phenanthroline complex. *Biochem Pharmacol* 31: 2801-2805.

[186] Gutteridge JM, Toeg D (1982) Iron-dependent free radical damage to DNA and deoxyribose. Separation of TBA-reactive intermediates. *Int J Biochem* 14: 891-893.

[187] Emerit J, Beaumont C, Trivin F (2001) Iron metabolism, free radicals, and oxidative injury. *Biomed Pharmacother* 55: 333-339.

[188] Raynal P, Pollard HB (1994) Annexins: the problem of assessing the biological role for a gene family of multifunctional calcium- and phospholipid-binding proteins. *Biochim Biophys Acta* 1197: 63-93.

[189] Anthony RS, McKelvie ND, Cunningham AJ, Craig JI, Rogers SY, et al. (1998) Flow cytometry using annexin V can detect early apoptosis in peripheral blood stem cell harvests from patients with leukaemia and lymphoma. *Bone Marrow Transplant* 21: 441-446.

[190] Kylarova D, Prochazkova J, Mad'arova J, Bartos J, Lichnovsky V (2002) Comparison of the TUNEL, lamin B and annexin V methods for the detection of apoptosis by flow cytometry. *Acta Histochem* 104: 367-370.

[191] Goldberg JE, Sherwood SW, Clayberger C (1999) A novel method for measuring CTL and NK cell-mediated cytotoxicity using annexin V and two-color flow cytometry. *J Immunol Methods* 224: 1-9.

[192] Watson DJ, Walton RM, Magnitsky SG, Bulte JW, Poptani H, et al. (2006) Structure-specific patterns of neural stem cell engraftment after transplantation in the adult mouse brain. *Hum Gene Ther* 17: 693-704.

[193] Pawelczyk E, Frank JA (2008) Transferrin receptor expression in iron oxide-labeled mesenchymal stem cells. *Radiology* 247: 913; author reply 914-915.

[194] Stauber RH, Afonina E, Gulnik S, Erickson J, Pavlakis GN (1998) Analysis of intracellular trafficking and interactions of cytoplasmic HIV-1 Rev mutants in living cells. *Virology* 251: 38-48.

[195] Stauber RH, Pavlakis GN (1998) Intracellular trafficking and interactions of the HIV-1 Tat protein. *Virology* 252: 126-136.

[196] Okon E, Pouliquen D, Okon P, Kovaleva ZV, Stepanova TP, et al. (1994) Biodegradation of magnetite dextran nanoparticles in the rat. A histologic and biophysical study. *Lab Invest* 71: 895-903.

[197] Pouliquen D, Lucet I, Chouly C, Perdrisot R, Le Jeune JJ, et al. (1993) Liver-directed superparamagnetic iron oxide: quantitation of T2 relaxation effects. *Magn Reson Imaging* 11: 219-228.

[198] Josephson L, Kircher MF, Mahmood U, Tang Y, Weissleder R (2002) Near-infrared fluorescent nanoparticles as combined MR/optical imaging probes. *Bioconjug Chem* 13: 554-560.

[199] Kircher MF, Mahmood U, King RS, Weissleder R, Josephson L (2003) A multimodal nanoparticle for preoperative magnetic resonance imaging and intraoperative optical brain tumor delineation. *Cancer Research* 63: 8122-8125.

[200] Josephson L, Mahmood U, Wunderbaldinger P, Tang Y, Weissleder R (2003) Pan and sentinel lymph node visualization using a near-infrared fluorescent probe. Molecular Imaging: *Official Journal of the Society for Molecular Imaging* 2: 18-23.

[201] Sosnovik DE, Schellenberger EA, Nahrendorf M, Novikov MS, Matsui T, et al. (2005) Magnetic resonance imaging of cardiomyocyte apoptosis with a novel magneto-optical nanoparticle. *Magn Reson Med* 54: 718-724.

[202] Xie J, Chen K, Huang J, Lee S, Wang J, et al. (2010) PET/NIRF/MRI triple functional iron oxide nanoparticles. *Biomaterials*.

[203] Meng XX, Wan JQ, Jing M, Zhao SG, Cai W, et al. (2007) Specific targeting of gliomas with multifunctional superparamagnetic iron oxide nanoparticle optical and magnetic resonance imaging contrast agents. *Acta Pharmacol Sin* 28: 2019-2026.

[204] Tung CH, Zeng Q, Shah K, Kim DE, Schellingerhout D, et al. (2004) In vivo imaging of beta-galactosidase activity using far red fluorescent switch. *Cancer Research* 64: 1579-1583.

[205] Ntziachristos V, Bremer C, Graves EE, Ripoll J, Weissleder R (2002) In vivo tomographic imaging of near-infrared fluorescent probes. Molecular Imaging: Official *Journal of the Society for Molecular Imaging* 1: 82-88.

[206] Ntziachristos V, Bremer C, Tung C, Weissleder R (2002) Imaging cathepsin B up-regulation in HT-1080 tumor models using fluorescence-mediated molecular tomography (FMT). *Academic Radiology* 9 Suppl 2: S323-325.

[207] Sosnovik DE, Nahrendorf M, Deliolanis N, Novikov M, Aikawa E, et al. (2007) Fluorescence tomography and magnetic resonance imaging of myocardial macrophage infiltration in infarcted myocardium in vivo. *Circulation* 115: 1384-1391.

[208] Lin Y, Weissleder R, Tung CH (2002) Novel near-infrared cyanine fluorochromes: synthesis, properties, and bioconjugation. *Bioconjugate Chemistry* 13: 605-610.

[209] Giepmans BN, Adams SR, Ellisman MH, Tsien RY (2006) The fluorescent toolbox for assessing protein location and function. *Science* 312: 217-224.

[210] Sutton EJ, Henning TD, Pichler BJ, Bremer C, Daldrup-Link HE (2008) Cell tracking with optical imaging. *Eur Radiol* 18: 2021-2032.

[211] Mothe AJ, Tator CH (2005) Proliferation, migration, and differentiation of endogenous ependymal region stem/progenitor cells following minimal spinal cord injury in the adult rat. *Neuroscience* 131: 177-187.

[212] Yamauchi M, Yamaguchi T, Kaji H, Sugimoto T, Chihara K (2005) Involvement of calcium-sensing receptor in osteoblastic differentiation of mouse MC3T3-E1 cells. *Am J Physiol Endocrinol Metab* 288: E608-616.

[213] Shichinohe H, Kuroda S, Lee JB, Nishimura G, Yano S, et al. (2004) In vivo tracking of bone marrow stromal cells transplanted into mice cerebral infarct by fluorescence optical imaging. *Brain Res Brain Res Protoc* 13: 166-175.

[214] Buursma AR, Rutgers V, Hospers GA, Mulder NH, Vaalburg W, et al. (2006) 18F-FEAU as a radiotracer for herpes simplex virus thymidine kinase gene expression: in-vitro comparison with other PET tracers. *Nucl Med Commun* 27: 25-30.

[215] Cao F, Lin S, Xie X, Ray P, Patel M, et al. (2006) In vivo visualization of embryonic stem cell survival, proliferation, and migration after cardiac delivery. *Circulation* 113: 1005-1014.

[216] Tjuvajev JG, Doubrovin M, Akhurst T, Cai S, Balatoni J, et al. (2002) Comparison of radiolabeled nucleoside probes (FIAU, FHBG, and FHPG) for PET imaging of HSV1-tk gene expression. *J Nucl Med* 43: 1072-1083.

[217] Yaghoubi SS, Barrio JR, Namavari M, Satyamurthy N, Phelps ME, et al. (2005) Imaging progress of herpes simplex virus type 1 thymidine kinase suicide gene therapy in living subjects with positron emission tomography. *Cancer Gene Ther* 12: 329-339.

[218] Bouchentouf M, Benabdallah BF, Dumont M, Rousseau J, Jobin L, et al. (2005) Real-time imaging of myoblast transplantation using the human sodium iodide symporter. *Biotechniques* 38: 937-942.

[219] Miyagawa M, Beyer M, Wagner B, Anton M, Spitzweg C, et al. (2005) Cardiac reporter gene imaging using the human sodium/iodide symporter gene. *Cardiovasc Res* 65: 195-202.

[220] Dwyer RM, Bergert ER, O'Connor MK, Gendler SJ, Morris JC (2006) Sodium iodide symporter-mediated radioiodide imaging and therapy of ovarian tumor xenografts in mice. *Gene Ther* 13: 60-66.

[221] Lee WW, Moon DH, Park SY, Jin J, Kim SJ, et al. (2004) Imaging of adenovirus-mediated expression of human sodium iodide symporter gene by 99mTcO4 scintigraphy in mice. *Nucl Med Biol* 31: 31-40.

[222] Niu G, Anderson RD, Madsen MT, Graham MM, Oberley LW, et al. (2006) Dual-expressing adenoviral vectors encoding the sodium iodide symporter for use in noninvasive radiological imaging of therapeutic gene transfer. *Nucl Med Biol* 33: 391-398.

[223] Rad AM, Iskander AS, Janic B, Knight RA, Arbab AS, et al. (2009) AC133+ progenitor cells as gene delivery vehicle and cellular probe in subcutaneous tumor models: a preliminary study. *BMC Biotechnol* 9: 28.

[224] Wu B, Yin J, Texier C, Roussel M, Tan KS (2010) Blastocystis legumain is localized on the cell surface, and specific inhibition of its activity implicates a pro-survival role for the enzyme. *J Biol Chem* 285: 1790-1798.

[225] Stern L, Perry R, Ofek P, Many A, Shabat D, et al. (2009) A Novel Antitumor Prodrug Platform Designed to Be Cleaved by the Endoprotease Legumain. *Bioconjug Chem*.

[226] Suwa T, Ozawa S, Ueda M, Ando N, Kitajima M (1998) Magnetic resonance imaging of esophageal squamous cell carcinoma using magnetite particles coated with anti-epidermal growth factor receptor antibody. *Int J Cancer* 75: 626-634.

[227] Ozawa S, Imai Y, Suwa T, Kitajima M (2000) What's new in imaging? New magnetic resonance imaging of esophageal cancer using an endoluminal surface coil and antibody-coated magnetite particles. *Recent Results Cancer Res* 155: 73-87.

[228] Yang C, Rait A, Pirollo KF, Dagata JA, Farkas N, et al. (2008) Nanoimmunoliposome delivery of superparamagnetic iron oxide markedly enhances targeting and uptake in human cancer cells in vitro and in vivo. *Nanomedicine* 4: 318-329.

[229] Montet X, Montet-Abou K, Reynolds F, Weissleder R, Josephson L (2006) Nanoparticle imaging of integrins on tumor cells. *Neoplasia* 8: 214-222.

[230] Liong M, Lu J, Kovochich M, Xia T, Ruehm SG, et al. (2008) Multifunctional inorganic nanoparticles for imaging, targeting, and drug delivery. *ACS Nano* 2: 889-896.

[231] Zhang Y, Kohler N, Zhang M (2002) Surface modification of superparamagnetic magnetite nanoparticles and their intracellular uptake. *Biomaterials* 23: 1553-1561.

[232] Hong G, Yuan R, Liang B, Shen J, Yang X, et al. (2008) Folate-functionalized polymeric micelle as hepatic carcinoma-targeted, MRI-ultrasensitive delivery system of antitumor drugs. *Biomed Microdevices* 10: 693-700.

[233] Leuschner C, Kumar CS, Hansel W, Soboyejo W, Zhou J, et al. (2006) LHRH-conjugated magnetic iron oxide nanoparticles for detection of breast cancer metastases. *Breast Cancer Res Treat* 99: 163-176.

[234] Van Kroonenburgh MJ, Pauwels EK (1988) Human immunological response to mouse monoclonal antibodies in the treatment or diagnosis of malignant diseases. *Nucl Med Commun* 9: 919-930.

[235] Muto MG, Finkler NJ, Kassis AI, Lepisto EM, Knapp RC (1990) Human anti-murine antibody responses in ovarian cancer patients undergoing radioimmunotherapy with the murine monoclonal antibody OC-125. *Gynecol Oncol* 38: 244-248.

[236] Legouffe E, Liautard J, Gaillard JP, Rossi JF, Wijdenes J, et al. (1994) Human anti-mouse antibody response to the injection of murine monoclonal antibodies against IL-6. *Clin Exp Immunol* 98: 323-329.

[237] Brochier J, Legouffe E, Liautard J, Gaillard JP, Mao LQ, et al. (1995) Immunomodulating IL-6 activity by murine monoclonal antibodies. *Int J Immunopharmacol* 17: 41-48.

[238] Otsuji E, Yamaguchi T, Yamaoka N, Kotani T, Kato M, et al. (1995) Biodistribution of murine and chimeric Fab fragments of the monoclonal antibody A7 in human pancreatic cancer. *Pancreas* 10: 265-273.

[239] Bond CJ, Marsters JC, Sidhu SS (2003) Contributions of CDR3 to V H H domain stability and the design of monobody scaffolds for naive antibody libraries. *J Mol Biol* 332: 643-655.

[240] Duan J, Wu J, Valencia CA, Liu R (2007) Fibronectin type III domain based monobody with high avidity. *Biochemistry* 46: 12656-12664.

[241] Hogemann-Savellano D, Bos E, Blondet C, Sato F, Abe T, et al. (2003) The transferrin receptor: a potential molecular imaging marker for human cancer. *Neoplasia* (New York) 5: 495-506.

[242] Deans AE (2006) Cellular MRI contrast via coexpression of transferrin receptor and ferritin. *Magn Reson Med* 56: 51-59.

[243] Ichikawa T, Hogemann D, Saeki Y, Tyminski E, Terada K, et al. (2002) MRI of transgene expression: correlation to therapeutic gene expression. *Neoplasia* 4: 523-530.

[244] Koretsky AP, Lin YJ, Schorle H, Jaenisch R (1996) Genetic control of MRI contrast by expression of the transferrin receptor. *Proc Int Soc for Magn Reson in Med* 4: 69.

[245] Martin SA, McCarthy A, Barber LJ, Burgess DJ, Parry S, et al. (2009) Methotrexate induces oxidative DNA damage and is selectively lethal to tumour cells with defects in the DNA mismatch repair gene MSH2. *EMBO Mol Med* 1: 323-337.

[246] Mathers JC (2009) Folate intake and bowel cancer risk. *Genes Nutr* 4: 173-178.

[247] Moon WK, Lin Y, O'Loughlin T, Tang Y, Kim DE, et al. (2003) Enhanced tumor detection using a folate receptor-targeted near-infrared fluorochrome conjugate. *Bioconjugate Chemistry* 14: 539-545.

[248] Zhang J, Rana S, Srivastava RS, Misra RD (2008) On the chemical synthesis and drug delivery response of folate receptor-activated, polyethylene glycol-functionalized magnetite nanoparticles. *Acta Biomater* 4: 40-48.

[249] Zhang Y, Zhang J (2005) Surface modification of monodisperse magnetite nanoparticles for improved intracellular uptake to breast cancer cells. *J Colloid Interface Sci* 283: 352-357.

[250] Tung CH, Lin Y, Moon WK, Weissleder R (2002) A receptor-targeted near-infrared fluorescence probe for in vivo tumor imaging. *Chembiochem* 3: 784-786.

[251] Sundaram S, Trivedi R, Durairaj C, Ramesh R, Ambati BK, et al. (2009) Targeted drug and gene delivery systems for lung cancer therapy. *Clin Cancer Res* 15: 7299-7308.

[252] Sundaram S, Durairaj C, Kadam R, Kompella UB (2009) Luteinizing hormone-releasing hormone receptor-targeted deslorelin-docetaxel conjugate enhances efficacy of docetaxel in prostate cancer therapy. *Mol Cancer Ther* 8: 1655-1665.
[253] Nasongkla N, Bey E, Ren J, Ai H, Khemtong C, et al. (2006) Multifunctional polymeric micelles as cancer-targeted, MRI-ultrasensitive drug delivery systems. *Nano Lett* 6: 2427-2430.
[254] Reardon DA, Nabors LB, Stupp R, Mikkelsen T (2008) Cilengitide: an integrin-targeting arginine-glycine-aspartic acid peptide with promising activity for glioblastoma multiforme. *Expert Opin Investig Drugs* 17: 1225-1235.
[255] Weibo C, Xiaoyuan C (2006) Anti-Angiogenic Cancer Therapy Based on Integrin avb3 Antagonism. *Anti-Cancer Agents in Medicinal Chemistry* 6: 407-428.
[256] Reardon DA, Fink KL, Mikkelsen T, Cloughesy TF, O'Neill A, et al. (2008) Randomized phase II study of cilengitide, an integrin-targeting arginine-glycine-aspartic acid peptide, in recurrent glioblastoma multiforme. *J Clin Oncol* 26: 5610-5617.
[257] Hotfilder M, Nowak-Gottl U, Wolff JE (1997) Tumorangiogenesis: a network of cytokines. *Klin Padiatr* 209: 265-270.
[258] Liekens S, De Clercq E, Neyts J (2001) Angiogenesis: regulators and clinical applications. *Biochem Pharmacol* 61: 253-270.
[259] Marme D (2001) [Tumor angiogenesis: new approaches to cancer therapy]. *Onkologie* 24 Suppl 1: 1-5.
[260] Mitra S, Gaur U, Ghosh PC, Maitra AN (2001) Tumour targeted delivery of encapsulated dextran-doxorubicin conjugate using chitosan nanoparticles as carrier. *J Control Release* 74: 317-323.
[261] Brannon-Peppas L, Blanchette JO (2004) Nanoparticle and targeted systems for cancer therapy. *Adv Drug Deliv Rev* 56: 1649-1659.
[262] Reddy LH (2005) Drug delivery to tumours: recent strategies. *J Pharm Pharmacol* 57: 1231-1242.
[263] Conti M, Tazzari V, Baccini C, Pertici G, Serino LP, et al. (2006) Anticancer drug delivery with nanoparticles. *In Vivo* 20: 697-701.
[264] Greish K (2007) Enhanced permeability and retention of macromolecular drugs in solid tumors: a royal gate for targeted anticancer nanomedicines. *J Drug Target* 15: 457-464.
[265] Cho K, Wang X, Nie S, Chen ZG, Shin DM (2008) Therapeutic nanoparticles for drug delivery in cancer. *Clin Cancer Res* 14: 1310-1316.
[266] Raghavan R, Brady ML, Rodriguez-Ponce MI, Hartlep A, Pedain C, et al. (2006) Convection-enhanced delivery of therapeutics for brain disease, and its optimization. *Neurosurg Focus* 20: E12.
[267] Saito R, Bringas JR, Panner A, Tamas M, Pieper RO, et al. (2004) Convection-enhanced delivery of tumor necrosis factor-related apoptosis-inducing ligand with systemic administration of temozolomide prolongs survival in an intracranial glioblastoma xenograft model. *Cancer Res* 64: 6858-6862.
[268] Perlstein B, Ram Z, Daniels D, Ocherashvilli A, Roth Y, et al. (2008) Convection-enhanced delivery of maghemite nanoparticles: Increased efficacy and MRI monitoring. *Neuro Oncol* 10: 153-161.

[269] Arbab AS, Jordan EK, Wilson LB, Yocum GT, Lewis BK, et al. (2004) In vivo trafficking and targeted delivery of magnetically labeled stem cells. *Hum Gene Ther* 15: 351-360.

[270] Song M, Kim YJ, Kim YH, Roh J, Kim SU, et al. (2010) *Targeted delivery of ferumoxide-labeled human neural stem cells using neodymium magnet in rat model of focal cerebral ischemia.* Hum Gene Ther.

[271] Hamasaki T, Tanaka N, Kamei N, Ishida O, Yanada S, et al. (2007) Magnetically labeled neural progenitor cells, which are localized by magnetic force, promote axon growth in organotypic cocultures. *Spine* (Phila Pa 1976) 32: 2300-2305.

[272] Sugioka T, Ochi M, Yasunaga Y, Adachi N, Yanada S (2008) Accumulation of magnetically labeled rat mesenchymal stem cells using an external magnetic force, and their potential for bone regeneration. *J Biomed Mater Res* A 85: 597-604.

[273] Shimizu K, Ito A, Arinobe M, Murase Y, Iwata Y, et al. (2007) Effective cell-seeding technique using magnetite nanoparticles and magnetic force onto decellularized blood vessels for vascular tissue engineering. *J Biosci Bioeng* 103: 472-478.

[274] Nakashima Y, Deie M, Yanada S, Sharman P, Ochi M (2005) Magnetically labeled human natural killer cells, accumulated in vitro by an external magnetic force, are effective against HOS osteosarcoma cells. *Int J Oncol* 27: 965-971.

[275] Chertok B, Moffat BA, David AE, Yu F, Bergemann C, et al. (2008) Iron oxide nanoparticles as a drug delivery vehicle for MRI monitored magnetic targeting of brain tumors. *Biomaterials* 29: 487-496.

[276] Chertok B, David AE, Huang Y, Yang VC (2007) Glioma selectivity of magnetically targeted nanoparticles: a role of abnormal tumor hydrodynamics. *J Control Release* 122: 315-323.

[277] Polyak B, Friedman G (2009) Magnetic targeting for site-specific drug delivery: applications and clinical potential. *Expert Opin Drug Deliv* 6: 53-70.

[278] Dandamudi S, Patil V, Fowle W, Khaw BA, Campbell RB (2009) External magnet improves antitumor effect of vinblastine and the suppression of metastasis. *Cancer Sci* 100: 1537-1543.

[279] Schulze K, Koch A, Petri-Fink A, Steitz B, Kamau S, et al. (2006) Uptake and biocompatibility of functionalized poly(vinylalcohol) coated superparamagnetic maghemite nanoparticles by synoviocytes in vitro. *J Nanosci* Nanotechnol 6: 2829-2840.

[280] Bonnemain B (2008) [Nanoparticles: the industrial viewpoint. Applications in diagnostic imaging]. *Ann Pharm Fr* 66: 263-267.

[281] Hofmann-Amtenbrink M, von Rechenberg B, Hofmann A (2009) *Superparamagnetic nanoparticles for biomedical applications*; Tan MC, editor. Kerala, India: Transworld Research Network.

[282] Butoescu N, Seemayer CA, Palmer G, Guerne PA, Gabay C, et al. (2009) Magnetically retainable microparticles for drug delivery to the joint: efficacy studies in an antigen-induced arthritis model in mice. *Arthritis Res Ther* 11: R72.

[283] Butoescu N, Seemayer CA, Foti M, Jordan O, Doelker E (2009) Dexamethasone-containing PLGA superparamagnetic microparticles as carriers for the local treatment of arthritis. *Biomaterials* 30: 1772-1780.

[284] van der Zee J (2002) Heating the patient: a promising approach? *Ann Oncol* 13: 1173-1184.

[285] van der Zee J, Koper PC, Lutgens LC, Burger CW (2002) Point-counterpoint: what is the optimal trial design to test hyperthermia for carcinoma of the cervix? Point: addition of hyperthermia or cisplatin to radiotherapy for patients with cervical cancer; two promising combinations--no definite conclusions. *Int J Hyperthermia* 18: 19-24.

[286] Hildebrandt B, Wust P, Ahlers O, Dieing A, Sreenivasa G, et al. (2002) The cellular and molecular basis of hyperthermia. *Crit Rev Oncol Hematol* 43: 33-56.

[287] Tseng HY, Lee GB, Lee CY, Shih YH, Lin XZ (2009) Localised heating of tumours utilising injectable magnetic nanoparticles for hyperthermia cancer therapy. *IET Nanobiotechnol* 3: 46-54.

[288] Le Renard PE, Buchegger F, Petri-Fink A, Bosman F, Rufenacht D, et al. (2009) Local moderate magnetically induced hyperthermia using an implant formed in situ in a mouse tumor model. *Int J Hyperthermia* 25: 229-239.

[289] Kikumori T, Kobayashi T, Sawaki M, Imai T (2009) Anti-cancer effect of hyperthermia on breast cancer by magnetite nanoparticle-loaded anti-HER2 immunoliposomes. *Breast Cancer Res Treat* 113: 435-441.

[290] Thiesen B, Jordan A (2008) Clinical applications of magnetic nanoparticles for hyperthermia. *Int J Hyperthermia* 24: 467-474.

[291] van Landeghem FK, Maier-Hauff K, Jordan A, Hoffmann KT, Gneveckow U, et al. (2009) Post-mortem studies in glioblastoma patients treated with thermotherapy using magnetic nanoparticles. *Biomaterials* 30: 52-57.

[292] Jordan A (2009) Hyperthermia classic commentary: 'Inductive heating of ferrimagnetic particles and magnetic fluids: Physical evaluation of their potential for hyperthermia' by Andreas Jordan et al., International Journal of Hyperthermia, 1993;9:51-68. *Int J Hyperthermia* 25: 512-516.

[293] Mitsumori M, Hiraoka M, Shibata T, Okuno Y, Nagata Y, et al. (1996) Targeted hyperthermia using dextran magnetite complex: a new treatment modality for liver tumors. *Hepatogastroenterology* 43: 1431-1437.

[294] Mitsumori M, Hiraoka M, Shibata T, Okuno Y, Masunaga S, et al. (1994) Development of intra-arterial hyperthermia using a dextran-magnetite complex. *Int J Hyperthermia* 10: 785-793.

[295] Melnikov OV, Gorbenko OY, Markelova MN, Kaul AR, Atsarkin VA, et al. (2009) Ag-doped manganite nanoparticles: new materials for temperature-controlled medical hyperthermia. *J Biomed Mater Res* A 91: 1048-1055.

[296] Atsarkin VA, Levkin LV, Posvyanskiy VS, Melnikov OV, Markelova MN, et al. (2009) Solution to the bioheat equation for hyperthermia with La(1-x)Ag(y)MnO(3-delta) nanoparticles: the effect of temperature autostabilization. *Int J Hyperthermia* 25: 240-247.

[297] Jordan A, Scholz R, Maier-Hauff K, van Landeghem FK, Waldoefner N, et al. (2006) The effect of thermotherapy using magnetic nanoparticles on rat malignant glioma. *J Neurooncol* 78: 7-14.

[298] Yanase M, Shinkai M, Honda H, Wakabayashi T, Yoshida J, et al. (1997) Intracellular hyperthermia for cancer using magnetite cationic liposomes: ex vivo study. *Jpn J Cancer Res* 88: 630-632.

[299] Kawai N, Futakuchi M, Yoshida T, Ito A, Sato S, et al. (2008) Effect of heat therapy using magnetic nanoparticles conjugated with cationic liposomes on prostate tumor in bone. *Prostate* 68: 784-792.

[300] Viroonchatapan E, Ueno M, Sato H, Adachi I, Nagae H, et al. (1995) Preparation and characterization of dextran magnetite-incorporated thermosensitive liposomes: an on-line flow system for quantifying magnetic responsiveness. *Pharm Res* 12: 1176-1183.

[301] Purushotham S, Ramanujan RV (2010) Thermoresponsive magnetic composite nanomaterials for multimodal cancer therapy. *Acta Biomater* 6: 502-510.

[302] Bhattarai SR, Kim SY, Jang KY, Lee KC, Yi HK, et al. (2008) Laboratory formulated magnetic nanoparticles for enhancement of viral gene expression in suspension cell line. *J Virol Methods* 147: 213-218.

[303] Scherer F, Anton M, Schillinger U, Henke J, Bergemann C, et al. (2002) Magnetofection: enhancing and targeting gene delivery by magnetic force in vitro and in vivo. *Gene Ther* 9: 102-109.

[304] Plank C, Scherer F, Schillinger U, Bergemann C, Anton M (2003) Magnetofection: enhancing and targeting gene delivery with superparamagnetic nanoparticles and magnetic fields. *J Liposome Res* 13: 29-32.

[305] Soeren W. Gersting USJLPNCRCPDRJR (2004) Gene delivery to respiratory epithelial cells by magnetofection. *The Journal of Gene Medicine* 6: 913-922.

[306] Stephanie Huth JLSWGCRCPUWJR (2004) Insights into the mechanism of magnetofection using PEI-based magnetofectins for gene transfer. *The Journal of Gene Medicine* 6: 923-936.

[307] Mykhaylyk O, Antequera YS, Vlaskou D, Plank C (2007) Generation of magnetic nonviral gene transfer agents and magnetofection in vitro. *Nat Protocols* 2: 2391-2411.

[308] Krotz F, Sohn HY, Gloe T, Plank C, Pohl U (2003) Magnetofection potentiates gene delivery to cultured endothelial cells. *J Vasc Res* 40: 425-434.

[309] Ko BS, Chang TC, Shyue SK, Chen YC, Liou JY (2009) An efficient transfection method for mouse embryonic stem cells. *Gene Ther* 16: 154-158.

[310] Buerli T, Pellegrino C, Baer K, Lardi-Studler B, Chudotvorova I, et al. (2007) Efficient transfection of DNA or shRNA vectors into neurons using magnetofection. *Nat Protoc* 2: 3090-3101.

[311] Kronick P, Gilpin RW (1986) Use of superparamagnetic particles for isolation of cells. *J Biochem Biophys Methods* 12: 73-80.

[312] Molday RS, Yen SP, Rembaum A (1977) Application of magnetic microspheres in labelling and separation of cells. *Nature* 268: 437-438.

[313] Molday RS, MacKenzie D (1982) Immunospecific ferromagnetic iron-dextran reagents for the labeling and magnetic separation of cells. *J Immunol Methods* 52: 353-367.

[314] MacKenzie D, Molday RS (1982) Organization of rhodopsin and a high molecular weight glycoprotein in rod photoreceptor disc membranes using monoclonal antibodies. *J Biol Chem* 257: 7100-7105.

[315] Molday RS, Molday LL (1984) *Separation of cells labeled with immunospecific iron dextran microspheres using high gradient magnetic chromatography.* FEBS Lett 170: 232-238.

[316] Miltenyi S, Muller W, Weichel W, Radbruch A (1990) High gradient magnetic cell separation with MACS. *Cytometry* 11: 231-238.

[317] Gao L, Zhuang J, Nie L, Zhang J, Zhang Y, et al. (2007) Intrinsic peroxidase-like activity of ferromagnetic nanoparticles. *Nat Nanotechnol* 2: 577-583.

[318] Yu F, Huang Y, Cole AJ, Yang VC (2009) The artificial peroxidase activity of magnetic iron oxide nanoparticles and its application to glucose detection. *Biomaterials* 30: 4716-4722.

[319] Rostovshchikova TN, Smirnov VV, Tsodikov MV, Bukhtenko OV, Maksimov YV, et al. (2005) Catalytic conversions of chloroolefins over iron oxide nanoparticles 1. Isomerization of dichlorobutenes in the presence of iron oxide nanopaticles immobilized on silicas with different structures. *Russian Chemical Bulletin, International Edition, 54*: 1418-1424.

[320] Rostovshchikova TN, Korobov MS, Pankratov DA, Yurkov GY, Gubin SP (2005) Catalytic conversions of chloroolefins over iron oxide nanoparticles 2.* Isomerization of dichlorobutenes over iron oxide nanoparticles stabilized on the surface of ultradispersed poly(tetraf luoroethylene). *Russian Chemical Bulletin, International Edition* 54: 1425-1432.

[321] Horv´ath D, Toth L, Guczi L (2000) Gold nanoparticles: effect of treatment on structure and catalytic activity of Au/Fe2O3 catalyst prepared by co-precipitation. *Catalysis Letters* 67: 117-128.

[322] Herzing AA, Kiely CJ, Carley AF, Landon P, Hutchings GJ (2008) Identification of active gold nanoclusters on iron oxide supports for CO oxidation. *Science* 321: 1331-1335.

[323] Brown VJ (2006) Fe-TAML catalyst for clean up. *Environmental Health Perspectives* 114: A656-A659.

INDEX

A

absorption, 24, 73, 111, 126, 165, 188
accessibility, 127
accuracy, 148
acetic acid, 144
acid, viii, 19, 20, 21, 23, 24, 25, 37, 43, 72, 73, 74, 75, 80, 81, 82, 85, 87, 91, 92, 103, 104, 105, 106, 107, 109, 120, 134, 135, 137, 138, 143, 150, 155, 175, 176, 183, 207
acidity, 16, 51
acromegaly, 99
acrylic acid, 92, 102, 103
active transport, 143, 181
acylation, 100
adenovirus, 159, 205
adhesion, 92, 107, 154
adhesive properties, 76, 109, 130
adhesives, 107
adipose, 107
adipose tissue, 107
adsorption, 80, 81, 85, 101, 120
adverse event, 168
aflatoxin, 124
AFM, 106
aggregates, 30, 46, 72, 86, 145
aggregation, 2, 72, 89, 94, 165
aggregation process, 72
AIDS, 75
alanine, 101, 150
alanine aminotransferase, 150
albumin, 72, 93, 99, 107, 109, 144, 158
alcohol, 17, 35, 37, 70, 86, 90, 91, 94, 101, 102, 130, 151, 153
alcohols, 26, 44
algorithm, 117
alloys, 25, 115
ALT, 150
alternative hypothesis, 52

aluminum, 24, 30
amines, 47
ammonium, 19, 75, 114, 134, 172
amplitude, 146
anatase, 157
anemia, 170, 198
aneurysm, 196, 197
angiogenesis, 184, 199, 207
angiography, 168, 170, 197, 198, 199
annealing, 25
ANOVA, 57, 58, 59, 61, 62, 65, 66
antibody, 85, 89, 121, 140, 143, 146, 156, 173, 183, 205, 206
anticancer activity, 147
anticancer drug, 147
anticoagulation, 173
antigen, 85, 121, 183, 186, 201, 208
antiphospholipid antibodies, 49
antisense, 155
antitumor, 147, 205, 208
apoptosis, 128, 145, 146, 150, 159, 177, 179, 203, 207
applications, vii, xxvii, xxviii, 2, 3, 4, 5, 11, 18, 19, 21, 28, 33, 35, 39, 40, 41, 44, 48, 51, 69, 71, 73, 74, 75, 80, 81, 84, 88, 89, 94, 95, 96, 102, 105, 106, 107, 108, 111, 114, 120, 123, 125, 127, 147, 150, 160, 162, 172, 181, 189, 191, 207, 208, 209
aqueous solutions, 44, 80, 91, 165
aqueous suspension, 72, 74, 86, 100
arginine, 103, 173, 207
argon, 14, 30, 32, 86, 87, 146
arteriography, 194
artery, 162, 197
arthritis, 186, 208
arthroplasty, 115
articular cartilage, 105
ascorbic acid, 75
ASI, 35
aspartate, 150
assessment, 127, 132, 148, 162, 192, 195, 196

Index

assumptions, 65, 67
astrocytes, 137, 154
asymmetry, 33
atherosclerosis, 196
atherosclerotic plaque, 168, 196
ATP, 139, 155
attachment, 73, 86, 123, 145, 172, 173, 179
Au nanoparticles, 85, 110, 118
Australia, 167
authors, vii, 1, 21, 24, 26, 27, 28, 30, 32, 33, 80, 150, 168, 184, 187

B

back pain, 147, 168
background, 109, 181
bacteria, 3, 33, 49
bacterial infection, 149
bacterium, 33, 127
band gap, 1
barium, 45, 94
BBB, 74, 81, 184
behavior, xxvii, 3, 4, 24, 25, 34, 36, 44, 105, 111, 116, 117, 125, 127, 149, 171, 179
benzene, 48, 174
bias, 66, 113
binding, 80, 85, 111, 123, 139, 150, 178, 182, 183, 203
biocatalysts, 87, 119
biocompatibility, 5, 51, 71, 72, 73, 76, 80, 81, 83, 84, 106, 125, 126, 137, 143, 147, 150, 151, 161, 162, 208
biodegradability, 81, 82, 83
biological media, 81, 84, 86, 131
biological responses, 15
biological systems, 3
bioluminescence, 139
biomaterials, 104
biomedical applications, vii, xxvii, xxviii, 4, 11, 19, 21, 27, 35, 38, 42, 71, 72, 73, 74, 76, 80, 82, 85, 89, 94, 100, 105, 108, 121, 125, 151, 152, 208
biosensors, vii
bioseparation, 113
biotin, 89, 124
blood, 11, 15, 52, 71, 73, 74, 81, 85, 88, 148, 159, 166, 168, 170, 171, 174, 177, 182, 184, 186, 190, 194, 196, 198, 199, 200, 208
blood flow, 148
blood stream, 73
blood vessels, 184, 208
blood-brain barrier, 190, 196
bloodstream, 74
body fluid, 73
Boltzman constant, 12

bonds, 26, 74, 79
bone, 74, 101, 105, 125, 126, 204, 208, 209
bone marrow, 125, 204
bone marrow transplant, 125
bowel, 165, 206
brain, 74, 75, 81, 99, 100, 104, 147, 150, 152, 154, 160, 161, 168, 170, 171, 184, 186, 196, 199, 201, 203, 207, 208
brain structure, 152, 171
brain tumor, 82, 104, 147, 152, 160, 170, 184, 186, 196, 203, 208
breast cancer, 74, 75, 82, 108, 137, 138, 181, 205, 206, 209
bromine, 142
bulk materials, xxvii

C

cadmium, 6, 7, 72, 156
calcium, 45, 94, 108, 203, 204
calibration, 117
Canada, viii
cancer, 11, 75, 78, 88, 98, 102, 118, 120, 123, 124, 126, 147, 149, 152, 153, 154, 157, 159, 160, 174, 181, 183, 186, 199, 205, 206, 207, 209, 210
cancer cells, 78, 98, 102, 120, 124, 152, 153, 154, 157, 159, 174, 186, 205
candidates, 46, 74, 83, 125, 148, 193
CAP, 111
capillary, 28
carbohydrate, 122, 168
carbon, 32, 47, 48, 72, 82, 84, 105, 107, 110, 113, 114, 115, 116, 117, 118, 120, 121, 123, 139, 158
carbon dioxide, 82, 114
carbon monoxide, 115
carbon nanotubes, 32, 48, 105, 110, 114, 117, 120, 123, 158
carbonization, 84
carboxylic acids, 26, 44
carcinoma, 95, 130, 131, 138, 142, 165, 168, 187, 195, 197, 205, 209
cardiac muscle, 202
cardiac output, 148
caries, 84
carrier, xxvii, 7, 30, 32, 73, 85, 92, 95, 96, 97, 100, 207
cartilage, 74, 100
catalysis, 113
catalyst, 82, 211
catalytic activity, 189, 190, 211
catalytic properties, 189
cation, 80, 123, 172
CD8+, 201

cell culture, 81, 128, 131, 135, 144, 155, 171, 172, 174
cell cycle, 91, 143, 145, 158
cell death, 141, 144, 153
cell killing, 187
cell line, 74, 84, 101, 129, 130, 131, 134, 138, 140, 142, 144, 155, 156, 172, 174, 178, 188, 210
cell lines, 74, 84, 101, 130, 131, 134, 155, 156
cell membranes, 83, 142
cell metabolism, 139
cell surface, 152, 171, 173, 178, 205
cellulose, 72
ceramic, 28, 32, 46, 47, 49, 114, 117
ceruloplasmin, 130, 132, 152
charge density, 71
chelates, 163, 200
chemical properties, 11, 151
chemical reactivity, 44
chemical vapor deposition, 32, 48, 116
chemical vapour deposition, 48
chemiluminescence, 139
chemotherapeutic agent, 183, 184
chemotherapy, 105
China, 34, 37, 92, 116, 118
chitin, 95
cholangiocarcinoma, 195
chondrocyte, 149, 155
chromatography, 2, 93, 101, 189, 210
circulation, 5, 71, 73, 81, 88, 90, 99, 148, 179, 186
classification, 195
clinical diagnosis, 147
clinical oncology, 121
clinical trials, 140, 147, 150, 170, 171, 184
clusters, 1, 6, 26, 47, 48, 71, 72, 97, 170, 189
CMC, 138
CNS, 158, 196
CO_2, 32, 48, 104, 107, 114
coagulation, 90
coatings, xxvii, 5, 15, 44, 51, 69, 71, 72, 73, 74, 84, 88, 91, 112, 115, 126, 127, 164, 165
color, 1, 137, 145, 203
communication, 156
components, 74, 117, 127, 130, 132, 167
composites, 24, 97, 100, 101, 115, 116, 117, 118, 123, 124
composition, 3, 5, 14, 15, 25, 26, 30, 33, 75, 89, 102, 115, 127, 129, 131, 158, 164, 171, 184, 188, 190
compounds, 1, 24, 25, 32, 39, 48, 74, 75, 96, 127, 131
compressibility, 80
computed tomography, viii, 167, 195
computer science, 126
computer simulations, 149

condensation, 24, 30, 31, 75, 86, 111
conjugation, 72, 81, 88, 142, 179
connective tissue, 130
contamination, 29, 143
control, 1, 2, 12, 19, 20, 28, 35, 39, 44, 52, 53, 71, 80, 83, 93, 97, 103, 109, 112, 119, 128, 130, 131, 134, 135, 137, 139, 140, 142, 143, 146, 147, 178, 184, 186, 206
control group, 52, 134, 186, 187
conversion, vii, 131, 140, 141
cooling, 25, 30
copolymers, 80, 96, 97, 103, 105, 107, 124
coronary arteries, 148
correlation, 200, 206
cost, xxvii, 5, 57, 127
cost constraints, 57
cost effectiveness, 127
coupling, 74, 87, 123
covalent bond, 74, 78, 86
covalent bonding, 86
CPC, 31
critical value, 55
crystal growth, 15
crystal structure, 14, 16
crystalline, 18, 24, 25, 26, 40, 79, 97, 113, 115
crystallinity, 20, 24, 25, 32, 82, 102
crystallites, 3, 6
crystallization, 3, 40
crystals, 2, 33, 130, 164, 165
CT scan, 197
culture, 99, 105, 106, 119, 128, 137, 140, 170, 173, 176
culture media, 170, 176
CVD, 115
cytochrome, 139
cytokines, 207
cytometry, 142, 145, 146, 157, 177, 203
cytoplasm, 141, 171, 187
cytoskeleton, 130, 132
cytotoxicity, vii, xxviii, 84, 91, 96, 100, 101, 119, 128, 129, 131, 133, 135, 137, 138, 139, 140, 141, 142, 143, 144, 145, 150, 152, 153, 154, 155, 156, 157, 173, 180, 188, 203

D

damages, 74, 129, 170
data analysis, 52
death, 126, 128
decision making, 51
decomposition, 4, 20, 25, 26, 41, 101
deficiency, 170, 198
degradation, 18, 71, 73, 75, 82, 86, 105, 107, 108, 150, 202

degradation process, 75
delivery, xxvii, 11, 36, 69, 73, 75, 81, 82, 84, 88, 96, 97, 100, 104, 106, 108, 109, 111, 119, 120, 121, 122, 125, 140, 147, 148, 149, 155, 156, 159, 161, 180, 182, 183, 184, 185, 187, 188, 202, 204, 205, 206, 207, 208, 210
dendritic cell, viii, 148, 158, 159, 161, 174, 177, 201
density, 1, 2, 30, 82, 116, 131, 147, 186, 196
deoxyribose, 202, 203
deposition, 30, 38, 47, 196
derivatives, 73, 142
detection, 49, 55, 71, 74, 85, 98, 102, 110, 111, 118, 121, 122, 124, 130, 138, 141, 143, 147, 151, 159, 160, 167, 170, 173, 177, 179, 191, 193, 194, 195, 196, 200, 203, 205, 206, 211
deviation, 33, 54, 55, 66, 67
diabetes, 142, 172, 201
diacrylates, 75
dialysis, 75, 105
differentiation, 101, 104, 172, 173, 178, 199, 201, 204
diffusion, 15, 30, 82
digestion, 41, 74, 147
diisocyanates, 38
diluent, 75
dimethacrylate, 75
direct measure, 145
disease model, 163
diseases, xxvii, 74, 170, 180, 190, 206
dispersion, 25, 32, 43, 72, 78, 83, 87, 109
dispersity, 17
dissociation, 147
distilled water, 175
distribution, 1, 4, 5, 11, 12, 14, 15, 19, 20, 21, 28, 32, 33, 34, 37, 45, 51, 53, 54, 61, 69, 73, 83, 85, 87, 92, 94, 152, 161, 162, 180, 184, 187, 196
division, 145
DNA, xxvii, 84, 85, 96, 98, 110, 111, 112, 126, 131, 139, 142, 143, 144, 145, 146, 155, 156, 159, 177, 178, 180, 183, 187, 202, 203, 206, 210
DNA damage, 144, 145, 146, 159, 177, 206
DNA lesions, 145, 146
DNA repair, 145
DNA sequencing, xxvii
docetaxel, 207
dosage, 130, 133, 144
double bonds, 74
drug carriers, 118
drug delivery, 4, 38, 51, 52, 70, 73, 74, 75, 80, 81, 82, 83, 88, 89, 90, 91, 94, 96, 99, 100, 103, 106, 107, 108, 109, 112, 119, 125, 147, 149, 152, 160, 161, 180, 182, 183, 184, 186, 187, 205, 206, 207, 208

drug discovery, 149, 161
drug release, 51, 82, 101, 107
drug therapy, 140
drug use, 75
drugs, 73, 88, 108, 122, 149, 183, 184, 186, 205, 207
duodenum, 165
dyes, 87, 143, 157, 180
dynamics, 7, 39, 115, 119

E

E.coli, 189
EAE, 201
electrocatalysis, 95
electrochemical deposition, 14, 26
electrochemistry, 44, 95, 114
electrolysis, 26
electrolyte, 26, 27
electromagnetic, 81, 155
electromagnetic field, 81, 155
electromagnetic fields, 155
electron, 1, 4, 15, 18, 26, 37, 86, 98, 136
electron diffraction, 26
electron microscopy, 37
electron paramagnetic resonance, 26
electrons, 15, 71, 78, 164
electrophoresis, 145, 146
electroporation, 170
electrospinning, 104
ELISA, 147, 160
emboli, 167, 186
embolism, 125
embolization, 167, 186
embryonic stem cells, 188, 210
emission, 138, 141, 145
EMMA, 97
emulsion polymerization, 43, 81
emulsions, 18, 37, 148
encapsulation, 33, 75, 82, 83, 86, 93, 97, 105, 126, 173, 179
encephalomyelitis, 170
endothelial cells, 74, 138, 155, 157, 158, 178, 184, 210
energy, 1, 12, 13, 14, 32, 49, 72, 78, 86, 98, 139, 153
engineering, xxvii, xxviii, 38, 82, 95, 100, 102, 104, 105, 106, 107, 113, 125, 151, 157, 208
environment, 3, 14, 71, 74, 84, 88, 127, 128, 129, 164
enzymatic activity, 97
enzyme immobilization, 49
enzyme-linked immunosorbent assay, 147
enzymes, 87, 137, 141, 180, 183
EPC, 178, 181
ependymal, 204

epithelial cells, 137, 188, 210
epithelium, 155
equilibrium, 11, 12, 48, 170, 198
erythrocytes, 72
esophageal cancer, 205
ester, 96, 108
ethanol, 20, 33, 76, 85, 87, 143, 146
ethylene, 24, 32, 47, 48, 72, 74, 75, 91, 96, 99, 100, 101, 112, 122, 129, 134, 137, 153, 154
ethylene glycol, 24, 74, 75, 91, 96, 99, 100, 101, 112, 122, 131, 134, 137, 153, 155
ethylene oxide, 75, 100
EU, 40
Europe, 150, 166, 167
evaporation, 82, 93, 138
EXAFS, 113
examinations, 127, 179
excitation, 1, 138, 141, 145
exclusion, 1, 93, 137, 141, 142, 143, 144, 152, 177
excretion, 73, 126, 165
experimental condition, 21, 25, 32
experimental design, 65
exposure, 18, 74, 117, 126, 127, 129, 130, 141, 143, 145, 146, 148, 155, 156, 186

F

fabrication, 1, 110, 116
fatty acids, 20
FDA, 73, 81, 82, 125, 142, 147, 150, 154, 166, 168, 170, 173, 175, 181, 200
FDA approval, 147
ferrite, 3, 4, 38, 39, 161
ferritin, 28, 44, 45, 46, 122, 167, 178, 206
ferrous ion, 2
fiber, 18, 105, 116
fibers, 106, 117
fibrinogen, 81
fibroblast growth factor, 107
fibroblasts, 91, 93, 99, 119, 129, 132, 153, 158, 174
fibrosis, 170, 190
fibrous cap, 126
film formation, 78
films, 1, 3, 30, 46, 47, 107, 122
fine tuning, 3
finite element method, 159
fixation, 143, 146, 147
fluid, 11, 28, 37, 43, 186
fluidized bed, 115, 116
fluorescence, 141, 142, 143, 145, 146, 157, 177, 179, 204, 206
folate, 91, 121, 183, 206
folic acid, 74, 122, 183
food, 19, 165, 166

food additives, 165
formula, 3, 4, 15, 52, 61, 164
France, 37, 69
free energy, 11, 12, 13, 19, 30
free radicals, 177, 203
FTIR, 24, 26, 78, 80, 114, 189
FTIR spectroscopy, 26
fuel, 114
functional analysis, 159
functionalization, 35, 74, 88, 89, 92, 96, 114, 122, 123, 124, 154
fusion, 99, 173, 189

G

gadolinium, 163, 164, 165, 167, 170, 184, 193, 195, 200
gamma rays, 86
gases, 30, 47
gasification, 116
gel, 17, 19, 24, 34, 42, 86, 87, 94, 102, 103, 109, 145, 146, 189
gel formation, 17
gelation, 93, 95, 103
gene, xxvii, 11, 84, 85, 96, 98, 100, 109, 110, 111, 120, 140, 148, 153, 155, 156, 159, 170, 178, 180, 181, 187, 191, 199, 203, 204, 205, 206, 210
gene expression, 155, 191, 204, 206, 210
gene targeting, 100
gene therapy, 11, 84, 148, 199, 204
gene transfer, 96, 109, 153, 187, 205, 210
generation, 6, 26, 45, 105, 138, 140, 200
genes, 140, 181, 188
germanium, 110
Germany, 36, 49, 92, 99, 150, 187
glia, 174, 202
glioblastoma, 89, 207, 209
glioblastoma multiforme, 207
glioma, 168, 169, 177, 180, 184, 187, 196, 199, 202, 209
glucose, 167, 211
glutamic acid, 101, 106
glycine, 103, 108, 189, 207
glycol, 24, 25, 72, 73, 74, 75, 91, 99, 100, 123, 152, 168, 206
gold, 30, 34, 72, 73, 84, 85, 86, 96, 98, 109, 110, 159, 190, 211
gold nanoparticles, 30, 86, 96, 98
groups, 2, 24, 26, 52, 53, 73, 78, 80, 81, 87, 88, 103, 120, 123, 128, 130, 140, 143, 146, 165, 172, 173, 178, 180, 186, 188, 189
growth, xxvii, 2, 3, 6, 11, 12, 21, 22, 25, 30, 33, 34, 35, 37, 39, 41, 47, 49, 79, 80, 83, 91, 97, 99, 100,

110, 112, 125, 132, 151, 154, 175, 176, 183, 184, 205, 208
growth factor, 100, 175, 176, 183, 184, 205
growth hormone, 99
growth mechanism, 154
growth rate, 132
guidance, 80, 148

H

half-life, 71, 73, 166, 168, 170, 180, 182, 184
harvesting, 49, 140
head and neck cancer, 196
heat, 20, 24, 75, 125, 143, 186, 190, 209
heating, 2, 21, 24, 28, 40, 81, 155, 189, 209
hemangioma, 169
hematopoietic stem cells, 178, 188
hemochromatosis, 167
hemodialysis, 198
hepatocellular carcinoma, 101, 166, 167, 194, 195, 200
hepatocytes, 120, 138, 155
hepatoma, 156
hepatotoxicity, 159
heptane, 19, 115
herpes, 181, 204
herpes simplex, 181, 204
herpes simplex virus type 1, 181, 204
heterogeneity, 92
histology, 107, 148, 184, 187
HIV, 111, 172, 200, 203
homogeneity, 28
Honda, 49, 121, 209
hormone, 184, 207
host, 72, 112, 126, 127, 171, 183
human brain, 161
human immunodeficiency virus, 172
human leukemia cells, 131
Hunter, 99
hybrid, 82, 108
hybridization, 86
hydrazine, 87
hydrocarbons, 116
hydrogels, 91, 94, 100, 101, 102, 105, 106, 107, 109
hydrogen, 6, 45, 80, 81, 86, 111, 165, 189
hydrogen atoms, 86
hydrogen bonds, 80, 81, 165
hydrolysis, 14, 24, 26, 36, 86, 87
hydrophilicity, 71, 76, 82
hydrothermal process, 40
hydrothermal synthesis, 20, 39, 40
hydroxide, 19, 24, 42, 76, 101, 103
hydroxyapatite, 105
hydroxyl, 71, 73, 78, 81, 86, 134, 165, 188, 202

hydroxyl groups, 71, 78, 81, 86, 165, 188
hyperplasia, 166, 193, 195
hypersensitivity, 167, 170
hyperthermia, 5, 11, 21, 34, 49, 69, 89, 95, 121, 122, 125, 146, 157, 186, 209
hypothalamus, 74
hypothesis, 51, 52, 54, 57
hypothesis test, 51, 52, 57
hypoxia, 184, 196

I

ideal, 33, 86, 198
identification, 66, 91, 148, 153, 154
IL-6, 206
image, 77, 83, 142, 164, 165, 167, 169, 181
images, 21, 22, 77, 78, 89, 161, 163, 164, 165, 166, 167, 168, 169, 177, 179, 180, 185, 186
imaging modalities, 180
imaging systems, 155
immobilization, 85, 105, 112, 118
immune response, 171
immune system, 71, 127
immunoglobulin, 85, 99
immunohistochemistry, 171
immunomodulatory, 201
immunosuppression, 156
in vitro, xxviii, 70, 84, 91, 93, 96, 101, 108, 109, 119, 120, 125, 127, 128, 131, 137, 138, 140, 142, 144, 146, 147, 148, 150, 153, 154, 155, 158, 170, 180, 185, 186, 187, 191, 205, 208, 210
in vivo, xxviii, 4, 5, 46, 51, 71, 73, 80, 81, 84, 88, 90, 92, 93, 96, 100, 102, 106, 109, 120, 125, 127, 147, 148, 149, 150, 153, 155, 158, 159, 160, 170, 171, 172, 178, 179, 180, 184, 188, 190, 191, 192, 200, 201, 202, 204, 205, 206, 210
incubation time, 129, 131, 138, 139, 171, 175
independence, 65, 66, 67
India, 208
indication, 131, 133, 150
induction, 85, 121, 153, 186
industrial sectors, xxvii
industry, xxvii, 11
infection, 127, 161
inflammation, 137, 158, 170, 186, 198
inhibition, 134, 158, 205
insight, 106, 125, 170
insulin, 119, 129, 152
integrated circuits, xxvii, 1
integrin, 88, 90, 184, 207
integrity, 74, 128, 130, 141, 177, 178
interaction, 57, 58, 59, 60, 61, 62, 65, 75, 78, 88, 89, 93, 99, 102, 126, 127, 131, 152, 159, 163, 165, 183, 190

interaction effect, 58, 59, 60, 62, 65
interaction effects, 63, 65
interactions, 5, 57, 62, 63, 72, 74, 78, 80, 84, 86, 91, 94, 102, 104, 114, 127, 131, 139, 163, 172, 203
interface, 6, 33, 91, 102, 104, 116, 117
interfacial reactivity, 115
intermetallic compounds, 29
internalization, 131, 154, 171, 202
interneurons, 171
intravenously, 147, 166, 170, 181
ion channels, viii
ion-exchange, 2
ionic strength, 14, 16, 51
ionization, 122
ions, 4, 14, 15, 25, 26, 27, 72, 73, 83, 84, 85, 94, 131, 164, 177, 186
IR spectroscopy, 120
Iran, 91
irradiation, 30, 75, 98, 120
ischemia, 208
isolation, 157, 161, 188, 210
issues, xxviii, 175

J

Japan, 6, 43, 101, 116, 150, 166, 167, 173, 180, 201
joints, 186
Jordan, 152, 155, 187, 199, 200, 201, 202, 208, 209

K

keratinocytes, 98
kerosene, 117
kidney, 88, 130, 150, 170, 198, 199
kidneys, 73, 167
killing, 187
kinetics, 27, 28, 73, 89, 90, 95, 101, 114
KOH, 140
Krebs cycle, 82

L

labeling, viii, 4, 74, 83, 89, 97, 112, 121, 125, 141, 142, 143, 146, 148, 149, 154, 155, 156, 157, 158, 170, 171, 172, 173, 174, 175, 176, 177, 178, 180, 188, 190, 191, 199, 200, 201, 210
labeling procedure, 171, 172, 177
lactate dehydrogenase, 139
lactic acid, 82, 105, 107
lactoferrin, 130, 132
laser ablation, 4, 11, 28, 29, 30, 31, 34, 46, 47
lattices, 24, 25
lesions, 138, 144, 160, 165, 166, 167, 169, 170, 180, 192, 193, 194, 195, 196
leukemia, 153, 188

Lewis acids, 71, 78
ligament, 104, 105
ligand, 26, 119, 140, 152, 183, 207
light scattering, 15, 43, 71, 89, 90, 93
line, 20, 57, 75, 138, 144, 210
linear model, 65
linkage, 2, 179, 183
lipid peroxidation, 147
lipids, 180, 187
liposomes, 33, 49, 72, 103, 158, 173, 184, 186, 209, 210
liquid phase, 107
liquid-phase synthesis, 28
liquids, xxvii, 18, 45, 46
lithium, 40, 117
liver, 81, 94, 124, 130, 138, 147, 150, 152, 160, 161, 162, 166, 167, 172, 185, 192, 193, 194, 195, 200, 209
liver cancer, 94, 124, 147, 162
liver cells, 130, 152
liver metastases, 192
local order, 114
localization, 122, 201
low temperatures, 4, 18, 25
luciferase, 139
luciferin, 139
luminescence, 139, 155
lung cancer, 168, 206
lung disease, 149
Luo, 34, 95, 98, 110, 116, 117, 118, 123
lymph, 121, 148, 160, 168, 179, 196, 197, 200, 203
lymph node, 121, 148, 168, 179, 196, 197, 203
lymphocytes, 141, 172, 175
lymphoid, 158
lymphoma, 165, 203
lysine, 83, 94, 130, 132
lysis, 146
lysozyme, 97

M

macromolecules, 106, 126, 172, 184
macrophages, 74, 81, 92, 131, 137, 153, 154, 156, 171, 174, 178, 180, 202
magnesium, 114
magnet, 113, 143, 180, 185, 189, 190, 208
magnetic field, 4, 5, 72, 84, 88, 125, 146, 147, 163, 164, 186, 187, 188, 189, 190, 210
magnetic field effect, 4
magnetic materials, 19, 71, 84
magnetic moment, 4, 85, 164, 186, 188
magnetic particles, 43, 49, 98, 113, 123, 125, 130, 139, 145, 148, 151, 160, 165, 187, 192, 201

magnetic properties, 5, 11, 14, 15, 30, 32, 40, 41, 42, 44, 48, 71, 84, 88, 92, 95, 97, 110, 111, 115, 118, 122
magnetic relaxation, 187
magnetic resonance, viii, xxvii, 4, 5, 21, 32, 35, 80, 94, 119, 125, 140, 148, 153, 154, 156, 157, 158, 159, 160, 161, 179, 182, 190, 191, 192, 193, 195, 196, 197, 199, 200, 201, 202, 203, 204, 205
magnetic resonance imaging, viii, xxvii, 4, 5, 21, 32, 35, 80, 94, 119, 125, 140, 148, 154, 156, 157, 158, 159, 160, 179, 182, 190, 191, 192, 193, 195, 196, 197, 199, 200, 201, 202, 203, 204, 205
magnetic sensor, 18
magnetization, 2, 4, 5, 15, 21, 24, 25, 26, 32, 52, 54, 72, 84, 119, 150, 164, 188
magnetoresistance, 69, 91
major histocompatibility complex, 121
majority, 19, 21, 25, 27, 28, 67, 188
malignant tumors, 183, 184, 190
manganese, 163, 164
manipulation, 109, 180
manufacturing, xxvii, xxviii, 25
mapping, 104, 159, 172
mass spectrometry, 101, 122
material surface, 126
materials science, xxvii, xxviii, 52
matrix, 2, 3, 26, 42, 59, 75, 78, 87, 94, 97, 108, 109, 113, 116, 117, 123, 189
matrix metalloproteinase, 123
measurement, 58, 66, 126, 136, 145, 147
measures, 140
mechanical properties, 116, 117, 118
media, 2, 5, 14, 16, 19, 25, 38, 51, 139, 171, 172, 175, 176
melanoma, 78, 130, 132, 145, 148, 161, 172
membranes, vii, 3, 19, 101, 106, 141, 210
memory, xxvii
mesenchymal stem cells, 104, 130, 140, 157, 159, 173, 177, 178, 185, 191, 201, 202, 203, 208
mesothelioma, 131
meta analysis, 168
meta-analysis, 197
metabolic pathways, 167
metabolism, 73, 126, 139, 178, 191, 193, 203
metabolites, 126
metal nanoparticles, 45, 46, 142
metal oxides, 26, 152
metal salts, 21, 24, 87
metals, 25, 26, 29, 39, 109
metastasis, 166, 168, 169, 208
mice, 81, 89, 138, 148, 149, 156, 159, 204, 205, 208
micelles, 1, 3, 6, 18, 19, 38, 87, 101, 103, 107, 124, 207

microemulsion, 11, 18, 19, 21, 34, 36, 39, 86, 87, 95, 97, 109, 121
microgels, 80
micrometer, 32, 69, 91
microscope, 15, 141, 143
microscopy, viii, 142, 143, 145, 161, 177, 184, 196, 199, 201
microspheres, 81, 103, 106, 107, 113, 123, 153, 159, 210
microstructure, 44, 49, 104, 116, 117, 118
migration, 15, 148, 154, 178, 179, 181, 182, 183, 184, 191, 199, 201, 202, 204
mixing, 25, 28
MMP, 123
MOCVD, 7
model, viii, 5, 12, 19, 45, 51, 59, 61, 63, 64, 65, 66, 67, 69, 81, 91, 92, 105, 107, 126, 137, 150, 154, 159, 168, 171, 179, 181, 186, 187, 190, 197, 199, 202, 207, 208, 209
modeling, viii, 46, 70, 116
models, 51, 65, 148, 172, 173, 175, 204, 205
mole, 33
molecular biology, 120, 126, 185
molecular structure, 80, 103
molecular weight, 19, 81, 84, 96, 109, 145, 172, 210
molecular weight distribution, 172
molecules, viii, 6, 18, 19, 35, 71, 72, 73, 74, 78, 79, 86, 105, 126, 163, 165, 172, 184, 188
monoclonal antibody, 172, 189, 206
monolayer, 19, 44, 86, 113
monomers, 21, 75, 82
Moon, 38, 93, 160, 195, 205, 206
morality, 59
morbidity, 128
morphology, 3, 5, 14, 20, 24, 28, 45, 83, 85, 118, 142, 143, 184
MTS, 129, 136, 137, 139
multicellular organisms, 144
multilayer films, 120
multiple sclerosis, 170, 198, 202
muscle stem cells, 174
myocardial infarction, 179, 202
myocardium, 170, 204

N

nanocomposites, xxvii, 92, 96, 111, 113, 119, 151
nanocrystals, 6, 7, 23, 39, 40, 41, 91, 114, 157, 191
nano-crystals, 20
nanodots, 35
nanofibers, 153
nanomaterials, xxvii, 5, 39, 71, 98, 123, 125, 127, 210
nanomedicine, xxvii

nanometer, 6, 27, 32, 35, 45, 47, 72
nanorods, 18, 19, 34, 79, 109, 111
nanostructured materials, xxvii, 25, 32
nanostructures, 47, 98, 112, 113
nanotechnology, vii, xxvii, 3, 127, 161
nanotube, 48, 114, 123
nanotube films, 48
nanowires, vii, 44
narcotic, 104
National Institutes of Health, 184
NATO, 35, 115
natural killer cell, 208
natural polymers, 81, 95
NCA, 30
Nd, 29
necrosis, 145, 146, 179, 187
neodymium, 208
nerve, 106
nervous system, 96
Netherlands, 35
network, 74, 75, 83, 84, 86, 102, 207
neuroblasts, 171
neuronal stem cells, 171, 178
neurons, 171, 188, 210
neurotransmitter, 139
New England, 160
niobium, 30
nitrates, 14
nitric oxide, 130
nitrides, 32
nitrogen, 14, 48
nitrogen gas, 48
NMR, 36, 48, 104, 151, 163, 172, 173, 191, 195, 198, 201
nonionic surfactants, 19, 36, 97
normal distribution, 52, 65, 67
North America, 151
nuclear magnetic resonance, 163, 172
nucleation, 2, 3, 11, 12, 13, 14, 21, 28, 33, 83
nuclei, 11, 12, 13
nucleic acid, 141, 142, 178
nucleus, 12, 145, 170, 179, 188
null hypothesis, 52, 53

O

oil, 19, 34, 37, 38, 39, 87, 95
oligodendrocytes, 172
opportunities, 88, 125
optical density, 129
optical microscopy, 131
optical properties, 1, 6, 7, 98, 109
optimization, viii, 5, 17, 20, 27, 29, 51, 65, 83, 197, 207

order, 11, 12, 15, 17, 19, 20, 24, 25, 26, 27, 28, 30, 32, 46, 59, 65, 66, 71, 72, 73, 74, 79, 80, 82, 84, 85, 87, 131, 139, 140, 141, 143, 144, 150, 181
organ, 25, 41, 121, 126, 140, 159, 168, 171, 178, 183, 191, 199
organelles, 74, 130
organic matter, 89
organic solvents, 20
outliers, 56, 65, 66, 67
oxidation, 2, 4, 15, 19, 20, 26, 41, 48, 75, 84, 117, 144, 190, 211
oxidative stress, 150
oxides, 3, 4, 5, 20, 24, 26, 29, 30, 35, 44, 49, 92, 104, 112, 114, 127, 141, 158, 164, 170, 175, 190, 196, 200, 202
oxygen, 15, 30, 32, 33, 48, 105, 178
ozonation, 113

P

PAA, 72, 80, 92, 137
paclitaxel, 101, 183
pain, 148, 167
pancreas, 165, 191, 201
pancreatic cancer, 183, 206
parameter, 5, 17, 18, 177
parameters, 4, 5, 14, 15, 20, 21, 24, 25, 27, 28, 29, 30, 51, 65, 83, 87, 88, 128, 129, 131, 148, 149, 164, 166, 177
pathways, 46, 167, 171, 172
peptides, 123, 172, 173, 179, 183, 188
performance, 20, 51, 87, 119, 168, 197
perfusion, 105, 168, 198
peripheral blood, 203
permeability, 17, 155, 156, 184, 207
permission, iv, 22, 23, 27, 29, 31, 77, 78, 79, 89, 173, 185
PET, 168, 179, 180, 197, 204
pH, 4, 14, 15, 24, 26, 37, 80, 81, 85, 86, 99, 101, 109, 127, 128, 164, 189
phagocytosis, 74, 171
pharmaceuticals, 4, 99
pharmacokinetics, viii, 150, 161, 193
phase transformation, 33, 49
phosphatidylserine, 143
photoluminescence, 6
photolysis, 47
photonics, 45
physical properties, 72, 125
physicochemical properties, 171
physics, xxvii, 43, 51, 90
physiology, 126, 173, 179
plasma, 81, 84, 101, 105, 141, 168, 178, 182, 194
plasma membrane, 141, 178

plasma proteins, 81
plasmid, 96, 155, 156, 187
PMMA, 113
polarization, 27
poly(methyl methacrylate), 113
polydispersity, 15, 83
polyesters, 74, 75, 106, 108
polyether, 73
polymer, viii, 2, 6, 17, 18, 37, 45, 51, 54, 55, 72, 75, 78, 79, 80, 81, 82, 83, 84, 90, 91, 94, 101, 102, 103, 106, 107, 108, 109, 112, 113, 121, 123, 130, 132, 151, 152, 153, 173
polymer blends, 107
polymer chains, 18, 82
polymer composites, viii, 2, 6
polymer films, 107
polymer matrix, 2, 78, 83
polymer molecule, 18, 79
polymer networks, 75, 83, 101, 109
polymer structure, 152
polymeric gels, 83
polymeric materials, 51
polymerization, 36, 38, 43, 75, 82
polymers, 1, 2, 29, 72, 73, 80, 107, 108, 109, 119, 122, 172
polypeptide, 73
polystyrene, 2, 24, 45, 83, 97, 122, 165, 188
polystyrene latex, 97
polyvinyl acetate, 76
polyvinyl alcohol, 69, 72, 76, 91, 92, 158, 159
polyvinylalcohol, 102
population, 52, 53, 54, 55, 171, 178, 189
porosity, 88, 190
Portugal, 42
positron, 179, 204
power, 29, 30, 32, 73, 125
precipitation, 11, 14, 15, 17, 18, 28, 35, 36, 37, 69, 83, 85, 87, 211
precursor cells, 174
prediction, 59, 63, 64, 65, 67
prediction models, 64
pressure, 20, 25, 30, 80, 184
probability, 14, 61, 67, 68
probe, 12, 120, 127, 128, 129, 139, 143, 160, 171, 178, 179, 181, 203, 205, 206
production, xxviii, 2, 14, 15, 45, 46, 48, 84, 125, 128, 131, 137, 139, 140, 144, 156, 167, 178
program, 197
programming, 37, 151
proliferation, 128, 129, 134, 138, 139, 140, 144, 145, 149, 158, 172, 173, 178, 204
propane, 116

properties, xxvii, 1, 3, 5, 6, 17, 32, 35, 36, 37, 40, 41, 44, 46, 47, 49, 69, 73, 78, 79, 80, 81, 83, 84, 88, 92, 94, 95, 99, 102, 103, 104, 109, 110, 112, 114, 115, 117, 147, 152, 158, 163, 164, 170, 171, 178, 186, 188, 190, 194, 195, 199, 204
prostate, 154, 156, 160, 168, 197, 200, 207, 209
prostate cancer, 154, 156, 160, 168, 197, 200, 207
proteins, 33, 81, 99, 112, 130, 131, 145, 173, 180, 185, 186, 189, 203
proteolytic enzyme, 73
proteome, 81
proteomics, 92
protocol, 85, 146, 188, 193, 195
protocols, 130, 153, 158, 188, 189
protons, 163, 164
pruritus, 168
psychiatric disorders, 74
pulse, 29, 30, 137, 164, 178, 194, 195
pumps, 125
purification, 111, 188, 189
purity, 15, 19, 148, 189
PVA, 5, 17, 18, 20, 36, 37, 39, 54, 72, 76, 78, 79, 86, 87, 102, 106, 119, 130, 131, 132, 133, 145, 146, 186
PVP, 26, 42, 72
pyrimidine, 181
pyrolysis, 24, 28, 32, 37, 45, 47, 48, 84, 93, 114, 116, 118

Q

quantum confinement, 1, 3
quantum dot, 6, 7, 26, 44, 97, 114, 122
quantum dots, 6, 7, 26, 44, 97, 114, 122
quartz, 124

R

radiation, 91, 156, 186
Radiation, 95, 98, 101, 114
radiation therapy, 156, 186
radical formation, 139
radical polymerization, 103
radicals, 86, 139, 202
radio, viii, 88, 155, 181
radius, 12, 13, 164
Raman spectra, 47
Raman spectroscopy, 26
range, 14, 17, 19, 24, 30, 32, 45, 66, 67, 88, 97, 104, 130, 137, 138, 143, 164, 170, 186, 189
reactants, 20, 21
reaction mechanism, 41
reaction temperature, 4, 5, 16, 20
reaction time, 20, 28

reactions, 2, 11, 20, 28, 46, 47, 131, 167, 170
reactive oxygen, 26, 144, 156
reagents, 28, 29, 180, 210
receptors, 93, 129, 152, 172, 183
recognition, 3, 57, 89, 113, 157
recommendations, iv, 57, 65
recovery, 19, 111, 163, 202
red blood cells, 147, 167
regeneration, 106, 208
region, 87, 198, 204
regression model, 64, 65
regulation, 125, 204
reinforcement, 117
rejection, 53, 121, 158, 168, 171, 191, 197, 199
relationship, 33, 126, 127
relaxation, 7, 30, 124, 163, 164, 167, 168, 172, 203
relaxation properties, 124, 167, 168
relaxation rate, 172
renal medulla, 121
repair, 74, 107, 145, 190, 206
replacement, 117, 198
replication, 65, 143
residuals, 65, 66, 67, 68
resistance, 117, 159, 180
resolution, 83, 125, 180, 196, 201
respect, 52, 173, 187
respiratory, 188, 195, 198, 210
responsiveness, 5, 210
retention, 149, 167, 185, 186, 187, 207
risk, xxviii, 53, 125, 126, 127, 172, 190, 206
risk assessment, xxviii, 125, 127
RNA, 146, 187
room temperature, 2, 4, 25, 85, 87

S

safety, xxviii, 125, 148, 160, 170, 192, 193, 194, 199
salt, 14, 51, 140, 141
salts, 3, 14, 15, 18, 19, 20, 85, 164
sample mean, 52, 55
saturation, 2, 5, 14, 15, 17, 21, 32, 54, 84, 167
scale system, 90
scattering, 86, 108, 111, 180
screening, 5, 51, 70
seeding, 118, 208
selected area electron diffraction, 117
selectivity, 81, 94, 147, 208
selenium, 72
self-assembly, 18, 26, 42, 75, 97, 113
semiconductor, 6, 38
semiconductors, 1, 109
sensing, 33, 48, 49, 110, 204
sensitivity, 81, 113, 138, 167, 168, 180

separation, xxvii, 6, 11, 12, 20, 21, 25, 85, 86, 89, 104, 107, 111, 112, 118, 125, 148, 151, 179, 185, 188, 189, 210
serum, 139, 150, 167, 170, 175, 176
serum albumin, 139
serum iron level, 150
shape, xxvii, 2, 4, 11, 12, 14, 15, 18, 19, 24, 30, 33, 40, 46, 78, 79, 83, 87, 98, 110, 130, 133, 164, 171, 172
Si_3N_4, 32
side effects, xxvii, 84, 127, 167
signals, 172, 181
signal-to-noise ratio, 52
significance level, 53, 67
signs, 148
silane, 47, 86, 113
silica, 19, 24, 36, 42, 72, 73, 84, 86, 87, 95, 96, 97, 98, 105, 108, 110, 111, 112, 113, 118, 119, 121, 122, 123, 124, 131, 148, 153, 187, 189
silicon, 6, 30, 32, 47, 110
silicon nanocrystals, 47
silk, 105
silver, 102, 110, 186
simulation, viii, 28, 51, 149
SiO_2, 38, 42, 87, 97, 110, 111, 112, 113, 118, 119, 120
siRNA, 187
skin, 98, 153
Slovakia, 97
smooth muscle cells, 139, 192
sodium, 18, 23, 25, 87, 94, 129, 132, 175, 176, 181, 205
sodium dodecyl sulfate (SDS), 25
soil, 152
solar cells, vii
sol-gel, 1, 11, 14, 24, 42, 86, 87, 112, 113, 119
solid phase, 12
solid solutions, 33
solid tumors, 101, 161, 207
solubility, 14, 20, 73, 138, 153
solvents, 20, 24, 25, 74
space, 3, 18, 63, 164
species, 11, 30, 33, 83, 88, 144, 156, 178
spectroscopy, 36
spectrum, xxvii, 79, 136
speed, 83, 188
spin, 15, 36, 82, 163, 164, 169, 185, 192, 193
spinal cord, 202, 204
spinal cord injury, 204
spleen, 81, 147, 148, 150, 166, 172, 192
squamous cell carcinoma, 205
stability, 15, 37, 43, 44, 69, 71, 72, 73, 74, 80, 81, 82, 85, 94, 123, 206

stabilization, 5, 35, 51, 72, 90, 122, 127, 160
standard deviation, 52, 54, 55, 66
standard error, 52
starch, 72, 84, 95, 96, 159
statistics, 54, 61, 69
stem cells, viii, 137, 140, 144, 146, 148, 154, 171, 173, 174, 175, 176, 177, 178, 190, 199, 200, 201, 208
storage, 1, 15, 165, 178
strategies, 3, 57, 87, 171, 173, 207
strategy, 86, 171, 173, 179, 180, 183, 184
strength, 21, 83, 175, 187, 194
striatum, 74, 199
stromal cells, 101, 204
strontium, 94
substrates, 26, 38, 105, 106
sucrose, 39
sugar, 120
suicide, 204
Sun, 8, 37, 39, 40, 41, 43, 80, 97, 99, 101, 103, 113, 122, 153, 157, 161, 200, 201
superiority, 167
supervision, vii
suppression, 29, 140, 198, 208
surface energy, 12, 71
surface modification, vii, xxviii, 41, 88, 112, 114, 139, 191
surface properties, xxvii, xxviii, 119
surface reactions, 116
surfactant, viii, 4, 17, 18, 19, 26, 36, 46, 87, 99, 113
susceptibility, 2, 21, 32, 73, 83, 159, 164, 194
suspensions, 32, 71, 120, 137
synthetic polymers, 172

T

T cell, 140, 156, 172, 200
T lymphocytes, 140
tamoxifen, 75, 77
targets, 30, 123, 183
TEM, 2, 15, 18, 21, 22, 77, 166, 189
TEOS, 24, 86, 87
test statistic, 54
testicular cancer, 197
testing, 5, 126, 152, 157
tetraethoxysilane, 24
tetrahydrofuran, 102
TGA, 114, 189
therapeutic agents, 183, 184
therapeutic approaches, 170
therapeutics, 88, 207
therapy, 5, 18, 110, 121, 123, 140, 148, 150, 151, 154, 155, 157, 160, 161, 173, 181, 186, 190, 192, 198, 199, 202, 205, 206, 207, 209, 210

thermal decomposition, 14, 20, 21, 23, 25, 41, 48, 84, 87
thermal energy, 4
thermal properties, 79
thermodynamics, 14
thermogravimetric analysis, 114
thin films, 34, 38, 112
threshold, 57, 126, 141
thrombin, 124
tissue, 74, 80, 82, 88, 92, 94, 95, 100, 101, 102, 104, 105, 107, 126, 140, 150, 157, 162, 163, 164, 165, 170, 178, 179, 180, 186, 190, 208
tissue plasminogen activator, 80, 92
titania, 30, 46
titanium, 30
toxic effect, 131, 141, 144, 148, 150, 175
toxicity, xxviii, 34, 51, 69, 70, 82, 91, 98, 125, 126, 127, 128, 129, 130, 131, 132, 133, 134, 138, 139, 140, 141, 144, 146, 147, 149, 150, 151, 152, 153, 155, 156, 158, 160, 161, 175, 184, 188, 193, 196
toxicology, 125, 126, 149
toxin, 126
tracking, xxvii, 83, 148, 149, 159, 161, 171, 172, 178, 180, 190, 191, 201, 202, 204
transfection, viii, 5, 84, 89, 96, 113, 140, 146, 152, 155, 172, 173, 175, 177, 181, 187, 199, 200, 201, 210
transferrin, 46, 119, 167, 178, 183, 206
transformation, 33, 49, 66
transformation processes, 49
transformations, 33
transfusion, 148
transgene, 206
transition, 24, 80, 112, 114, 116
transition metal, 112, 114
transmission, 2, 15, 140
transmission electron microscopy, 2, 140
Transmission Electron Microscopy, 7
transplantation, 140, 142, 148, 156, 158, 172, 201, 203, 205
transport, 4, 123, 126, 171, 172
trapezium, 117
trial, 57, 147, 160, 168, 175, 192, 195, 197, 209
tricarboxylic acid, 82
tricarboxylic acid cycle, 82
triglycerides, 19
troubleshooting, viii
trypsin, 179
tumor, viii, 46, 75, 78, 88, 90, 92, 102, 125, 146, 147, 149, 159, 161, 168, 174, 179, 180, 182, 184, 185, 186, 190, 194, 196, 199, 202, 204, 205, 206, 207, 208, 209

tumor cells, 46, 75, 78, 88, 90, 92, 125, 148, 159, 161, 184, 187, 190, 202, 205
tumor necrosis factor, 207
tumors, 82, 89, 121, 147, 161, 165, 182, 183, 184, 186, 194, 201, 209
tumours, 161, 192, 196, 207, 209
two sample t-test, 52, 53, 55

U

Ukraine, 115
ultrasonography, viii
ultrasound, 25, 26, 84, 91, 148
United States, 96, 121, 155, 157, 166, 201
UV, 75, 136, 141, 158

V

vacuum, 113, 125
Valencia, 206
vapor, 11, 25, 116, 117
variance, 52, 53, 54, 57, 61, 65, 66, 68
variations, 5, 148
vasculature, 148
vector, 96, 156, 187, 201
vehicles, 36, 73, 109, 180, 184
vesicle, 6, 126, 144
vessels, 182, 184
vinyl monomers, 75
viral vectors, 181

W

white blood cells, 74
white matter, 198
wound healing, 83, 106

X

xenografts, 205
X-ray, viii, 15, 24, 26, 79, 102, 108, 116
X-ray analysis, 26
X-ray diffraction, 24, 79
X-ray diffraction data, 24
XRD, 15, 59

Y

yttrium, 45

Z

zeolites, 1, 7
zinc oxide, 45
zirconia, 116
ZnO, 26, 46, 157